W9-AET-098

Advances in

INORGANIC CHEMISTRY

AND

RADIOCHEMISTRY

———

Volume 25

CONTRIBUTORS TO THIS VOLUME

B. J. Aylett

Kailash C. Dash

Peter P. Edwards

J. B. Farmer

Hubert Schmidbaur

G. E. Toogood

M. G. H. Wallbridge

Advances in
INORGANIC CHEMISTRY
AND
RADIOCHEMISTRY

EDITORS

H. J. EMELÉUS

A. G. SHARPE

University Chemical Laboratory
Cambridge, England

VOLUME 25

1982

ACADEMIC PRESS
A Subsidiary of Harcourt Brace Jovanovich, Publishers

New York London
Paris San Diego San Francisco São Paulo Sydney Tokyo Toronto

ACADEMIC PRESS, INC.
111 Fifth Avenue, New York, New York 10003

United Kingdom Edition published by
ACADEMIC PRESS, INC. (LONDON) LTD.
24/28 Oval Road, London NW1 7DX

LIBRARY OF CONGRESS CATALOG CARD NUMBER: 59–7692

ISBN 0–12–023625–7

PRINTED IN THE UNITED STATES OF AMERICA

82 83 84 85   9 8 7 6 5 4 3 2 1

CONTENTS

LIST OF CONTRIBUTORS. vii

Some Aspects of Silicon–Transition-Metal Chemistry

B. J. AYLETT

I. Introduction 1
II. Preparative Routes to Molecular Compounds 3
III. Reactivity of Molecular Silicon–Transition-Metal Compounds. . . . 39
IV. Information from Physical Methods 82
V. Topics of Special Interest 107
  Appendix 118
  References 120

The Electronic Properties of Metal Solutions in Liquid Ammonia and Related Solvents

PETER P. EDWARDS

I. Introduction 135
II. The Isolated Solvated Electron in Dilute Solutions 138
III. Electronic Properties of Dilute Solutions 142
IV. Concentrated Solutions and the Nonmetal-to-Metal Transition . . . 168
V. Concluding Remarks 178
  Appendix 178
  References 180

Metal Borates

J. B. FARMER

I. Introduction 187
II. Nomenclature 188
III. Structural Characteristics of Metal Borates 188
IV. Aqueous Solutions of Metal Borates 200
V. Preparation and Properties of Metal Borates 211
VI. Concluding Remarks 225
  References 225

Compounds of Gold in Unusual Oxidation States

HUBERT SCHMIDBAUR AND KAILASH C. DASH

I. Introduction 239
II. Compounds of Gold in Oxidation State −1. 240
III. Compounds of Gold in Fractional Oxidation States: Gold Clusters . . . 243
IV. Compounds of Gold in Oxidation State +2. 249
V. Compounds of Gold in Oxidation State +5. 257
   References 262

Hydride Compounds of the Titanium and Vanadium Group Elements

G. E. TOOGOOD AND M. G. H. WALLBRIDGE

I. Titanium, Zirconium, and Hafnium Hydrides 267
II. Vanadium, Niobium, and Tantalum Hydrides 305
   References 334

INDEX 341

CONTENTS OF PREVIOUS VOLUMES 350

LIST OF CONTRIBUTORS

Numbers in parentheses indicate the pages on which the authors' contributions begin.

B. J. AYLETT (1), *Department of Chemistry, Westfield College, University of London, London NW3 7ST, England*

KAILASH C. DASH* (239), *Anorganisch-Chemisches Institut der Technischen Universität München, D-8046 Garching, Federal Republic of Germany*

PETER P. EDWARDS (135), *University Chemical Laboratory, Cambridge CB2 1EW, England*

J. B. FARMER (187), *Borax Research Limited, Chessington, Surrey KT9 1SJ, England*

HUBERT SCHMIDBAUR (239), *Anorganisch-Chemisches Institut der Technischen Universität München, D-8046 Garching, Federal Republic of Germany*

G. E. TOOGOOD (267), *Department of Chemistry, University of Waterloo, Waterloo, Ontario, Canada N2L 3G1*

M. G. H. WALLBRIDGE (267), *Department of Chemistry and Molecular Sciences, University of Warwick, Coventry CV4 7AL, England*

* Permanent address: Department of Chemistry, Utkal University, Vani Vihar, Bhubaneswar 751004, India.

SOME ASPECTS OF SILICON–TRANSITION-METAL CHEMISTRY

B. J. AYLETT

Department of Chemistry, Westfield College, University of London, London, England

I.	Introduction	1
II.	Preparative Routes to Molecular Compounds	3
	A. Reaction of a Transition-Metal Anion with a Silicon Halide	3
	B. Reaction of a Thallium Metal Carbonylate with a Silicon Halide	11
	C. Reaction of a Silyl Alkali-Metal Compound with a Metal Halide	11
	D. Reactions with Mercury Silyl Compounds	11
	E. Reactions of Transition-Metal Carbonyls with Si–H Compounds Giving H_2	13
	F. Oxidative Addition and Oxidative Addition–Elimination Reactions	21
	G. Other Methods	30
III.	Reactivity of Molecular Silicon–Transition-Metal Compounds	39
	A. General	39
	B. Cleavage of the Silicon–Transition-Metal Bond	39
	C. Ligand Exchange at Silicon	58
	D. Adducts with Lewis Bases at Silicon	63
	E. Substitution of Ligands at the Transition Metal	70
	F. Reactions at the Metal Carbonyl Group	76
	G. Disproportionation, Reductive Elimination, and Other Reactions	79
IV.	Information from Physical Methods	82
	A. Electron and X-Ray Diffraction Studies	82
	B. Vibrational Spectroscopy	91
	C. NMR Spectroscopy	93
	D. Mass Spectrometry and Appearance Potentials	98
	E. Optical Activity in Silicon–Transition-Metal Compounds	100
	F. Other Techniques	106
V.	Topics of Special Interest	107
	A. Pyrolysis and Chemical Vapor Deposition: Metal Silicides	107
	B. Silicon–Transition-Metal Cluster Compounds	111
	C. Metal-Complexed Silylene Derivatives	116
	Appendix	118
	References	120

I. Introduction

Compounds with silicon–transition-metal bonds fall into two distinct classes, which until recently showed no points of connection. The first class comprises the metal silicides: typical binary examples are FeSi,

1

$CoSi_2$, and V_3Si, and some have been known for more than 100 years (59). Other more complex types containing three or more components, such as Mn_5Si_3C, are also known. They all form three-dimensional giant lattices, often of an unusual and complicated kind, and can best be considered as intermediate in nature between alloy systems and macromolecular covalent compounds. The study of silicides has considerably intensified in the past decade, since it has been found that some show promise as electronic materials, while at the same time they are strong and highly resistant to chemical attack. A number of books and general reviews may be noted (18, 257, 318, 340, 360, 423).

The second class consists of molecular compounds containing one or more silicon–transition-metal bonds. The first example, $Me_3SiFe(CO)_2$-$(\eta^5$-$C_5H_5)$, was prepared in 1956 (359), but nearly 10 years then elapsed before other compounds were described (26, 94). These heralded many more, and now examples are known in which silicon is bonded to almost every transition metal (Fig. 1). Curiously, no Si–Ag compounds have been described; there are also no reports of derivatives of lanthanides or actinides. Most work has involved Fe, Co, Pt, Mn, Re, Mo, Ru, and Ni, in roughly decreasing order of frequency. Almost all well-characterized molecular silicon–transition-metal compounds known at present are diamagnetic; some possible exceptions are noted in Section II,F. The most recent comprehensive reviews of the area were published in 1973 (134) and 1974 (235), covering the literature until 1971 and 1972, respectively; these contain details of earlier reviews. Other surveys of certain aspects have also appeared, two of them very recently (24, 25, 201).

In this article the intention is to place the main emphasis on studies of molecular silicon–transition-metal compounds, in particular those from the author's laboratory and those that have been reported since the comprehensive reviews noted above. As, however, it is now known that these molecular compounds can act as convenient precursors for certain transition-metal silicides, aspects of the properties of silicides will also be outlined, and relationships between the two classes of compounds examined. Compounds with bonds between transition metals

Sc	Ti	V	Cr	Mn	Fe	Co	Ni	Cu	Zn
Y	Zr	Nb	Mo	Tc	Ru	Rh	Pd	Ag	Cd
La	Hf	Ta	W	Re	Os	Ir	Pt	Au	Hg
Ac									

FIG. 1. Transition metals known to form molecular compounds with bonds to silicon are shown in bold-face type; dubious examples are in italics. No derivative of any lanthanide or actinide element is known.

and other group IV elements, namely C, Ge, Sn, and Pb, will be referred to only in so far as they provide illuminating parallels or contrasts with the corresponding silicon derivatives.

II. Preparative Routes to Molecular Compounds

The most important synthetic routes continue to be (1) the elimination of an alkali halide between the salt of a transition-metal anion and a silicon halide, and (2) oxidative addition and addition–elimination reactions. The present position regarding the scope and limitations of these and other routes is outlined in this section.

A. REACTION OF A TRANSITION-METAL ANION WITH A SILICON HALIDE

Table I shows recent applications of this method together with some earlier results of particular interest.

1. The Anion

Considering first the transition-metal-containing anion (174), it is clear that the most used are $M(CO)_3(Cp)^-$ (M = Cr, Mo, W; Cp = η^5-C_5H_5), $Mn(CO)_5^-$, $Re(CO)_5^-$, $Fe(CO)_2(Cp)^-$ and $Co(CO)_4^-$; in the case of doubly charged species, $Fe(CO)_4^{2-}$ has been employed with limited success, but no use has been made of anions of higher charge such as $Co(CO)_3^{3-}$ (175, 176). Neither have successful applications of binuclear or polynuclear anions yet been reported, although a low yield of a compound thought to be $SiH_3Fe(CO)_4Fe(CO)_4SiH_3$ resulted from the reaction of $[Fe_2(CO)_8]^{2-}$ and H_3SiI in ether (80). Reactions involving anions in which one or more carbonyl groups are replaced by tertiary phosphines have also proved unrewarding except in the case of $[Co(CO)_3(PPh_3)]^-$ (138). Although steric factors are clearly significant (see below), there seems no reason why systems of this kind should not be more generally applicable, given a sensible choice of silicon halide and solvent.

The relative electronegativities of some of these groups appear to increase in the order (147, 174)

$$Fe(CO)_4 < Fe(CO)_2(Cp) < Re(CO)_5 < Mn(CO)_5 < Co(CO)_4$$

while the nucleophilicities of the derived anions correspondingly decrease in the same order: It is interesting that a recent kinetic study of

TABLE I

REACTIONS OF ANIONS WITH SILICON HALIDES

Entry	Anion[a]	Silicon halide	Solvent[b]	Success[c]	Remarks	Ref.
1	$V(CO)_6^-$	H_3SiI	Ether, isopentane	✓	—	(8)
2		$MePh_2SiCl$	THF	✗	—	(273)
3	$Cr(CO)_3(Cp)^-$	R_3SiCl	Ether	✗	Similarly Mo, W	(357)
4		H_3SiBr	None	✓	Similarly Mo, W	(214)
5		R_3SiX	Nonpolar solvents	✓	PR_4^+ cation	(302)
6	$Cr_2(CO)_{10}^{2-}$	R_2SiX_2	THF	✗	Succeeded with Ge, Sn	(315)
7	$Mo(CO)_3(Cp)^-$	Me_3SiBr	Cyclohexane	✓	[d]	(306)
		$MeHSiCl_2$	Cyclohexane	✓	Forms ClMeHSi–Mo deriv.[d]	(306)
		$RSiCl_3(R = H, Me)$	Cyclohexane	✓	Forms Cl_2RSi–Mo deriv.[d]	(306)
		$SiX_4(X = Cl, Br)$	Cyclohexane	✓	Forms X_3Si–Mo deriv.[d]; similarly Cr, W	(306)
8	$Mo(CO)_3(Cp)^-$	R_3SiCl	Cyclohexane or methylcyclohexane	✓	Similarly W; compounds labile in polar solvents	(310)
9	$Mo(CO)_3(cp)^-$	$R_2(CH_2{=}CH)SiCl$	Not stated	✓	Similarly W	(308)
10		R_5Si_2X	Cyclohexane	✓	Similarly W[d]	(300)
		$XSiR_2SiR_2X$	Cyclohexane	✓	Forms $XSiR_2SiR_2$–Mo deriv.; similarly W[d]	(300)
11	$Mo(CO)_3(C_7H_7)^-$	Ph_3SiCl	THF	✗	Succeeded with Ge, Sn	(255)
12	$Mo(CO)_2(PPh_3)(Cp)^-$	Ph_3SiCl	THF	✗	—	(138)
13	$Mn(CO)_5^-$	H_3SiI	Ether, isopentane	✓	Similarly Re	(7, 28, 29)
14		H_2SiI_2	Ether, isopentane	✓	Only disubstd. deriv. found; similarly Re	(32)
15		H_2SiCl_2	n-Hexane	✓	Mono- and disubstd. derivs. found	(2)
16		Me_3SiBr	Pentane	✓	—	(194)
17		Me_3SiBr	Cyclohexane	✓	[d]	(306)
18		Me_3SiCl	None	✓	—	(62a)
19		$Me_{2n+1}Si_nX$ $(n = 1, 2)$	None	✓	—	(335)

4

No.	Reagent	Silicon compound	Solvent		Remarks	Ref.
20	$Mn(CO)_5^-$	Me_3SiCl	THF	×	PPh_3 also present	(146)
21		$\underline{CH_2CH_2CH_2}Si(R)Cl$	n-Hexane	√	Low yield, much Si–O–Si	(136)
22		R_3SiCl	Ether, CH_2Cl_2	×	$Ph_3PNPPh_3^+$ cation, succeeded with Ge, Sn	(151)
23	$Mn(CO)_4(PPh_3)^-$	Ph_3SiCl	THF	×	Similarly Re	(138)
		Ph_3SiCl	THF	×	—	(138)
24	$Mn(CO)_3L_2^-$	R_3SiX	THF	×	Similarly Ge, but Sn successful	(350)
25	$Fe(CO)_4^{2-}$	H_3SiI	Me_2O, isopentane	√	Some $SiH_3FeH(CO)_4$ also produced	(38, 39)
26		Me_3SiI	THF	×	Si–O–C derivative formed	(54, 331)
27		$ClMe_2Si(CH_2)_3Cl$	Ether	√	Cyclic $CH_2(CH_2)_2SiMe_2Fe(CO)_4$ (I) produced	(132)
28	$Fe(CO)_2(Cp)^-$	Me_3SiCl	THF	√	—	(359)
29		R_3Si^*Cl	THF	√	Optically active MePhNpSi deriv. (Np = 1-naphthyl)	(92)
30		H_2SiI_2	Benzene, isopentane	√	Only disubstd. deriv. found	(32)
31		$R_2(CH_2{=}CH)SiCl$	Not stated, THF	√		(123, 308)
32		$ClRHSiFe(CO)_2(Cp)$	Methylcyclohexane	√	$HRSi[Fe(CO)_2Cp]_2$ produced	(309)
33		$\underline{CH_2(CH_2)_3}Si(R)Cl$	THF	√	—	(136)
34	$Fe(CO)_2(Cp)^-$	$[ClMe_2SiCH_2{-}]_2$	THF	√	$[{-}CH_2SiMe_2Fe(CO)_2(Cp)]_2$ produced	(271, 272)
		Me_2HSiCl	THF	√	Disubstd. deriv. formed.	(271, 272)
		$Me_3Si(SiMe_2)_nCl$ $(n = 1{-}3)$	THF	√		(271, 272)
		$Cl(SiMe_2)_nCl$ $(n = 2, 3)$	THF	√	$n = 3(i\text{-}); n = 4(t\text{-})$isomer	(271, 272)
35		$Me_{2n+1}Si_nCl$ $(n = 1{-}4)$	THF	√	$SiISiMe_2Fe(CO)_2(Cp)]_4$ produced	(335)
		$(ClMe_2Si)_4Si$	THF	√	$XMe_2SiSiMe_2Fe(CO)_2(Cp)$ formed[a]	(335)
36		$XSiMe_2SiMe_2X$	Cyclohexane	√		(300)
37		$RSiCl_3$ (R = H, Me)	Cyclohexane	√	Cl_2RSi–Fe deriv. formed[a]	(306)
		Me_3SiBr	Cyclohexane	√	[a]	(306)
		$RMeSiCl_2$ (R = H, Me)	Cyclohexane	√	ClRMeSi–Fe deriv. formed[a]	(306)

(table continues)

5

TABLE I (Continued)

Entry	Anion[a]	Silicon halide	Solvent[b]	Success[c]	Remarks	Ref.
38	$Co(CO)_4^-$	H_3SiI	Ether, Me_2O	✓	—	(26, 29)
39		H_2SiI_2	Ether	✓	Disubstd. deriv. only	(29)
40	$Co(CO)_4^-$	H_2SiCl_2	Ether	✓	Mono- and disubstd. derivs.	(2)
		$HSiCl_3$	Ether	✓	Monosubst. deriv. only	(2)
		Me_2HSiCl	Ether	✓	—	(2)
41		Ph_3SiCl	THF	✗	$(Ph_3Si)_2O$ produced	(138)
42		Me_3SiCl	Ether	✓	—	(336)
		Me_2SiCl_2	Ether	✗	No isolable Si–Co product	(336)
		$RSiCl_3$ (R = Me, Cl)	Ether	✗	$Cl_2RSiOCCo_3(CO)_9$ formed	(336)
		Me_nSiCl_{4-n} (n = 0–3)	THF	✗	$Cl_{3-n}Me_nSiO(CH_2)_4(CO)Co(CO)_4$ formed	(336)
43		SiI_4	n-Hexane	✓	$(OC)_4CoSiCo_3(CO)_9$ produced	(396)
44		[cyclic silicon–Br structure]	Methylcyclohexane	✓	[cyclic Si–Co(CO)₄ structure] formed, also di- and trisubstd. analogs[e]	(186)
45	$Co(CO)_3(PPh_3)^-$	Ph_3SiCl	THF	✓	$trans\text{-}Ph_3SiCo(CO)_3(PPh_3)$ formed	(138)
46	$Co_3(CO)_{10}^-$	R_3SiCl (R = Me, Ph)	Ether	✗	$Me_3SiOCCo_3(CO)_9$ produced	(312)
47	$Co[P(OR)_3]_4^-$	R_3SiCl	Not stated	✗	—	(325)
48	$Ir(CO)_4^-$	H_3SiI	Ether	✓	—	(37)
49	$M(PF_3)_4^-$ (M = Rh, Ir)	R_3SiCl	Ether	✗	Ge analog similar, but Sn analog successful	(51)

[a] Cp = η^5-C_5H_5.
[b] THF = tetrahydrofuran
[c] Refers to formation (✓) or nonformation (✗) of silicon–metal derivative.
[d] Products decompose in THF, yielding Si–O–Si derivatives.
[e] Similar derivatives prepared from $M(CO)_3(Cp)^-$ (M = Cr, Mo, W), $Fe(CO)_2(Cp)^-$, and $Mn(CO)_5^-$.

the reaction of benzyl chloride with some of these anions showed that the rates fell in the order (322)

$$Fe(CO)_4^{2-} > Mn(CO)_5^- > Co(CO)_4^-$$

and it was concluded that nucleophilic attack by $Co(CO)_4^-$ in this system was unimportant; the extent of ion pairing and the ease of removal of Cl^- were, however, highly significant. Nearly all studies involving silicon halides have used Na^+ or, less commonly, K^+ as the counterion, and there are no systematic studies of the effects of ion pairing. Two reactions involving large cations have been reported: the first (302) was successful,

$$Et_3P{=}CHMe + HCr(CO)_3(Cp) \longrightarrow Et_4\overset{+}{P}[Cr(CO)_3Cp]^- \tag{1}$$

$$\downarrow \begin{array}{l} R_3SiCl \\ \text{nonpolar solvent} \end{array}$$

$$R_3SiCr(CO)_3(Cp)$$

$$(R_3 = Me_3, Cl_2H, ClMeH)$$

but the second (151) was not.

$$(PNP)^+Mn(CO)_5^- + R_3SiCl \xrightarrow[\text{or CH}_2\text{Cl}_2]{\text{ether}} [PNP]^+[Mn(CO)_4Cl_2]^- + \cdots \tag{2}$$

In the latter case, where $(PNP)^+ = (Ph_3P{=}N{=}PPh_3)^+$, it was shown in a separate experiment that $Me_3SiMn(CO)_5$ quickly reacts with $(PNP)^+Cl^-$ to give $(PNP)^+[Mn(CO)_4Cl_2]^-$ as a major product, so that any silicon–metal compound formed initially would be rapidly destroyed.

There have been no successful syntheses using carbonylate anions with halide or pseudohalide substituents. Indeed, it is known that cyanide-containing anions react with organosilicon halides to give isonitrile derivatives (50), e.g.,

$$Na[(\eta^3\text{-}C_3H_5)Mn(CO)_3CN] + R_3SiX \rightarrow (\eta^3\text{-}C_3H_5)Mn(CO)_3(CNSiR_3) \tag{3}$$

2. The Silicon Halide

Several effects are related to the nature of the silicon halide. Hydrido halides usually react readily with a range of anions, while organosilicon halides may not; for instance, H_2SiCl_2 reacts with $Co(CO)_4^-$ in ether giving good yields of silicon–cobalt derivatives (2), while Me_2SiCl_2 does not (336). This is probably related in the main to the electron-accepting properties of hydridosilicon halides, since the compounds SiH_3X behave as Lewis acids of modest strength (22), but their analogs R_3SiX are much weaker.

Bulky groups can be tolerated when only one metal group is to be attached to each silicon atom, as for example in the following successful reaction (92):

$$MePh(1\text{-naphthyl})SiCl + Fe(CO)_2(Cp)^- \xrightarrow{THF} MePh(1\text{-naphthyl})SiFe(CO)_2(Cp) \quad (4)$$

When attempts are made to attach two or more such groups, however, steric effects may become critical. Reactions of anions with di-, tri-, and tetrahalides usually result in monosubstituted products, e.g. ($2, 306$),

$$RSiCl_3 + Mo(CO)_3(Cp)^- \xrightarrow{cyclohexane} RCl_2SiMo(CO)_3(Cp) \quad (R = H,Me) \quad (5)$$

$$HSiCl_3 + Co(CO)_4^- \xrightarrow{ether} HCl_2SiCo(CO)_4 \quad (6)$$

Disubstituted products result only when dihydridosilicon halides are used or in the special case of reactions with $Fe(CO)_2(Cp)^-$ ($2, 29, 32, 309$), e.g.,

$$H_2SiI_2 + M(CO)_5^- \xrightarrow{ether} H_2Si[M(CO)_5]_2 \quad (M = Mn,Re) \quad (7)$$

$$MeHSiCl_2 + Fe(CO)_2(Cp)^- \xrightarrow{methylcyclohexane} HMeSi[Fe(CO)(Cp)]_2 \quad (8)$$

The sole reported example of a product with more than two metal atoms attached to silicon is the cluster derivative $(OC)_9Co_3SiCo(CO)_4$, made from SiI_4 and $Co(CO)_4^-$ in n-hexane (396); here the steric effects are reduced as a result of $Co–Co$ bond formation (see Section V,B).

The nature of the halogen attached to silicon also affects the reaction; where comparisons can be made, iodides react more rapidly and completely than chlorides, e.g. ($2, 29, 32$),

$$H_2SiI_2 + M(CO)_n^- \to H_2Si[M(CO)_n]_2 \quad (9)$$

$$H_2SiCl_2 + M(CO)_n^- \to H_2ClSi[M(CO)_n] + H_2Si[M(CO)_n]_2 \quad (10)$$

where $M(CO)_n^-$ is $Co(CO)_4^-$ or $Mn(CO)_5^-$. In the first case, only disubstituted products could be isolated. This parallels previous observations made on analogous germanium compounds (109): Ph_2GeBr_2 gave disubstituted products with a range of metal anions, whereas Ph_2GeCl_2 did not. All this is consistent with the idea noted earlier that the role of the leaving halide ion is important. Furthermore, there are a number of examples of cases in which organogermanium and organotin halides react successfully with transition-metal anions, but organosilicon halides do not (e.g., *138, 151, 255, 263, 315*). This may well be related to greater ionic character of Ge–X and Sn–X bonds as compared with Si–X, and a consequently enhanced tendency for halogen to leave as X^-.

Silicon compounds with both silicon–halogen and carbon–halogen bonds can react using both these functions; an interesting example is the synthesis of the ferrasilacyclopentane (**I**) (*132*):

$$\text{ClMe}_2\text{SiCH}_2\text{CH}_2\text{CH}_2\text{Cl} + \text{Fe(CO)}_4^{2-} \xrightarrow{\text{ether}} \underset{\textbf{(I)}}{\begin{array}{c}\diagup\text{SiMe}_2\\\mid\\\diagdown\text{Fe(CO)}_4\end{array}} \qquad (11)$$

3. The Solvent

In a few cases (Table I, entries 4, 18, and 19), successful syntheses have been carried out by direct reaction between a solid metal carbonylate and liquid silicon halide; usually, however, a solvent is employed.

Because transition-metal anions can be prepared conveniently in tetrahydrofuran solution, this cyclic ether was often used in earlier attempts to prepare silicon–metal compounds. It is now generally realized, however, that tetrahydrofuran can frustrate these attempts in two ways. First, it promotes electrophilic attack by the silicon compound on oxygen atoms of coordinated carbonyl groups; this leads to the formation of products with Si–O bonds (*54, 138, 262, 300, 306, 310, 336, 337*), e.g.,

$$\text{Ph}_3\text{SiCl} + \text{Mn(CO)}_5^- \xrightarrow{\text{THF}} (\text{Ph}_3\text{Si})_2\text{O} + \text{Mn}_3(\text{CO})_{14}^- + \cdots \qquad (12)$$

$$\text{Me}_3\text{SiI} + \text{Fe(CO)}_4^{2-} \xrightarrow{\text{THF}} (\text{Me}_3\text{SiOC})_4\text{Fe}_2(\text{CO})_6 + \cdots \qquad (13)$$

$$\text{Ph}_3\text{SiCl} + \text{Co(CO)}_4^- \xrightarrow{\text{THF}} (\text{Ph}_3\text{Si})_2\text{O} + \text{CoCl}_2 + \cdots \qquad (14)$$

In fact, preparations involving tetrahydrofuran are successful only when using the highly nucleophilic species $\text{Fe(CO)}_2(\text{Cp})^-$ and, in a solitary example, $\text{Co(CO)}_3(\text{PPh}_3)^-$ (*138*) [which is more nucleophilic than Co(CO)_4^-; note however that $\text{Mn(CO)}_4(\text{PPh}_3)^-$ did not yield Si–Mn products (*138*)]:

$$\text{Me}_3\text{SiCl} + \text{Fe(CO)}_2(\text{Cp})^- \xrightarrow{\text{THF}} \text{Me}_3\text{SiFe(CO)}_2(\text{Cp}) \qquad (15)$$

$$\text{Ph}_3\text{SiCl} + \text{Co(CO)}_3(\text{PPh}_3)^- \xrightarrow{\text{THF}} trans\text{-Ph}_3\text{SiCo(CO)}_3(\text{PPh}_3) \qquad (16)$$

$$\text{Ph}_3\text{SiCl} + \text{Mn(CO)}_4(\text{PPh}_3)^- \xrightarrow{\text{THF}} (\text{Ph}_3\text{Si})_2\text{O} + \text{ClMn(CO)}_3(\text{PPh}_3)_2 \\ + [\text{Mn(CO)}_4(\text{PPh}_3)]_2 + \cdots \qquad (17)$$

The role of tetrahydrofuran in facilitating silicon transfer from metal to oxygen is further discussed in Section III,F,2.

The second harmful effect of tetrahydrofuran is that it may undergo ring opening and insert into the silicon–transition-metal bond, again with the formation of a siloxy derivative (252, 336, 337); e.g.*

$$Me_nCl_{3-n}SiCl + Co(CO)_4^- \xrightarrow{THF} Me_nCl_{3-n}SiO(CH_2)_4COCo(CO)_4 \qquad (18)$$

This possibility had been recognized in the case of simple silyl compounds many years ago (20):

$$H_3SiI + \overline{CH_2(CH_2)_3O} \rightarrow H_3SiO(CH_2)_4I \rightarrow \text{decomposition products} \qquad (19)$$

Consequently, in the early work with hydridosilicon derivatives, diethyl ether was normally used; in the case of very volatile products, dimethyl ether offered some practical advantages. It was tacitly assumed that a polar solvent was essential in order to dissolve, at least partly, the transition-metal carbonyl derivative. More recently, however, it has become clear that alkanes, although nonpolar, provide a very suitable reaction medium, and can be used in cases where ethers, for example, are inimical to the products (8, 32, 306, 310). The hydrocarbon, besides acting as a diluent for the silicon halide, seems to assist the separation of alkali halide from the surface of the reacting transition-metal carbonyl salt.

Even with alkanes as the reaction medium, extensive formation of siloxanes can sometimes occur, as in the reaction between a silacyclobutane derivative and $Mn(CO)_5^-$ (136):

$$\text{⌐Si(R)Cl} + Mn(CO)_5^- \xrightarrow{n\text{-hexane}} \text{⌐Si(R)—O—(R)Si⌐} + \text{⌐Si(R)Mn(CO)_5} \qquad (20)$$

Major product　　　　　Minor product　　(20)

4. Conclusions

In summary, this procedure will be most effective with an anion of high nucleophilicity, an iodosilane of low steric requirements, and a nonpolar solvent. Reactions usually begin at low temperatures, although long shaking at room temperature may be needed for completion. The method can be capricious, and occasionally a familiar system may yield no product at all, with no obvious explanation. Nevertheless, it has considerable extra potential, particularly for the synthesis of silicon derivatives of polynuclear metal carbonyls.

---

* Additional CO insertion to give an acyl derivative has also occurred in this case.

B. REACTION OF A THALLIUM METAL CARBONYLATE WITH A SILICON HALIDE

This variant of the normal route using alkali metal salts (Section II,A) depends on the fact that metallic thallium reacts smoothly with some metal carbonyls to give $TlM(CO)_n$ species. These Tl(I) compounds are slightly soluble in hydrocarbons and react readily with iodosilanes (8, 121):

$$Tl + V(CO)_6 \xrightarrow{\text{benzene}} TlV(CO)_6 \xrightarrow[\text{isopentane}]{H_3SiI} H_3SiV(CO)_6 \qquad (21)$$

$$2Tl + Co_2(CO)_8 \rightarrow TlCo(CO)_4 \xrightarrow[\text{isopentane}]{H_3SiI} H_3SiCo(CO)_4 \qquad (22)$$

The method is limited by the tendency of some Tl(I) metal carbonylates to disproportionate, even at low temperatures, e.g. (79),

$$3TlMn(CO)_5 \xrightarrow{-40^\circ C} Tl[Mn(CO)_5]_3 + 2Tl \qquad (23)$$

C. REACTION OF A SILYL ALKALI-METAL COMPOUND WITH A METAL HALIDE

Table II lists preparations or attempted preparations using this route.

The method is of limited applicability, since although silyl alkali metal compounds R_3SiM are known when R_3Si is $H_{2n+1}Si_n$ ($n = 1–5$), $Me_{2n+1}Si_n$ ($n = 1, 3, 4$), and $Me_nPh_{3-n}Si$ ($n = 0–2$), and several dilithium derivatives $Li(SiPh_2)_nLi$ ($n = 4–6$) have also been described, all are rather difficult to prepare and handle (24). It will be seen that only some of them have yet been used to prepare silicon–transition-metal compounds.

The interesting group of anionic derivatives $R_3SiM(CO)_5^-$ (M = Cr, Mo, W) have been prepared in the following way (254):

$$R_3SiLi + Et_4N^+[ClM(CO)_5]^- \xrightarrow{THF} Et_4N^+[R_3SiM(CO)_5]^- \qquad (24)$$

D. REACTIONS WITH MERCURY SILYL COMPOUNDS

Some applications of the use of substituted disilyl mercury compounds, $Hg(SiX_3)_2$ (X = Me, Cl), are shown in Table III; there is an early review article (421). In most cases, the mercury compound reacts with a metal–halogen bond to give a silicon halide and mercury as by-product, e.g. (entry 1),

$$Mo(CO)_3(Cp)Cl + Hg(SiMe_3)_2 \rightarrow Me_3SiMo(CO)_3(Cp) + Me_3SiCl + Hg \qquad (25)$$

TABLE II

REACTIONS OF SILYL ALKALI METAL COMPOUNDS WITH METAL HALIDES

Entry	Silyl alkali metal derivative/solvent	Metal halide	Product	Ref.
1	$Si_4Ph_8Li_2$/THF	$(Cp)_2TiCl_2$	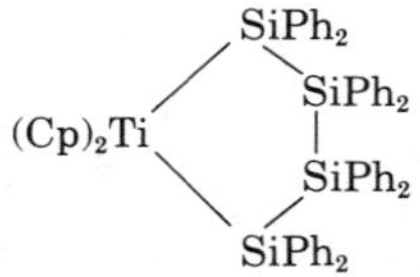 (Cp)₂Ti with SiPh₂ ring structure	(239)
2	H_3SiK/glyme	$(Cp)_2TiCl_2$	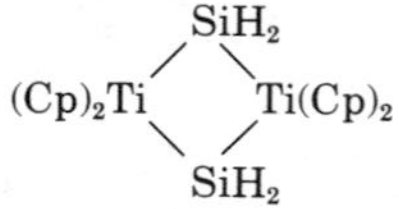 (Cp)₂Ti–Ti(Cp)₂ with SiH₂ bridge structure	(232)
3	Ph_3SiLi/THF	$(Cp)_2MCl_2(M = Zr, Hf)$	$Ph_3SiM(Cl)(Cp)_2$	(273)
4	Ph_3SiLi/THF	$(Cp)_2VCl_2$	No Si–V deriv. isolated	(273)
5	R_3SiLi/THF ($R_3 = Ph_3$, $MePh_2$)	$ClM(CO)_5^-$ (M = Cr, Mo, W)	$R_3SiM(CO)_5^-$; similarly, Ge, Sn	(254)
6	$i\text{-}Me_7Si_3Li$/THF	$BrMn(CO)_5$	$i\text{-}Me_7Si_3Mn(CO)_5$	(335)
	$t\text{-}Me_9Si_4Li$/THF	$BrMn(CO)_5$	$t\text{-}Me_9Si_4Mn(CO)_5$	(335)
	$t\text{-}Me_9Si_4Li$/THF	$BrMn(CO)_4(PPh_3)$	$t\text{-}Me_9Si_4Mn(CO)_4(PPh_3)$	(335)
7	H_3SiK	$BrFe(CO)_2(Cp)$	$H_3SiFe(CO)_2(Cp)$	(9)
8	R_3SiLi	$Cl_2Pt(PMe_2Ph_2)$	$(R_3Si)_2Pt(PMe_2Ph)_2$	(97, 98)
9	$PhMe_2SiLi$	CuX (X = I, CN)	$PhMe_2SiCu \cdot LiX$ $(PhMe_2Si)_2CuLi \cdot LiCN$ } not isolated	(6, 185)
10	Ph_3SiLi	$ClAu(PPh_3)$	$Ph_3SiAu(PPh_3)$	(43)

TABLE III

REACTIONS WITH MERCURY DISILYL DERIVATIVES

Entry	Mercurial	Metal compound	Silicon–metal product	Ref.
1	$Hg(SiMe_3)_2$	$Mo(CO)_3(Cp)Cl$	$Me_3SiMo(CO)_3(Cp)$; similarly W	(306)
2	$Hg(SiCl_3)_2$	$Mn_2(CO)_{10}$	$Cl_3SiMn(CO)_5$	(64)
3	$Hg(SiMe_3)_2$	$Fe(CO)_5{}^a$	cis-$(Me_3Si)_2Fe(CO)_4$	(262)
4	$Hg(SiMe_3)_2$	$Hg[Co(CO)_4]_2$	$Me_3SiCo(CO)_4$	(293)
5	$Hg(SiMe_3)_2$	$(Et_3P)_2Ir(CO)Cl$	$(Me_3Si)_2Ir(CO)(HgSiMe_3)(PEt_3)_2$	(241)
6	$Hg(SiMe_3)_2$	cis-$(Et_3P)_2PtCl_2$	$trans$-$(Me_3Si)PtCl(PEt_3)_2$	(196)
7	$Hg(SiMe_3)_2$	cis-$(diphos)PtCl_2{}^b$	cis-$(Me_3Si)PtCl(diphos)$ + cis-$(Me_3Si)_2Pt(diphos)$	(112, 113)

^{a} UV irradiation needed; $Fe_2(CO)_9$ and $Fe(CO)_4Br_2$ are also effective.
^{b} diphos = $Ph_2PCH_2CH_2PPh_2$.

When using the monodentate phosphine derivative shown in entry 6, only one Pt–Cl bond is broken; this is attributed to the high trans influence of the Me_3Si group in the monosubstituted and isomerized product. With a bidentate phosphine (entry 7), however, a change in stereochemistry is prevented and both mono- and disubstituted products are formed.

In the case of reaction with the Ir(I) compound in entry 5, the presumed initial product, $(Et_3P)_2Ir(CO)(SiMe_3)$, undergoes oxidative addition with further $Hg(SiMe_3)_2$ to give the Ir(III) final product shown.

The first of the two reactions using metal(0) starting materials (entry 2) is brought about by heating,

$$Hg(SiCl_3)_2 + Mn_2(CO)_{10} \rightarrow 2Cl_3SiMn(CO)_5 + Hg \qquad (26)$$

while the second (entry 3) is promoted by "intense Alberta sunlight." Finally, the novel reaction between two mercury derivatives (entry 4) should be noted; the yield is quantitative.

$$Hg[Co(CO)_4]_2 + Hg(SiMe_3)_2 \xrightarrow{cyclopentane} 2Me_3SiCo(CO)_4 + 2Hg \qquad (27)$$

E. REACTIONS OF TRANSITION-METAL CARBONYLS WITH Si–H COMPOUNDS GIVING H_2

Although some of these reactions could equally well be classified as oxidative addition or oxidative addition–elimination reactions, it is convenient to list them separately. There was an initial report by Chalk and Harrod (94) of the reaction

$$Co_2(CO)_8 + 2R_3SiH \rightarrow 2R_3SiCo(CO)_4 + H_2 \qquad (28)$$

and later work that established that metal carbonyl hydrides may act as intermediates, e.g. (*41, 42, 95*),

$$Co_2(CO)_8 + R_3SiH \rightarrow HCo(CO)_4 + R_3SiCo(CO)_4 \qquad (29)$$

$$HCo(CO)_4 + R_3SiH \rightarrow H_2 + R_3SiCo(CO)_4 \qquad (30)$$

Following these, this method has been widely used; Table IV lists more recent applications. Only those reactions that evolve dihydrogen are included here. Oxidative addition reactions leading to metal hydride derivatives are included in Table V, while processes involving elimination of dihydrogen between hydridosilanes and metal hydride derivatives containing no carbonyl groups are deferred until Section II,G,1.

Reactions involving $Co_2(CO)_8$ proceed very readily at or below room temperature, but carbonyl derivatives of the other metals listed must be heated, sometimes quite vigorously. Thus, long heating at 150°C in a sealed tube was needed to effect the reactions shown in entries 4 and 6, and quantities of siloxanes were obtained, no doubt as the result of attack on silicon by oxygen of coordinated carbonyl groups (*335*) (cf. Section III,F,2). Ultraviolet irradiation of some of these systems can lead to different products with metal–hydrogen bonds: for example, entry 35 may be compared with entry 27 of Table V. Reactivity tends to decrease on going down a transition-metal group; thus the reactions shown in entries 1 and 7 require temperatures of 50 and 150°C respectively. Curiously, in the latter case the diphenylphosphino group appears not to become coordinated to the metal.

Dihydridosilanes and hydridodisilanes often give rise to cluster compounds (e.g., entries 9, 11, 13, 16, 19, 20, 28). It is believed that these can arise as a result of extrusion of SiR_2 groups from the precursors, e.g.,

$$RSiMe_2SiMe_2H \xrightarrow{\Delta} RSiMe_2H + :SiMe_2 \qquad (31)$$

The silylene species may be coordinated to a metal center rather than being free (see Section V,C). When two Si–H groups are separated by —$(CH_2)_2$— links, either a cyclic (entry 8) or a linear (entry 26) compound may result: the latter is somewhat unstable thermally. Correspondingly, cyclic (entry 10) or linear (entry 27) species are formed from the dihydrido derivatives $(HSiMe_2)_2X$ (X = O,CH$_2$); the linear compound with two $Co(CO)_4$ groups is again thermally unstable.

In a similar way, two or more $Co(CO)_4$ groups attached to the *same*

TABLE IV

REACTIONS OF METAL CARBONYL DERIVATIVES WITH SILICON HYDRIDES

Entry	Metal carbonyl derivative	Silicon hydride	Treatment[a]	Product	Ref.
1	$Mn_2(CO)_{10}$	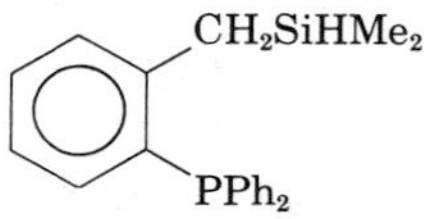 (benzene ring with CH_2SiHMe_2 and PPh_2)	Δ	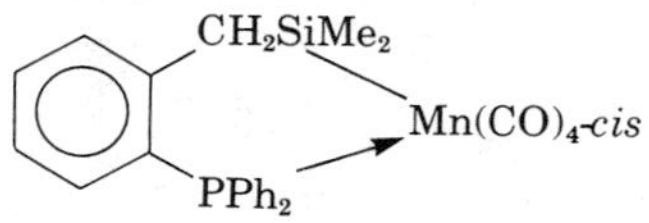 (CH_2SiMe_2—$Mn(CO)_4$-*cis*, benzene with PPh_2)	(13)
2		$HSiF_3$	Δ	$F_3SiMn(CO)_5$ (and Re analog)[b]	(373, 397)
3		$HSiPh_xCl_{3-x}$ ($x = 1, 2$)	Δ	$Ph_xCl_{3-x}SiMn(CO)_5$	(107)
4		HSi_2Me_5	Δ	$Me_5Si_2Mn(CO)_5$	(335)
		HSi_3Me_7-i		i-$Me_7Si_3Mn(CO)_5$	(335)
		HSi_4Me_9-t		t-$Me_9Si_4Mn(CO)_5$	(335)
5	$HMn(CO)_5$	H_2SiCl_2	Δ	$Cl_2Si[Mn(CO)_5]_2$	(2)
6	$[Mn(CO)_4(PPh_3)]_2$	$HSiMe_3$	Δ	$trans$-$Me_3SiMn(CO)_4(PPh_3)$[c]	(335)
7	$Re_2(CO)_{10}$	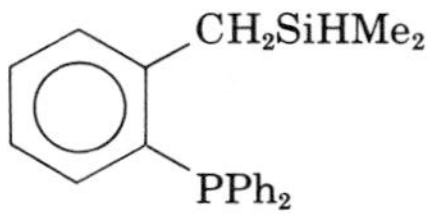 (benzene ring with CH_2SiHMe_2 and PPh_2)	Δ	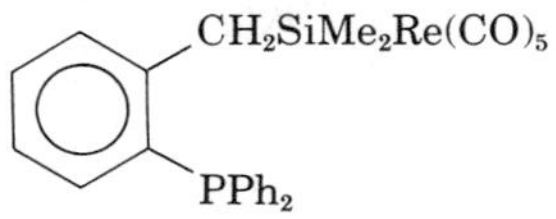 ($CH_2SiMe_2Re(CO)_5$, benzene with PPh_2)	(13)
8	$Fe(CO)_5$	$[HSiMe_2CH_2]_2$	hv	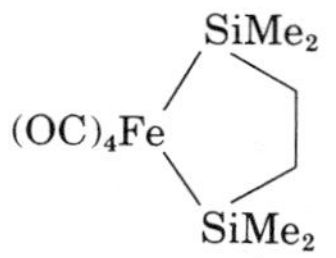 ($(OC)_4Fe$ with $SiMe_2$ ring) [Similarly Ru, Os analogs from $M_3(CO)_{12}$]	(418)

(*table continues*)

TABLE IV (Continued)

Entry	Metal carbonyl derivative	Silicon hydride	Treatment[a]	Product	Ref.
9	$Fe(CO)_5$	$HSiMe_2SiMe_2H$	$h\nu$ or Δ	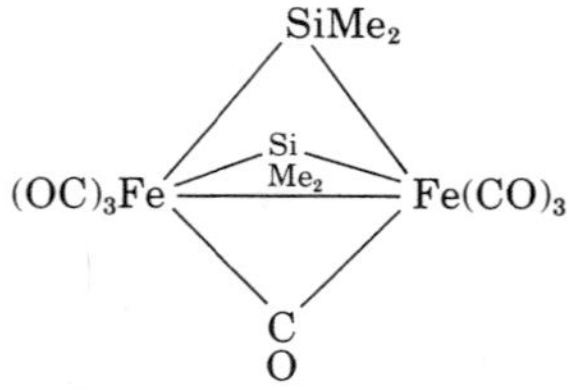 (OC)_3Fe—Fe(CO)_3 bridged by SiMe_2, SiMe_2 and CO	(289)
10		$(HSiMe_2)_2X$ (X = O, CH$_2$)		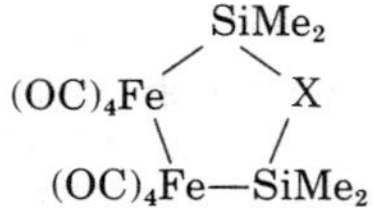 $(OC)_4Fe$ with SiMe_2—X—SiMe_2 bridge; $(OC)_4Fe—SiMe_2$ [Similarly Ru analog, from $Ru_3(CO)_{12}$]	(208)
10a		Ph_2SiH_2	$h\nu$	$[Ph_2SiFe(CO)_4]_2$	(124)
		R_2SiH_2 ($R_2 = PhMe$)	$h\nu$	$(OC)_3Fe(\mu\text{-}SiR_2)_2(\mu\text{-}CO)Fe(CO)_3$	(124)
11	$Fe_2(CO)_9$	$HSiMe_2SiMe_2H$		$(OC)_3Fe(\mu\text{-}SiMe_2)_2(\mu\text{-}CO)Fe(CO)_3$	(268)
12	$Fe_2(CO)_9$ or $Fe_3(CO)_{12}$	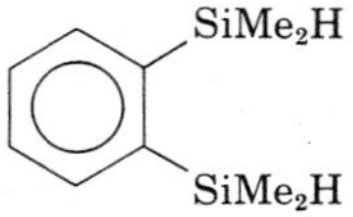 o-C$_6$H$_4$(SiMe$_2$H)$_2$		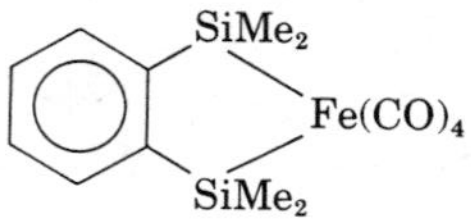 (benzo-ring) Fe(CO)_4 with two SiMe_2 bridges	(182)
		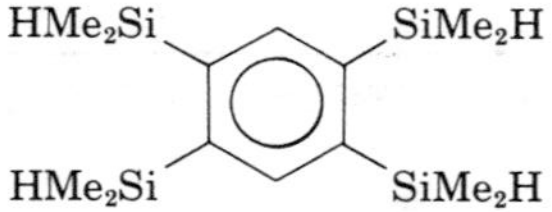 C$_6$H$_2$(SiMe$_2$H)$_4$		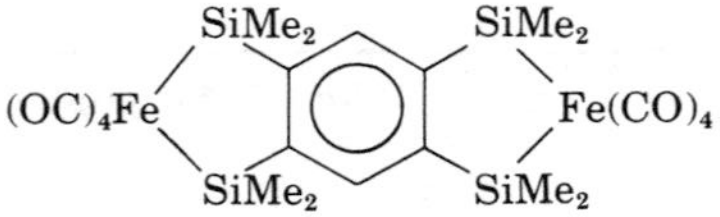 $(OC)_4Fe$ and $Fe(CO)_4$ bridged by SiMe_2 groups on benzene ring (Similarly Ru analogs)	(182)

13	Fe₃(CO)₁₂	H₂SiPh₂	Δ	(OC)₄Fe—SiPh₂—Fe(CO)₄	(77)
14	[Fe(CO)₂(Cp)]₂	HSiCl₃	Δ	Cl₃SiFe(CO)₂(Cp) (Cl₃Si)₂FeH(CO)(Cp)ᵃ	(260)
15		HSiPhₙCl₃₋ₙ (n = 1, 2)	Δ	(Cp)Fe(CO)₃⁺(Cl₃Si)₂Fe(CO)(Cp)⁻ PhₙCl₃₋ₙSiFe(CO)₂(Cp)	(107)
16	M₃(CO)₁₂(M = Ru, Os)	HSi₂Me₅	Δ or hv	(OC)₃(Me₃Si)M—M(SiMe₃)(CO)₃ (SiMe₂/SiMe₂ bridge)	(76)
17	Ru₃(CO)₁₂	(diphos)Pt[SiMe₂/SiMe₂] benzene structure		(diphos)Pt[Me₂Si/Me₂Si—C₆—Si Me₂/Si Me₂]Ru(CO)₄	(182)
18	Os₃(CO)₁₂	HSiMe₃	Δ	Me₃SiOsH(CO)₄ [Me₃SiOs(CO)₄]₂ (Me₃Si)₂Os(CO)₄	(243)
19	H₂Os(CO)₄	HSi₂Me₅	Δ	Os₃/SiMe₂ cluster structure	(76)
20	[Me₃SiRu(CO)₄]₂	HMe₂SiSiMe₂H		(OC)₃Ru(μ-SiMe₂)₃Ru(CO)₃	(76)
21	Co₂(CO)₈	HSiR₃		R₃SiCo(CO)₄	(323)
22		H₂SiR₂		HR₂SiCo(CO)₄ R₂SiCo₂(CO)₇ R₂[(OC)₄Co]SiOCCo₃(CO)₉	(181)

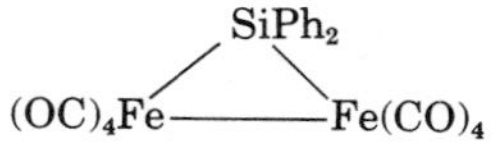

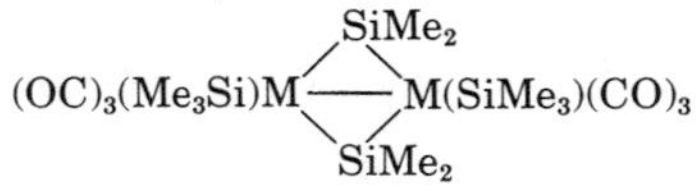

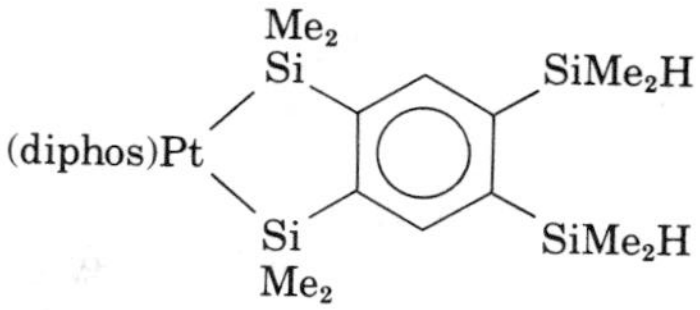

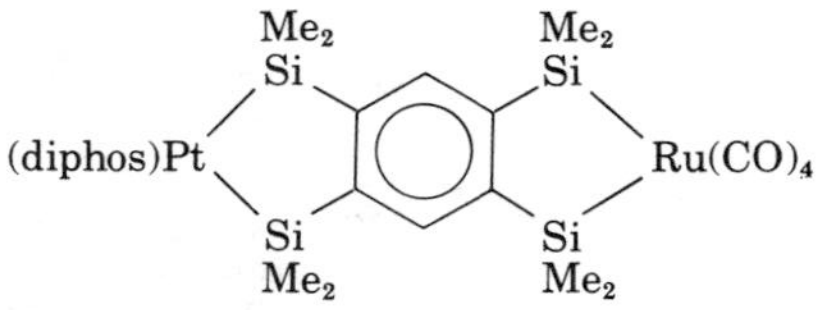

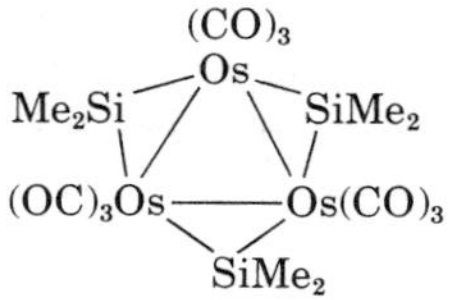

(table continues)

TABLE IV (Continued)

Entry	Metal carbonyl derivative	Silicon hydride	Treatment[a]	Product	Ref.
23	$Co_2(CO)_8$	H_2SiCl_2		$HCl_2SiCo(CO)_4$	(2)
				$Cl_2Si[Co(CO)_4]_2$	(2)
		$H_2ClSiMn(CO)_5$		$HClSi[Mn(CO)_5][Co(CO)_4]$	
				$ClSiMn(CO)_5][Co_2(CO)_7]$	
24	$Co_2(CO)_8$	$HSiR_3$		$R_3SiCo(CO)_4$	(73)
25		$HSi(C_6F_5)_3$	Δ	$(C_6F_5)_3SiCo(CO)_4$ [e]	(398)
26		$[HMe_2SiCH_2]_2$		$[(OC)_4CoSiMe_2CH_2—]_2$	(418)
27		$(HMe_2Si)_2X$ (X = O, CH₂)		$[(OC)_4CoSiMe_2]_2X$	(208)
28		$HMe_2SiSiMe_2H$		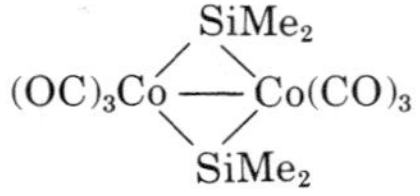 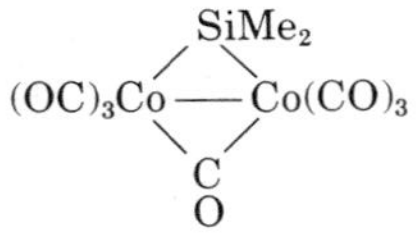 $HMe_2SiSiMe_2Co(CO)_4$ [f]	(268)
29	$Co_2(CO)_8$	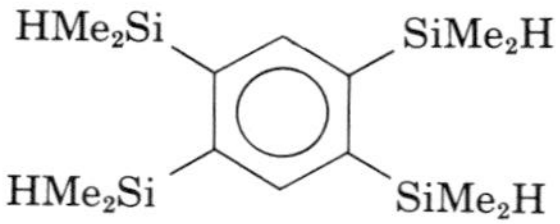		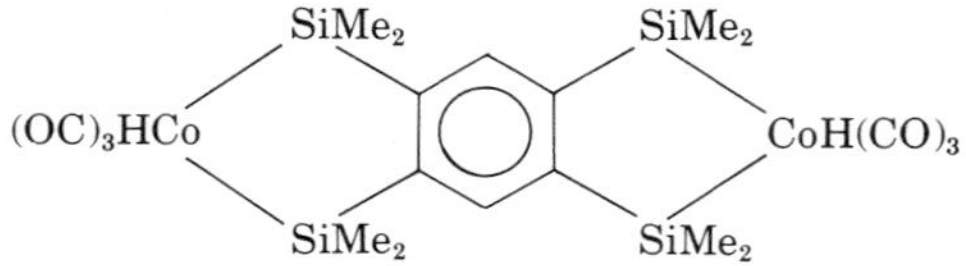	(182)

18

No.	Reactant	Reagent	Product	Ref.
30	(cyclic Si ring, Si–H)	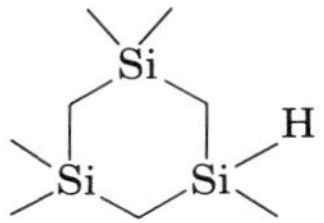 HSiF$_3$	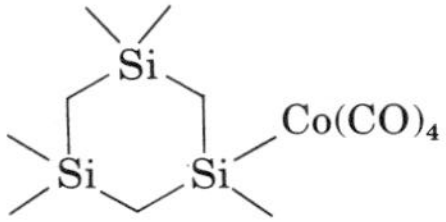 (cyclic Si ring)–Co(CO)$_4$	(186)
31	[Co(CO)$_3$L]$_2$ (L = PR$_3$, AsR$_3$)	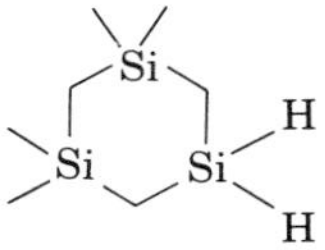 HSiF$_3$	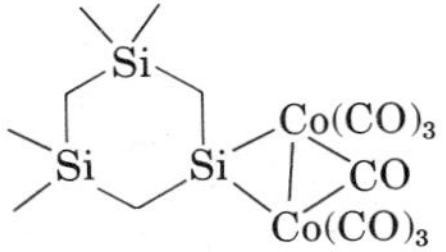 trans-F$_3$SiCo(CO)$_3$L	(227)
32	HCo(CO)$_3$(PPh$_3$)	HSiF$_3$	trans-F$_3$SiCo(CO)$_3$(PPh$_3$)	(227)
		HSiPh$_3$	Ph$_3$SiCoH$_2$(CO)$_2$(PPh$_3$) [h]	(227)
33	HCo(CO)$_2$(PPh$_3$)$_2$	HSiF$_3$	F$_3$SiCo(CO)$_2$(PPh$_3$)$_2$	(227)
34	HCo(CO)(PPh$_3$)$_3$	HSiF$_3$	F$_3$SiCo(CO)(PPh$_3$)$_3$	(227)
35	Rh(CO)$_2$(Cp)	HSiCl$_3$	(Cl$_3$Si)$_2$Rh(CO)(Cp)	(346)
36	M$_2$(PF$_3$)$_8$ (M = Rh, Ir) H(PF$_3$)$_4$ (M = Rh, Ir)	HSiR$_3$	R$_3$SiM(PF$_3$)$_4$	(51)

(Row 30 product also bears a bridged structure: Si bonded to Co(CO)$_3$, CO, Co(CO)$_3$.)

[a] Δ, Heating required; hv, irradiation required.

[b] Similar reactions occur on heating HSiF$_3$ with [Fe(CO)$_2$(Cp)]$_2$, [M(CO)$_3$(Cp)]$_2$ (M = Mo, W), and Mn$_2$(CO)$_8$(PPh$_3$)$_2$.

[c] Attempted preparation of Me$_5$Si$_2$Mn(CO)$_4$(PPh$_3$) from HSi$_2$Me$_5$ + [Mn(CO)$_4$(PPh$_3$)]$_2$ led to small yields of Me$_3$SiMn(CO)$_4$(PPh$_3$).

[d] Derived from the first product by oxidative addition–elimination with a second molecule of HSiCl$_3$; loss of H$^+$ gives the anion in the third product.

[e] Similar reactions also with [Fe(CO)$_2$(Cp)]$_2$, M$_2$(CO)$_{10}$ (M = Mn, Re), Mn$_2$(CO)$_8$(PPh$_3$)$_2$.

[f] Cobalt cluster derivatives also formed.

[g] With CO/benzene, gives trans-Ph$_3$SiCo(CO)$_3$(PPh$_3$).

[h] Both PPh$_3$ molecules probably in axial positions.

silicon atom often undergo condensation to yield clusters, e.g. (entry 22),

$$H_2SiPh_2 \xrightarrow{Co_2(CO)_8} HPh_2SiCo(CO)_4 \xrightarrow{Co_2(CO)_8} Ph_2Si[Co_2(CO)_7]$$
$$\textbf{(II)}$$

$$\downarrow Co_2(CO)_8$$

$$[(OC)_4Co]Ph_2SiOCCo_3(CO)_9$$
$$\textbf{(III)} \tag{32}$$

It can be seen that first a Co_2 cluster (II) is produced, then a Co_3 unit (III) (accompanied by migration of silicon from metal to oxygen). This topic is taken up further in Section V,B. Other examples of the formation of $Co_2(CO)_7$ groups are shown in entries 23, 28, and 30.

A different type of sequential process is illustrated in entry 18: $Os_3(CO)_{12}$ behaves as a source of "$Os(CO)_4$" groups, and initial insertion is followed by loss of dihydrogen and metal–metal bond formation. This Os–Os bond is then attacked by further Me_3SiH (243).

$$HSiMe_3 + \text{"}Os(CO)_4\text{"} \rightarrow Me_3SiOs(H)(CO)_4$$

$$\downarrow -H_2$$

$$[Me_3SiOs(CO)_4]_2$$

$$\downarrow HSiMe_3$$

$$Me_3SiOsH(CO)_4 + (Me_3Si)_2Os(CO)_4$$
$$\text{etc.} \tag{33}$$

Analogously, $Fe_3(CO)_{12}$, $Fe_2(CO)_9$, and irradiated $Fe(CO)_5$ can act as sources of "$Fe(CO)_4$" in similar reactions.

Steric factors are important in reactions of this type. The substituted cobalt carbonyl hydrides $HCo(CO)_{4-n}(PPh_3)_n$ react with increasing difficulty as n increases (entries 32–34). Another general effect is that a hydridosilane $HSiX_3$ will react more readily (and the product be more robust) as X becomes more electronegative (227, 230). This seems to be valid for all oxidative addition processes.

The metal carbonyl analogs $M_2(PF_3)_8$ and $HM(PF_3)_4$ (M = Rh,Ir) follow the same reaction pattern, but somewhat less readily; this is no doubt a result of steric factors, e.g. (entry 36),

$$Ir_2(PF_3)_8 + 2HSiR_3 \xrightarrow{\Delta} 2R_3SiIr(PF_3)_4 + H_2 \qquad (34)$$

F. Oxidative Addition and Oxidative Addition–Elimination Reactions

This category is taken to comprise those reactions in which the formal oxidation state of the metal increases by 2.* Most of the entries in Table V involve the cleavage of Si–H or Si–Si bonds, e.g. (entries 8 and 49),

$$[Re(I)](CO)_3(Cp) + HSiPh_3 \xrightarrow{h\nu} Ph_3Si[Re(III)]H(CO)_2(Cp) + CO \qquad (35)$$

$$[Pt(0)(PPh_3)_4 + Si_2Cl_6 \rightarrow (Cl_3Si)_2[Pt(II)](PPh_3)_2 + 2PPh_3 \qquad (36)$$

while the most common metal oxidation states in the starting materials are Mn(I), Fe(0), Co(I), Rh(I), Ir(I), and especially Pt(0). It is not always easy, however, to decide the final metal oxidation state (e.g., entries 9 and 47). The particular case of reactions between metal carbonyl derivatives and Si–H compounds with elimination of dihydrogen has already been dealt with in Section II,E; it may be noted that such reactions commonly involve an increase in metal oxidation state of less than 2.

In both the examples given above, there is concomitant loss of one or more neutral ligands. Elimination of CO is the rule in reactions of mononuclear metal carbonyls (e.g., entry 12) and cyclopentadienyl metal carbonyls (e.g., entry 4), but not those of polynuclear carbonyls (e.g., entry 16) or carbonyl halides (e.g., entry 33). Elimination of tertiary phosphines often occurs, especially when more than two molecules are present in the initial complex; however, this is not always the case (see entry 24). Clearly, steric requirements and the dictates of the 18-electron rule determine the composition of the product, and normally act in concert; when they conflict, as in the case of $R_3SiRuH_3(PR_3)_n$ ($n = 2$ or 3; entry 22), variable stoichiometry may result. Chelating diphosphines, with somewhat reduced steric requirements, are usually retained (e.g., entry 19), while complexed olefins are invariably lost; the bulky ligand P(cyclohexyl)$_3$ is associated with unusual products (entries 47 and 48). Particular mention may be made of the 17-electron species $Cl_3SiVH(Cp)_2$ and $(Cl_3Si)_2V(Cp)_2$ shown

* Two reactions are included (entries 3 and 17) that involve metal–metal bonded dimers as precursors and are difficult to rationalize in this way.

TABLE V

OXIDATIVE ADDITION (–ELIMINATION) REACTIONS

Entry	Metal derivative	Silicon compound	Treatment	Silicon–metal product	Ref.
1	$(Cp)_2Ti(CO)_2$	$HSiCl_3$	Δ	$Cl_3SiTi(CO)H(Cp)_2$ [a]	(88)
2	$(Cp)_2V$	$HSiCl_3$	Δ	$Cl_3SiVH(Cp)_2$ [b]; $(Cl_3Si)_2V(Cp)_2$ [b]	(88)
3	$[Mo(CO)_3(Cp)]_2$	Si_2Cl_6		$Cl_3SiMo(CO)_3(Cp)$	(197)
4	$(Cp)Mn(CO)_3$	$HSiR_3$	hv	$R_3SiMnH(CO)_2(Cp)$	(225)
5	$(Cp)Mn(CO)_3$	$R_3SiSiR_2R^1$		$(R_3Si)(R_2R^1Si)Mn(CO)_2(Cp)$ [c]	(400)
6	$(Me-Cp)Mn(CO)_3$	HSi^*R_3 $[Si^*R_3 = MePh(1\text{-naphthyl})Si]$	hv	$R_3SiMnH(CO)_2(Me\text{-}Cp)$	(117, 118)
7	$(Me-Cp)Mn(CO)_3$	$HSiR^1R^2X$	hv	$XR^1R^2SiMnH(CO)_2(Me\text{-}Cp)$	(119)
8	$(Cp)Re(CO)_3$	$HSiPh_3$	hv	$Ph_3SiReH(CO)_2(Cp)$	(249)
9	$Re_2(CO)_{10}$	$HSiCl_3$	hv	? $Cl_3SiRe(CO)_4\!-\!H\!-\!Re(CO)_5$ [d]	(248, 251)
10	$Fe(CO)_5$	$HSiR_3$	hv	$cis\text{-}R_3SiFeH(CO)_4$	(259)
11		Si_2Cl_6		$[Cl_3SiFe(CO)_4]_2$	(197)
12		$HSiR^1R^2X$	hv	$XR^1R^2SiFeH(CO)_4$	(395)
13		$HSiMe_2(NEt_2)$	hv	$(Et_2N)Me_2SiFeH(CO)_4$ [e]	(394)
14	$Fe_2(CO)_9$	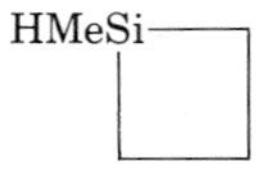 HMeSi (cyclobutane ring)		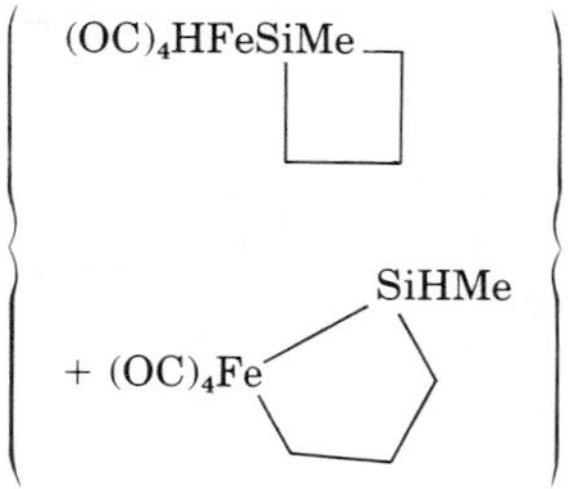 $(OC)_4HFeSiMe$ (cyclobutane) + $(OC)_4Fe$(tetrahydrosilole, SiHMe) (in 2:1 ratio)	(133)
15	$Fe_2(CO)_9$	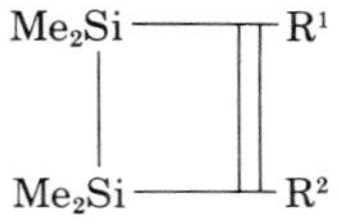 Me_2Si–Me_2Si with R^1, R^2		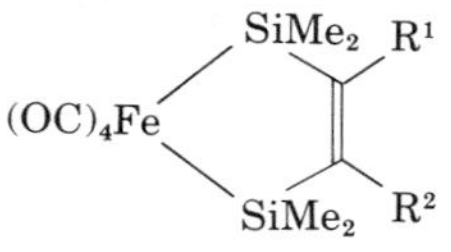 $(OC)_4Fe$(ferracycle: $SiMe_2R^1$, $SiMe_2R^2$)	(386)

22

No.	Substrate	Silane	Product	Conditions	Ref.
16	$Fe_3(CO)_{12}$	$HSiCl_3$	$cis\text{-}Cl_3SiFeH(CO)_4$ $cis\text{-}(Cl_3Si)_2Fe(CO)_4$ (Similarly Ru, Os)	hv	(361)
17	$[(Cp)Fe(CO)_2]_2$	$HSiCl_3$	$(Cl_3Si)_2FeH(CO)(Cp)$	hv	(259)
18	$Fe(II)H_2(PMePh_2)_4$	$HSiR_3$	$R_3SiFe(IV)H_3(PMePh_2)_3$		(228)
19	$(diphos)_2Fe(C_2H_4)$[f]	$HSiR_3$	$R_3SiFeH(diphos)_2$ $(R_3Si)_2Fe(diphos)_2$		(291, 355)
20	$RuCl_2L_3\ (L = PPh_3)$	$HSiR_3$	$R_3SiRuHL_2\ (R = Cl,\ OEt)$ $(R_3Si)_2RuL_2\ (R = Cl)$		(411) (411)
21	$RuH_2L_4\ (L = PPh_3)$ $RuHClL_3\ (L = PPh_3)$ $RuCl_2L_3\ (L = PPh_3)$ }	$HSiR_3$	$R_3SiRuH_3L_3$ $R_3SiRuHL_n\ (n = 2,\ 3)$		(284, 285) (284, 285)
22	$RuH_2L_4\ (L = PR_3)$ $RuHClL_3\ (L = PR_3)$ $RuCl_2L_3\ (L = PR_3)$ }	$HSiR_3$	$R_3SiRuHL_n\ (n = 2,\ 3)$ (not when R = Me, Et)		(226)
23	$Co(CO)_2(Cp)$	$HSiCl_3$	$Cl_3SiCoH(CO)(Cp)$	hv/Δ	(259)
24	$CoH(X_2)L_3\ (L = PPh_3,$ $X = H,\ N)$	$HSiR_3$	$R_3SiCoH_2L_3$		(15, 17)
25	$(diphos)_2CoH$[f]	$HSiCl_3$	$[(diphos)_2CoH_2]^+SiCl_3^-$		(291)
26	$(diphos)_2CoH$[f]	$HSiCl_3,\ Ph_3SiCl,\ SiCl_4$	None; $(diphos)_2CoCl$ and $[(diphos)_2Co(H)Cl]^+Cl^-$ produced		(16)
27	$Rh(CO)_2(Cp)$	$HSiR_3$	$R_3SiRhH(CO)(Cp)$	hv	(346)
28	$RhClL_3$ $(L = PPh_3,\ etc.)$	$HSiR_3$	$R_3SiRhH(Cl)L_2$		(230, 344)
29	$Rh(C_8H_{12})(Cp)$	$HSiCl_3$	$(Cl_3Si)_2Rh(C_8H_{14})(Cp)$		(195)
	$[Rh(C_8H_{12})Cl]_2$	$HSiCl_3$	$(Cl_3Si)_2Rh(C_8H_{12})Cl$		(195)
	$RhClL_3\ (L = PPh_3)$	$HMeSi(OSiMe_3)_2$	$Me(Me_3SiO)_2SiRhH(Cl)L_2$		(195)
30	$IrH(CO)L_3\ (L = PPh_3)$	$HSiR_3$	$R_3SiIrH_2(CO)L_2$		(179, 221, 223)
31		$DSiR_3$	$R_3SiIrHD(CO)L_2$		(180)

(table continues)

TABLE V (Continued)

Entry	Metal derivative	Silicon compound	Treatment	Silicon–metal product	Ref.
32	$IrH(CO)L_3$	$(HMe_2Si)_2X$ (X = O, CH_2)		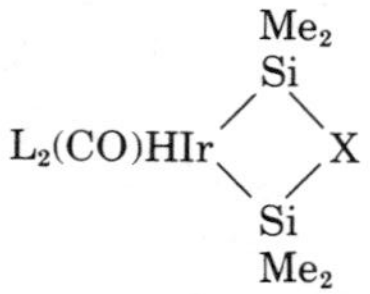 $L_2(CO)HIr$ with $\begin{smallmatrix}Me_2\\Si\end{smallmatrix}\!\!-\!\!X\!\!-\!\!\begin{smallmatrix}Si\\Me_2\end{smallmatrix}$ bridge	(139, 207)
33	$IrCl(CO)L_2$ (L = PPh_3)	$HSiR_3$		$R_3SiIrH(Cl)(CO)L_2$[g]	(70, 229)
34	$IrX(CO)L_2$ (L = PEt_3, PPh_3)	H_3SiY (Y = H, Cl, Br, I, Me, SiH_3)		$YH_2SiIrH(X)(CO)L_2$ (exchange between X and Y possible)	(165, 166)
35	$IrCl(CO)L_2$ (L = PPh_3)	Si_2Cl_6		$(Cl_3Si)_2IrCl(CO)L_2$	(197)
36	$IrCl(CO)L_2$ (L = PEt_3)	$Hg(SiMe_3)_2$		$(Me_3Si)_2Ir(HgSiMe_3)(CO)L_2$	(241)
37	$[Ir(diphos)_2]^+BPh_4^-$[f]	$HSiR_3$		$[R_3SiIrH(diphos)_2]^+BPh_4^-$	(222)
38	NiL_4 (L = PPh_3)	$HSiPh_3$		? $(Ph_3Si)_2NiL_2$[h]	(291)
		$HSiCl_3$		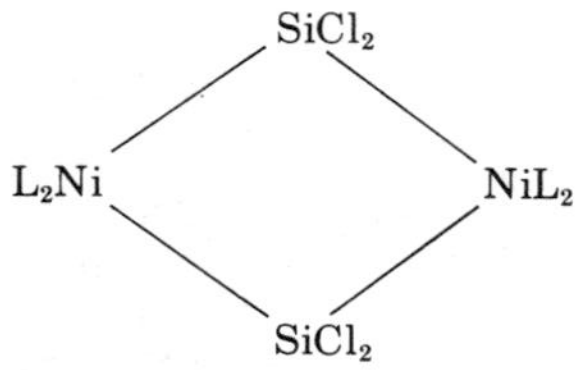 $L_2Ni\overset{SiCl_2}{\underset{SiCl_2}{\diamondsuit}}NiL_2$	(291)
39	$Ni(CO)_4$	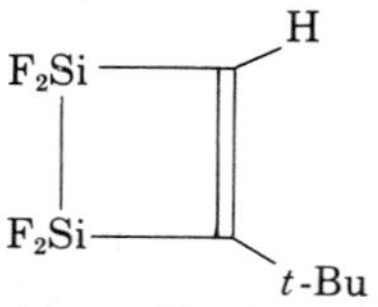 $F_2Si\overset{H}{\underset{t\text{-Bu}}{>}\!\!=\!\!<}$ (ring with two F_2Si)		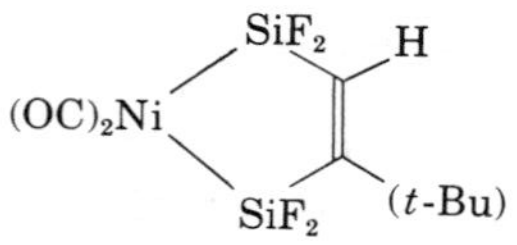 $(OC)_2Ni\overset{SiF_2}{\underset{SiF_2}{\diamondsuit}}\overset{H}{\underset{(t\text{-Bu})}{=}}$	(100, 297)

24

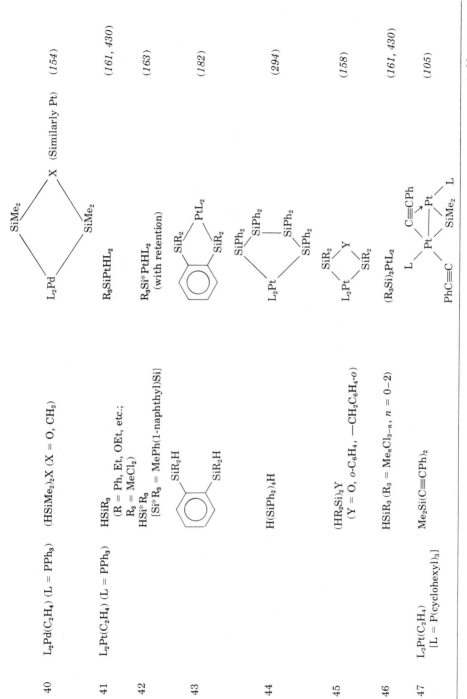

40	$L_2Pd(C_2H_4)$ ($L = PPh_3$)	$(HSiMe_2)_2X$ ($X = O$, CH_2)	(154)
41	$L_2Pt(C_2H_4)$ ($L = PPh_3$)	$HSiR_3$ ($R = Ph$, Et, OEt, etc.; $R_3 = MeCl_2$)	(161, 430)
42		HSi^*R_3 [$Si^*R_3 = MePh(1\text{-naphthyl})Si$]	(163)
43			(182)
44		$H(SiPh_2)_4H$	(294)
45		$(HR_2Si)_2Y$ ($Y = O$, $o\text{-}C_6H_4$, $-CH_2C_6H_{4\text{-}o}$)	(158)
46		$HSiR_3$ ($R_3 = Me_nCl_{3-n}$, $n = 0\text{-}2$)	(161, 430)
47	$L_2Pt(C_2H_4)$ [$L = P(\text{cyclohexyl})_3$]	$Me_2Si(C\equiv CPh)_2$	(105)

(table continues)

25

TABLE V (Continued)

Entry	Metal derivative	Silicon compound	Treatment	Silicon–metal product	Ref.
48	LPt(C₂H₄)₂ [L = P(cyclohexyl)₃]	HSiR₃		L(R₃Si)Pt——Pt(SiR₃)L (with bridging H groups)	(205, 206, 415)
49	PtL₄ (L = PPh₃)	Si₂Cl₆		(Cl₃Si)₂PtL₂	(197, 390)
50	PtL₄ (L = PPh₃)	HSiR₃ (R₃ = Cl₃, MeCl₂)		R₃SiPtHL₂ (R₃Si)₂PtL₂	(99, 183)
			?	L₂Pt ⟨Si(R₂)·Si(R₂)⟩ PtL₂ⁱ	
51	PtL₄ (L = PPh₂Me)	HSiCl₃		(Cl₃Si)₂PtL₂	(161)
52	Pt(diphos)₂ᶠ	R₃SiX		None [gives (diphos)PtX₂ and R₆Si₂]	(113)
53	L₂Pt(CO)₃ [L₂ = (PPhMe₂)₂ or diphosᶠ]	(HPh₂SiCH₂—)₂		L₂Pt ⟨SiPh₂—CH₂—SiPh₂⟩	(159)

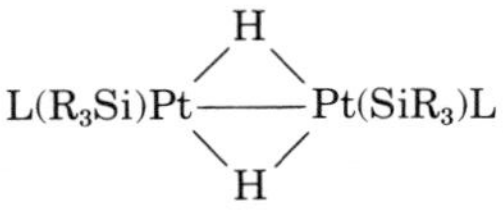

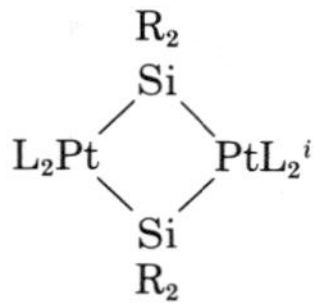

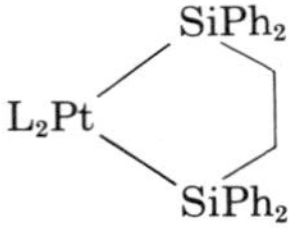

$$L(R_3Si)Pt \overset{H}{\underset{H}{\rule{1.5cm}{0.4pt}}} Pt(SiR_3)L$$

54	Pt₃L₆ (L = t-BuNC)	HSiR₃	$L(R_3Si)Pt\begin{smallmatrix}CH=N(t\text{-}Bu)\\ \diagdown\\ \diagup\\ t\text{-}BuN=CH\end{smallmatrix}Pt(SiR_3)L$	(104)
55	L₂PtI₂ (L = PEt₃)	H₃SiI	IH₂SiPtI₂(H)L₂	(56, 58)
56	L₂PtI₂ (L = PEt₃)	H₃SiC≡CX (X = H, CF₃)	XC≡CSiH₂PtI₂(H)L₂ᶜ	(10)

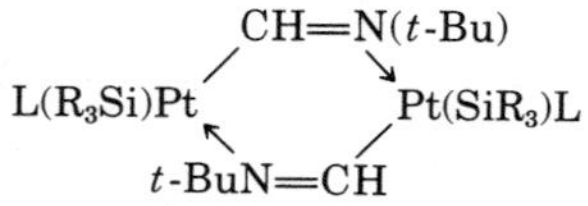

[a] Impure.
[b] Expected to be paramagnetic.
[c] Not isolated.
[d] Possible Re–H–Re bridge.
[e] Rearranges to $Et_2HN{\rightarrow}Me_2SiFe(CO)_4$ (see Section V,C).
[f] diphos = Ph₂PCH₂CH₂PPh₂.
[g] With extra HSiR₃, loses R₃SiCl by reductive elimination, then adds more HSiR₃ to give R₃SiIrH₂(CO)L₂.
[h] Composition uncertain: as shown, this compound is expected to be paramagnetic.
[i] Uncertain compound.

27

as products in entry 2. These are expected to be paramagnetic, since dimerization is highly unlikely; together with the less-certain examples in entry 38, they represent the first examples of such behavior in molecular silicon–transition-metal compounds.

Attempted oxidative addition reactions involving cleavage of silicon–halogen bonds have proved unsuccessful (see entries 26 and 52), and when Si–H and Si–halogen bonds are present in the same molecule, it is the former that cleaves, even in the case of iodine (entries 34 and 55). For example,

$$trans\text{-}(Et_3P)_2Ir(I)(CO)Cl + H_3SiCl \longrightarrow \underset{\substack{\displaystyle ClH_2Si \diagup \;\; | \;\; \diagdown PEt_3 \\ CO \\ \textbf{(IV)}}}{\overset{\substack{Cl \\ Et_3P \diagdown \; | \;\; \diagup H}}{Ir(III)}} \qquad (37)$$

When disilane, Si_2H_6, is used in this reaction, the Si–H linkage is broken in preference to the Si–Si bond (165). Silicon–carbon bonds are not usually susceptible to cleavage, although Si–alkynyl bonds provide an exception (entry 47). If both Si–H and Si–alkynyl bonds are present, however, it is the former that cleaves (entry 56). The special case of metal insertion into Si–C bonds of strained heterocycles is dealt with separately in Section II,G,2, while an isolated but interesting example of Si–Hg bond cleavage appears in entry 36.

There is abundant evidence that, in any particular system, the stability of adducts formed by oxidative addition using a series of silanes $HSiX_3$ increases as X becomes more electronegative; a typical order is

$$HSi(aryl)_3 \lesssim HSi(alkyl)_3 < HSi(OR)_3 < HSiCl_3 < HSiF_3$$

Thus, adducts of triarylsilanes frequently dissociate readily (i.e., undergo reductive elimination) or may exist only in solution in the presence of an excess of triarylsilane. Sometimes the only effect of the added silane may be to reduce metal ligands, e.g. (411),

$$(Ph_3P)_3RuCl_2 + Et_3SiH \rightarrow (Ph_3P)_3RuHCl + Et_3SiCl \qquad (38)$$

while in other cases, this reduction is followed by addition of further $HSiX_3$ to the hydrido complex so formed. For this particular system, there have been conflicting reports of the products from reactions between various Ru(II) species and silanes (entries 20–22). In the most recent study (226), it is persuasively argued that the normal product is the Ru(IV) species $R_3SiRuH_3(PR_3)_n$ ($n = 2,3$), in which the metal is six- or seven-coordinate. Another controversial reaction was that between a cobalt(I) hydride and trichlorosilane (entries 25 and 26). It was first supposed that the product was a novel salt of the anion $SiCl_3^-$ [Eq. (39a)], but it was later shown that not only $HSiCl_3$ but also chlorosilanes with no Si–H bonds underwent the same reaction, giving Co(I)–Cl compounds [Eq. (39b)].

$$(diphos)_2CoH \xrightarrow{HSiCl_3} [(diphos)_2CoH_2]^+SiCl_3^- \tag{39a}$$

$$(diphos)_2CoH \xrightarrow{HSiCl_3,\ R_3SiCl,\ SiCl_4} (diphos)_2CoCl + [(diphos)_2CoHCl]^+Cl^- \tag{39b}$$

The high formal oxidation states of metals in some of these adducts is noteworthy, e.g., Fe(IV) (entries 17 and 18), Ru(IV) (entries 21 and 22), and Pt(IV) (entries 55 and 56). Such adducts are important because they provide definite examples of species often postulated as intermediates in oxidative addition–reductive elimination processes (compare Section II,G,1) and in homogeneous catalysis (*134, 220a, 410a*). In the case of germanium, a tris(germyl) adduct of Pt(IV) has been described (*57*), but no more than two silyl groups per metal atom are known to result from oxidative addition.

In a few cases, the stereochemistry of adducts has been elucidated from their vibrational and NMR spectra. The arrangement in compound (**IV**) has already been displayed, the product of the reaction in entry 30 is known to be (**V**) (*223*) [and this is confirmed by an X-ray study of the GeMe$_3$ analog (*202*)], while the product in entry 33 is (**VI**) (*229*). The adduct $Ph_3SiMnH(CO)_2(Cp)$ (entry 4) adopts a piano-stool

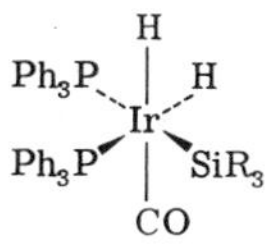

(**V**) 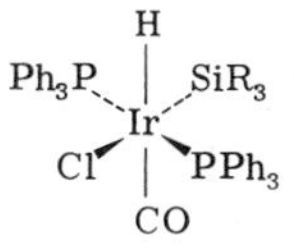   (**VI**) 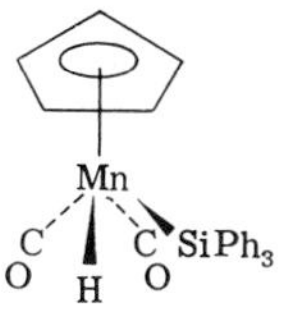   (**VII**)

arrangement with cis carbonyl groups (**VII**) (*225*), while the rhenium analog (entry 8) has both cis and trans isomers (*404*); reaction of $HSiR_3$ with $L_2Pt(olefin)$ complexes leads to formation of cis derivatives (entries 41–46).

Some details of the crystal structure of adducts referred to in entries 8, 10, 17, 28, 47, 48, and 54 will be found in Section IV,A (*404, 403, 313,*

327, 105, 205, 104 respectively), while various cluster compounds formed by reactions analogous to oxidative addition processes are discussed in Section V,B.

Besides these determinations of static structures, further studies of the kinetics of oxidative addition to Ir(I) complexes have been made. Beginning with $Ir(CO)H(PPh_3)_3$ (**VIII**), the first step is seen as dissociation to the four-coordinate planar intermediate (**IX**). This then deforms and undergoes concerted cis addition to yield the adduct (**V**) (*179, 224*). In a comparison of group IV hydrides, Ph_3MH, the rates increased in the order Si<Ge<Sn; this was attributed not so much to variations in M–H bond energy but to the effects of differing polarizability and solvation of M–Ir products on the activation energy for addition.

$$
\underset{\textbf{(VIII)}}{
\begin{array}{c}
H \\
| \quad \; \nearrow PPh_3 \\
Ph_3P-Ir \\
| \quad \searrow PPh_3 \\
CO
\end{array}}
\underset{+PPh_3}{\overset{-PPh_3}{\rightleftharpoons}}
\underset{\textbf{(IX)}}{
\begin{array}{c}
H \\
| \\
P-Ir-P \\
| \\
CO
\end{array}}
\xrightarrow{R_3SiH'}
\underset{\textbf{(V)}}{
\begin{array}{c}
H \\
P_{\diagdown} | _{\diagup} H' \\
Ir \\
P^{\diagup} | ^{\diagdown} SiR_3 \\
CO
\end{array}}
\qquad (40)
$$

This picture of the addition process receives support from a study of the same reaction using Ph_3SiD. The same product (**V**, H' = D) is formed initially, although H–D exchange rapidly scrambles the hydrogen positions (*180*).

G. OTHER METHODS

1. Elimination of Small Molecules

Table VI lists various ways in which the elimination of small molecules has been used to produce silicon–transition-metal bonds; most can be pictured as proceeding via consecutive processes of oxidative addition and reductive elimination. Dihydrogen may result from reaction between compounds with M–H and Si–H bonds (entries 1–10).

$$L_nMH + HSiR_3 \rightarrow R_3SiML_n + H_2 \qquad (41)$$

This is a good route to compounds that are functionally substituted at silicon (entries 6–10).

Alkanes can also be formed by the interaction of M–alkyl and Si–H species.

$$L_nMR + HSiR'_3 \rightarrow R'_3SiML_n + RH \qquad (42)$$

This method was unsuccessful when R was the bulky CH_2SiMe_3 group (entry 11). A six-coordinate M(IV) adduct is probable intermediate in entries 12–15, and elimination of groups other than R + H from this adduct provides a reasonable explanation for the volatile products observed in entry 12 and the formation of cis-$R_3SiPt(H)L_2$ in entry 15.

Elimination of hydrogen halide occurs both from M–H and Si–halogen (entry 18) and from M–halogen and Si–H systems (entries 17 and 19); a Lewis base must normally be added to drive the reaction from left to right (entries 17 and 18).

$$L_nM\text{—}X + H\text{—}SiR_3$$
$$\xrightarrow{\text{NEt}_3} R_3SiMl_n + NHEt_3X$$
$$L_nM\text{—}H + X\text{—}SiR_3 \tag{43}$$

By contrast, elimination of an amine to yield Si–M bonds has been achieved only between M–H and Si–N compounds (entries 20 and 21).

Elimination of alkenes has been already dealt with in Section II,F (oxidative addition–elimination), while other reactions involving elimination of dihydrogen may be found in that section and in Section II,E; this latter section includes many cases in which metal hydride species are believed to act as intermediates.

2. Addition to Si–C Bonds of Strained Heterocycles

These reactions, shown in Table VII, are chiefly a special type of oxidative addition process. Most important are those involving insertion of "$Fe(CO)_4$"; this is thought to occur via electrophilic Si–C cleavage in an essentially concerted process. Experiments with substituted silacyclobutanes have shown that the process is both regio- and stereospecific (135). Other metals seem reluctant to undergo this reaction: the manganese analog in entry 23 is thermally very unstable.

While silacyclopentanes and sitacyclohexanes do not appear to react in this way, it seems likely that the very strained three-membered ring in silacyclopropanes will.

3. Formation of Adducts with M→Si Bonds

In principle, electron donation could take place from filled orbitals of a transition metal into empty silicon 3d orbitals to give adducts. Fluorosilanes in particular are known to act as Lewis acids (23). The somewhat confused experimental situation is summarized in Table VIII.

Initial experiments (entry 29) suggested that a 1:1 complex was

TABLE VI

ELIMINATION OF SMALL MOLECULES

Entry	Metal compound	Silicon compound	Silicon–metal product	Ref.
(a) Elimination of H_2				
1	$(Cp)_2WH_2$	$HSiCl_3$	$Cl_3SiWH(Cp)_2$	(88)
2	$FeH_2(diphos)_2$ [a]	$HSiX_3$	$X_3SiFeH(diphos)_2$ $(X_3Si)_2Fe(diphos)_2$	(291, 355)
3	$FeHCl(diphos)_2$ [a]	$HSiX_3$	$X_3SiFeCl(diphos)_2$	(291)
4	CoH_3L_3 (L = PPh_3)	$HSiX_3$	$X_3SiCoH_2L_3$	(15, 17)
5	$RhHL_3$ (L = PPh_3)	$HSiCl_3$	$(Cl_3Si)_2RhL_2$	(195)
6	$PtH(X)L_2$ (L = PEt_3)	$H_{4-n}SiX'_n$ $(X, X' = Cl, Br, I; n = 0-3)$	$trans\text{-}X'_nH_{3-n}SiPtXL_2$	(58)
7	$PtH(X)L_2$ (L = PEt_3)	$H_3SiC{\equiv}CX$ (X = H, CF_3)	$trans\text{-}(XC{\equiv}CSiH_2)Pt(X)L_2$	(10)
8	$PtH(X)L_2$ (L = PR_3)	$(H_3Si)_2Y$ $(Y = O, NH, S, Se, Te)$	$trans\text{-}(SiH_3YSiH_2)Pt(X)L_2 + Y[H_2SiPt(X)L_2]_2$	(167, 168)
9	$PtH(X)L_2$ (L = PR_3)	$(H_3Si)_3Z$ (Z = N, P)	$trans\text{-}[(SiH_3)_2ZSiH_2]Pt(X)L_2$	(168)
10	PtH_2L_2 [L = P(cyclohexyl)$_3$]	H_3SiX (X = H, Cl, SiH_3)	$trans\text{-}(XSiH_2)Pt(H)L_2$	(169)
(b) Elimination of alkane				
11	$(Cp)_2V(CH_2SiR_3)$	$HSiEt_3$	Unsuccessful (successful with Ge,Sn)	(372)
12	$(bipy)NiEt_2$ [b]	$HSiX_3$	$(X_3Si)_2Ni(bipy)$ [c]	(274, 276)
13	$NiMe(Cp)L$ (L = PPh_3)	$HSiCl_3$	$Cl_3SiNi(Cp)L$	(276)
14	$PtMe_2L_2$ [$L_2 = (Ph_2P)_2CH_2$]	$HSiMe_3$	$Me_3SiPt(Me)L_2$	(200)
15	$cis\text{-}PtMe_2L_2$ (L = PMe_2Ph)	$HSiR_3$	$cis\text{-}R_3SiPtHL_2$ $cis\text{-}(R_3Si)_2PtL_2$	(162)
		H_2SiPh_2	$cis\text{-}(HPh_2Si)_2PtL_2$	(162)

32

(c) Elimination of hydrogen halide

16	$(Cp)_2TiCl_2$	$HSiCl_3$	Unsuccessful	(88)
17	$Mo(CO)_2(C_7H_7)Cl$	$HSiCl_3$ [d]	$Cl_3SiMo(CO)_2(C_7H_7)$	(255)
18	$HMn(CO)_5$	$Cl_3SiCH_2CH_2PMe_2$ [d]	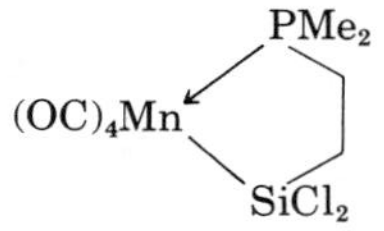	(211)
18a	$Fe_2(CO)_9$	$HSiMe_2Cl$ [d]	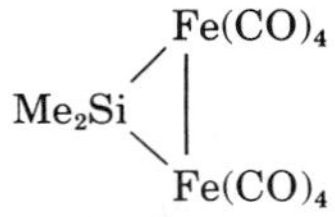	(68)
19	$Ni(Cp)_2 + NiCl_2L_2$ ($L = PPh_3$)	$HSiCl_3$	$Cl_3SiNi(Cp)L$	(199, 203)

(d) Elimination of amine

20	$HMo(CO)_3(Cp)$	Me_2NSiMe_3	$Me_3SiMo(CO)_3(Cp)$ (similarly W)	(86)
21	$HMn(CO)_5$	$Me_2NSiMe_2CH_2CH_2PMe_2$	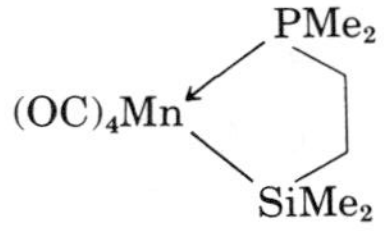	(211)

[a] diphos = $Ph_2PCH_2CH_2PPh_2$.
[b] bipy = 2,2'-bipyridyl.
[c] Evolves C_2H_6, also H_2, $EtSiCl_3$.
[d] NEt_3 added.

TABLE VII

ADDITION TO STRAINED HETEROCYCLIC RINGS

Entry	Metal compound	Silicon compound	Treatment	Silicon–metal product	Ref.
22	$TiCl_4$	$(Me_2SiCH_2)_2$	Δ	None (TiCSi formed)	(329)
23	$Mn(CO)_3(Cp)$	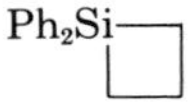 Ph_2Si □	hν	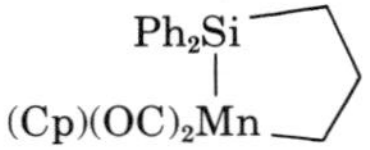 Ph_2Si—$(Cp)(OC)_2Mn$ [a]	(400)
24	$Fe_2(CO)_9$ [b]	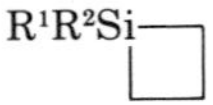 R^1R^2Si □		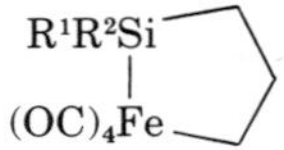 R^1R^2Si $(OC)_4Fe$	(131, 132, 135)
25	$Fe_2(CO)_9$	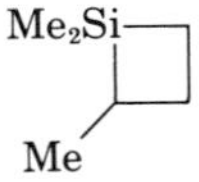 Me_2Si □ Me		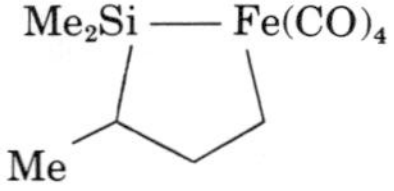 Me_2Si—$Fe(CO)_4$ Me	(135)

[a] Decomposes at −78°C.

[b] $Fe(CO)_5$ or $Fe_3(CO)_{12}$ with irradiation can also be used.

TABLE VIII

ADDUCTS WITH M→Si BONDS

Entry	Metal compound	Silicon compound	Silicon–metal complex	Ref.
26	$Fe_2(CO)_9$ or $Fe(CO)_5$	$(Me_2PCH_2CH_2SiF_3)_n$	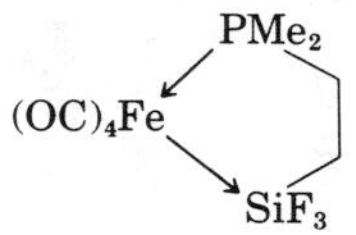 (**X**) [a]	(210)
27	$Ir(CO)H(PPh_3)_2$	SiF_4	None; $[Ir(CO)H_2(PPh_3)_3]^+SiF_5^-$ isolated	(69)
28	$trans$-$PdCl_2(PhCN)_2$	$(Me_2PCH_2CH_2SiF_3)_n$	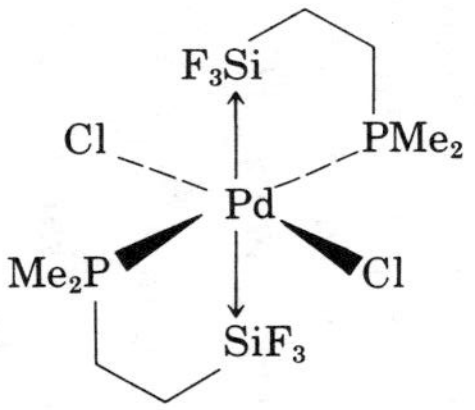 (**XI**)	(212)
29	$Pt(PPh_3)_4$	SiF_4	$(Ph_3P)_2Pt \cdot SiF_4$	(153)
30	$Pt(PPh_3)_3$	SiF_4	None; $[PtH(PPh_3)_3]^+SiF_5^-$ isolated	(69)

[a] Mainly present in solution at 25°C as $(OC)_4Fe \leftarrow PMe_2CH_2CH_2SiF_3$.

formed between the donor fragment $(Ph_3P)_2Pt$ and SiF_4; this complex in turn formed an adduct with 1 mol ammonia. However, later work (entries 27 and 30) indicated that the product in this system and that from $Ir(CO)H(PPh_3)_2$ and SiF_4 were ionic derivatives containing the SiF_5^- anion [1:1 adducts between SiF_5^- and ammonia are already known (108a)]. Evidently, reaction in glass vessels had led to production of hydrogen fluoride.

The interesting ligand molecule $SiF_3CH_2CH_2PMe_2$, which undergoes extensive self-association (209), attaches itself to iron or palladium centers via the phosphorus atom (entries 26 and 28). Further interaction then occurs between the metal atom and the SiF_3 group; this is weak in the former case but stronger in the latter, which has been described as an example of a three-center two-electron bond system. Electron donation is postulated from the filled Pd $4d_{z^2}$ orbital into 3d orbitals of both silicon atoms (212).

4. Rearrangements, Anion Attack, and Miscellaneous Reactions

a. Rearrangements. Table IXA shows three examples of rearrangements leading to silicon–metal bonds:

$$Si—C—M \rightarrow C—Si—M \quad \text{(entries 31 and 32)} \quad (44)$$

$$Si—Si—C—M \rightarrow Si—C—Si—M \quad \text{(entry 33)} \quad (45)$$

While it is tempting to ascribe these processes to an increase in bond energy on going from C–M to Si–M systems, entries 34 and 35 show

TABLE IXA

REARRANGEMENTS

Entry	Reaction	Ref.
31	$Me_2HSiCH_2Fe(CO)_2(Cp) \overset{\Delta}{\rightarrow} Me_3SiFe(CO)_2(Cp)$	(352)
32	$Me_2HSiCH_2Fe(CO)_2(Cp) \underset{PPh_3}{\overset{h\nu}{\longrightarrow}} Me_3SiFe(CO)(PPh_3)(Cp)$	(352)
33	$Me_3SiSiMe_2CH_2Fe(CO)_2(Cp) \overset{h\nu}{\rightarrow} Me_3SiCH_2SiMe_2Fe(CO)_2(Cp)$	(353)
34	$(ClCH_2)Cl_{2-x}Me_xSiFe(CO)_2(Cp) \overset{\Delta}{\rightarrow} Cl_{3-x}Me_xSiCH_2Fe(CO)_2(Cp)$ $(x = 1,2)$	(427, 428)
35	$(ClCH_2)Cl_2SiFe(CO)_2(Cp) \overset{AlCl_3}{\longrightarrow} Cl_3SiCH_2Fe(CO)_2(Cp)$	(427, 428)
35a	$(BrCH_2)Me_2SiFe(CO)_2(Cp) \overset{0°C}{\longrightarrow} BrMe_2SiCH_2Fe(CO)_2(Cp)$	(427, 428)

TABLE IXB

ANIONIC ATTACK

Entry	Metal compound	Silylanion/solvent	Silicon–metal compound	Ref.
36	$Fe(CO)_5$	Ph_3SiLi/THF	$Li[Ph_3SiFe(CO)_4]$	(288)
37	$Ni(CO)_4$	Ph_3SiLi/THF	$Li[Ph_3SiNi(CO)_3]^a$	(288)
38	$Ni(COD)_2$ [b]	Ph_3SiK/THF	$K_3[(Ph_3Si)_3Ni](THF)_3$	(417)
39	$Ni(COD)_2$ [b]	$Ph_3SiNa/glyme$	$Na_2[(Ph_3Si)_2Ni](glyme)_4$	(417)

[a] NMe_4^+ salt also isolated.
[b] Cyclo-octa-1,5-diene.

that under certain circumstances the rearrangement

$$Cl—C—Si—M \rightarrow Cl—Si—C—M \tag{46}$$

can occur, no doubt assisted by the high Si–Cl bond energy (see also entry 35a).

b. Silyl anion attack. To the extent that alkali metal organosilyl derivatives R_3SiM are ionic in nature (24), they can be regarded as sources of R_3Si^- anions, isoelectronic with tertiary phosphines. As Table IXB shows, they do indeed react with transition-metal(0) compounds to yield anionic complexes with silicon–metal bonds. As yet, little is known of the chemical behavior of these interesting species.

c. Miscellaneous reactions. Attempted reaction of the type:

$$\overset{\diagdown}{\diagup}Ti—Cl + Me_3SiSiMe_3 \nrightarrow \overset{\diagdown}{\diagup}Ti—SiMe_3 + Me_3SiCl \tag{47}$$

was unsuccessful (Table IXC, entry 40), but reaction of a hydridodisilane with $Fe_2(CO)_9$ led to extrusion of Me_2Si, which then coordinated to the metal; the remaining silane added oxidatively to the metal center, giving the compound $Me_2Si \rightarrow FeH(CO)_3SiMe_2R$ as product (entry 45). Another attempt to produce a silylene–metal species gave a product for which the dimeric structure (**XII**) was proposed (entry 41). Metal–stabilized silylene derivatives are discussed further in Section V,C.

Formal insertion of $Fe(CO)_x$ into the Si–Si bond of a vinyldisilane leads to the interesting silicon-substituted allyl derivative (**XIV**)*

* This reaction has now been reinterpreted in terms of an η^2-substituted vinyl $Fe(CO)_4$ product (see Appendix).

TABLE IXC

MISCELLANEOUS

Entry	Reaction	Ref.
40	$(Cp)_2TiCl_2 + Me_3SiSiMe_3 \rightarrow$ No Si–Ti products	(273)
41	$W(CO)_6 + SiI_4 \overset{h\nu}{\rightarrow} [(OC)_5WSiI]_2(\mu\text{-}I)_2{}^a$ (**XII**)	(392)
42	$Mn_2(CO)_{10} + Me_3Si(Cp) \rightarrow$ $(\eta^5\text{-}Me_3SiC_5H_4)Mn(CO)_3 + Me_3SiMn(CO)_5{}^b + Mn(CO)_3(Cp)^b$	(1)
43	$fac\text{-}[Ph_3SiC(O)]Re(CO)_3(diphos)^c(\textbf{XIII}) \overset{\Delta}{\rightarrow}$ $mer\text{-}Ph_3SiRe(CO)_3(diphos) + CO^d$	(14)
44	$Fe_2(CO)_9 + CH_2{=}CHSi_2Me_5 \rightarrow$ 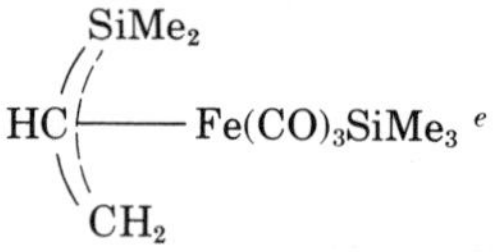 $\begin{array}{c} SiMe_2 \\ \diagup\diagup \\ HC{\longmapsto}Fe(CO)_3SiMe_3{}^e \\ \diagdown\diagdown \\ CH_2 \end{array}$ (**XIV**)	(382, 385)
45	$Fe_2(CO)_9 + HSiMe_2SiMe_2R \rightarrow Me_2Si{\rightarrow}FeH(CO)_3(SiMe_2R)$ $(R = H,Me)$	(384)

^{a} Dissociates in THF solution, giving $I_2SiW(CO)_5 \cdot$ THF; see also Section V,C.
^{b} Minor products.
^{c} diphos = $Ph_2PCH_2CH_2PPh_2$.
^{d} Irreversible process.
^{e} See footnote, p. 37, and Appendix.

shown in entry 44; although air-sensitive, it is stable thermally up to 80°C. Cleavage of a Si–(η^1-C_5H_5) bond and addition across a metal–metal bond is the minor reaction pathway in entry 42.

The great reluctance of silicon–metal compounds to undergo carbonyl insertion reactions, familiar in analogous C–metal chemistry, is discussed in Section III,B. Entry 43 shows the reverse process: a sila-acyl derivative (**XIII**), prepared by the reaction (14) of Eq. (48),

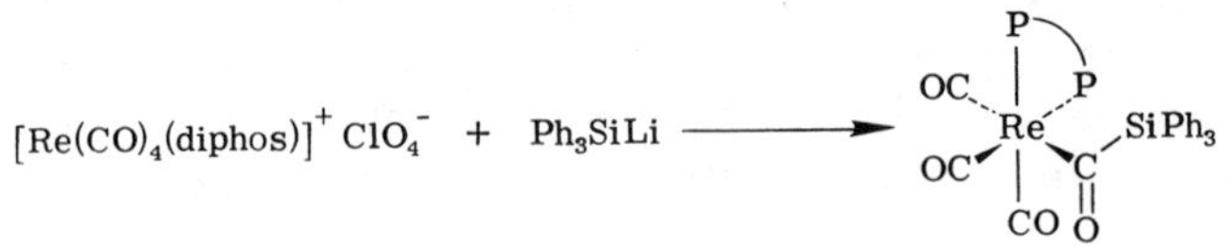

$$[Re(CO)_4(diphos)]^+ ClO_4^- + Ph_3SiLi \longrightarrow \underset{(\textbf{XIII})}{\begin{array}{c} OC{\diagdown}\overset{P{\frown}}{\underset{\underset{CO\ O}{|}}{\underset{|}{\overset{|}{Re}}}}\overset{P}{\diagup}{\diagdown}{\underset{C}{{\overset{SiPh_3}{\diagup}}}} \\ OC{\diagup} \end{array}} \qquad (48)$$

decomposes irreversibly above about 182°C to yield a silicon–rhenium compound.

III. Reactivity of Molecular Silicon–Transition-Metal Compounds

A. GENERAL

A pattern of reactivity for typical molecular silicon–transition-metal compounds can be summarized in terms of Fig. 2.
Here we see the following modes of behavior:

(1) Cleavage of the silicon–metal bond. This includes as important subdivisions: insertion reactions, exchange of one silicon-containing group for another, and exchange of one metal-containing group for another (mode 1′).

(2) Ligand exchange at silicon, while the Si–M bond remains intact.

(3) Formation of adducts by attack of Lewis bases at silicon.

(4) Ligand exchange at the metal, while the Si–M bonds remains intact. Two important types of ligand, hydrogen (mode 4b) and carbonyl (mode 4a) are differentiated from the rest (mode 4c).

(5) Reactions involving the carbonyl group, in which oxygen acts as a nucleophile and/or carbon acts as an electrophile.

(6) Disproportionation reactions, either at a silicon or a metal center, the former being more common.

Examples of all these processes are considered next.

B. CLEAVAGE OF THE SILICON–TRANSITION-METAL BOND

Cleavage reactions, classified according to the nature of the attacking group, are collected together in Table X. Besides these, adduct formation (mode 3) and reductive elimination reactions also involve cleavage of Si–M bonds, but it is convenient to treat these separately in

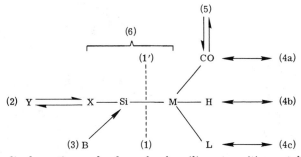

FIG. 2. Idealized reaction modes for molecular silicon–transition-metal compounds (see text).

TABLE X

CLEAVAGE REACTIONS OF SILICON–METAL COMPOUNDS

Entry	Silicon–metal compound/reagent	Success	Products	Ref.
Cleavage by dihydrogen				
1	cis-(Me$_3$Si)$_2$Pt(diphos)^a/H$_2$	✓	Si–H + M–H	(112, 113)
2	cis-(Me$_3$Si)PtH(diphos)^a/H$_2$	✓	Si–H + M–H	(112, 113)
3	cis-(Me$_3$Si(PtCl(diphos)^a/H$_2$	✓	Si–H + M–H	(111)
Cleavage by halogen compounds				
4	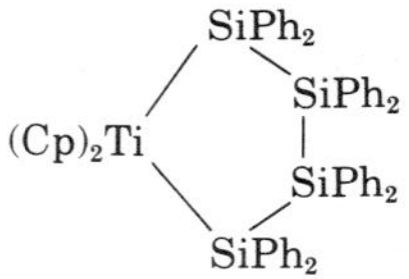 (Cp)$_2$Ti〈SiPh$_2$–SiPh$_2$–SiPh$_2$–SiPh$_2$〉 /CHCl$_3$/$h\nu$	✓	(Cp)$_2$TiCl$_2$, Si$_5$Ph$_{10}$, C$_2$Cl$_4$	(239)
5	(Ph$_3$Si)ZrCl(Cp)$_2$/HX	✓	Ph$_3$SiH + (Cp)$_2$ZrClX	(273)
	(Ph$_3$Si)ZrCl(Cp)$_2$/CDCl$_3$	✓	? + (Cp)$_2$ZrCl$_2$	(273)
6	H$_3$SiCr(CO)$_3$(Cp)/HCl (also Mo, W analogs)	✓	Si–Cl + H–M	(214)
7	R$_3$SiCr(CO)$_5^-$/HCl (also Mo, W analogs)	✓	Si–H + Cl–M	(254)
7a	(CH$_2$=CH)Me$_2$SiW(CO)$_3$(Cp)/HX (X = F, Br)	✓	Si–X + H–W	(308)
8	F$_3$SiMn(CO)$_5$/HBr (also Re analog)	✗	^{b}	(373)
9	Me$_3$SiMn(CO)$_5$/I$_2$	✓	Si–I + I–Mn	(102)
	(also Ge, Sn, Pb analogs)		Kinetic study: Si < Ge ≪ Sn < Pb	
10	Ph$_3$SiMn(CO)$_5$/X$_2$ (X = Cl, Br)	✓	Si–X + X–Mn	(101)
11	Ph$_3$SiMn(CO)$_5$/I$_2$	✗	—	(101)
12	R$_3$Si*MnH(CO)$_2$(Cp')^{c,d}/Cl$_2$	✓	R$_3$Si*H + Cl–Mn	(117, 118)
13	XRR'SiMnH(CO)$_2$(Cp')^c/HCl	✓	Si–H + Cl–Mn	(119)
14	Me$_3$SiFe(CO)$_2$(Cp)/HCl	✓	Si–Cl + H–Fe	(330)
15	Me$_3$SiFe(CO)$_2$(Cp)/Cl$_2$	✓	Si–Cl + Cl–Fe	(67)
16	Me$_3$SiFe(CO)$_2$(Cp)/HCl	✓	Si–Cl + H–Fe^e	(67)
17	Me$_3$SiFe(CO)$_2$(Cp)/ICl	✓	Si–I + I–Fe	(67)
18	Me$_3$SiFe(CO)$_2$(Cp)/CF$_3$I	✓	Si–F + I–Fe	(67)

19	R₃Si*Fe(CO)₂(Cp)^d/Cl₂	✓	Si–Cl + Cl–Fe	(91, 92)
20	R₃Si*Fe(CO)₂(Cp)^d/HCl	✗	—	(91, 92)
21	R₃Si*Fe(CO)₂(Cp)^d/KHF₂	✗	—	(91, 92)
22	(CH₂=CH)Me₂SiFe(CO)₂(Cp)/HX (X = F, Cl, Br, I, CF₃COO, etc.)	✓	Si–H + X–Fe	(308)
23	(CH₂=CH)Ph₂SiFe(CO)₂(Cp)/HCl	✗	—	(123)
24	(CH₂=CH)Ph₂SiFe(CO)₂(Cp)/Cl₂	✓	Si–Cl + Cl–Fe	(123)
25	(CH₂=CH)Ph₂SiFe(CO)₂(Cp)/KHF₂	✗	—	(123)
26	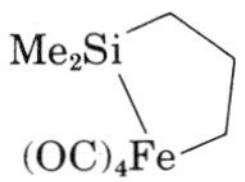 Me₂Si⌐(OC)₄Fe⌐ /HCl	✓	n-PrMe₂SiCl + FeCl₂ [f]	(132)
27	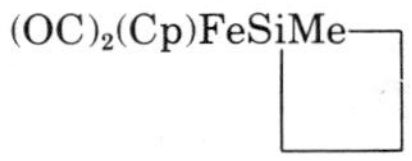 (OC)₂(Cp)FeSiMe⌐⌐ /HCl	✓	n-PrMeClSiFe(CO)₂(Cp) (70%) / Si–Cl + H–Fe (20%)	(136)
28	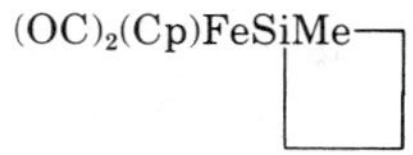 (OC)₂(Cp)FeSiMe⌐⌐ /Cl₂, Br₂	✓	Si–X + X–Fe	(136)
29	MeₓCl₃₋ₓSiCo(CO)₄/HX (x = 0–2)	✓ [g]	Si–X + H–Co	(215)
30	R₃Si*Co(CO)₄/X₂ (X = Cl, Br)	✓	Si–X + X–Co	(115)
31	R₃Si*PtCl(PPhMe₂)₂^d/X₂ (X = Br, I)	✓	Si–Cl + X–Pt	(163, 164)
32	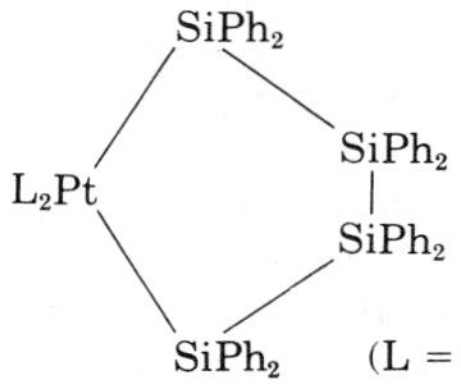 L₂Pt⌐SiPh₂–SiPh₂⌐SiPh₂ SiPh₂ /Br₂ (L = PPh₃)	✓	Si₄Ph₈Br₂ + L₂PtBr₂	(294)
Cleavage by oxygen and sulfur compounds				
33	H₃SiCr(CO)₃(Cp)/H₂O	✓	SiH₄ [h]	(214)
34	F₃SiMn(CO)₅/CO (also Re analog)	✗	[b]	(373)

(table continues)

41

TABLE X (Continued)

Entry	Silicon–metal compound/reagent	Success	Products	Ref.
35	F$_3$SiMn(CO)$_5$/SO$_2$	×	[b]	(373)
36	Me$_3$SiMn(CO)$_5$/MeOH	✓	Si–O + H–Mn	(193)
37	Me$_3$SiMn(CO)$_5$/PhCHO	✓	Si–O + Mn$_2$(CO)$_{10}$	(264)
38	R$_3$Si* MnH(CO)$_2$(Cp')[c,d]/H$_2$O	✓	Si*–O + ?	(117, 118)
39	R$_3$Si* MnH(CO)$_2$(Cp')[c,d]/MeOH	✓	Si*–O + ?	(117, 118)
40	XRR'SiMnH(CO)$_2$(Cp')[c]/H$_2$O	✓	Si–O + ?	(119)
41	XRR'SiMnH(CO)$_2$(Cp')[c]/MeOH	✓	Si–O + ?	(119)
42	Ph$_2$Si[ReH(CO)$_4$]$_2$/"H$_2$SiO$_3$"[i]	✓	? + H$_2$Re(CO)$_8$	(53)
43	Ph$_3$SiRe(CO)$_3$(diphos)[a]/CO	×	cf. Section II,G,4,c	(14)
44	Me$_3$SiFe(CO)$_2$(Cp)/SO$_2$	×	Insertion does occur with Ge, Sn analogs	(65)
45	R$_3$Si* Fe(CO)$_2$(Cp)[d]/H$_2$O	✓	Si–O + ?	(91, 92)
46	R$_3$Si* Fe(CO)$_2$(Cp)[d]/MeOH	×	—	(91, 92)
47	R$_3$Si* Fe(CO)$_2$(Cp)[d]/MeO⁻	×	—	(91, 92)
48	(CH$_2$=CH)Ph$_2$SiFe(CO)$_2$(Cp)/H$_2$O	×	—	(123)
49	(CH$_2$=CH)Ph$_2$SiFe(CO)$_2$(Cp)/EtO⁻	×	—	(123)
50	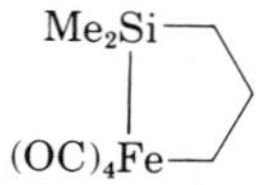 Me$_2$Si⟨⟩(OC)$_4$Fe /H$_2$O	✓	n-PrMe$_2$SiOH + ?	(132)
51	(Me$_3$Si)$_2$Fe(CO)$_4$/PhCHO	✓	Si–O + ?	(264)
52	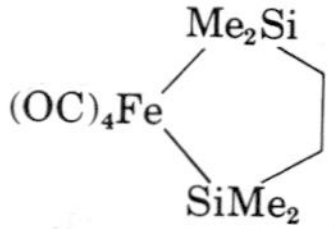 Me$_2$Si⟨⟩(OC)$_4$Fe SiMe$_2$ /PhCHO	✓	Si–O + ?[j]	(264)
53	Cl$_x$Me$_{3-x}$SiCo(CO)$_4$/H$_2$O (x = 1, 2)	✓	Si–O + ?	(215)
54	Cl$_x$Me$_{3-x}$SiCo(CO)$_4$/CO (x = 0–2)	×	Even at 90°C/4000 atm	(215)
55	R$_3$Si* Co(CO)$_4$[d]/H$_2$O	✓	Si*–O + ?	(115)
56	R$_3$Si* PtCl(PPhMe$_2$)$_2$[d]/PhSH	✓	Si*–H + S–Pt	(163, 164)
57	R$_3$Si* PtH(PPh$_3$)$_2$[d]/PhSH	✓	Si*–H + S–Pt	(163, 164)
58	R$_3$Si* PtH(PPh$_3$)$_2$[d]/PhCOCl	✓	Si*–H + Pt(Cl)OCPh	(163, 164)

Cleavage by nitrogen and phosphorus compounds

59	$H_3SiMo(CO)_3(Cp)/NHMe_2/70°C^k$	✓	$H_3SiNMe_2 + H–Mo$	(214)
60	$H_3SiM(CO)_3(Cp)/PF_5$ (M = Cr, Mo, W)	✓	$HSiF_3, PF_3, PF_3O$, etc.	(214)
61	$Me_5Si_2W(CO)_3(Cp)/Me_3P{=}CH_2$ (also Cr, Mo analogs)	✓	$Me_3P{=}CH(Si_2Me_5)^{l,m} + Me_4P^+W(CO)_3(Cp)^-$	(305)
62	$Ph_3SiMn(CO)_5/P(OR)_3/80°C$	✓	$(Ph_3SiO)_3PO + RCOMn(CO)_3[P(OR)_3]_2$	(376)
63	$R_3Si^*MnH(CO)_2(Cp')^{c,d}/PR_3$	✓	$R_3Si^*\ H + Mn–PR_3$	(117, 118)
64	$R_3Si^*Fe(CO)_2(Cp)/P(OPh)_3/h\nu$	✓	$R_3Si^*\ H + Fe–(o{-}C_6H_4O)–P$ (see text)	(93)
65	$X_3SiFe(CO)_2(Cp)/Me_3P{=}CH_2$	✓	$Me_3P{=}CH(SiX_3)^{m,n} + Me_4P^+Fe(CO)_2(Cp)^-$	(301, 303)
66	$Me_2Si{\rightarrow}FeH(CO)_3(SiMe_2R)/PPh_3$	✓	$Fe(CO)_3(PPh_3)_2 + ?$	(384)
66a	$(Cl_3Si)_2Fe(CO)_4/PPh_3$	✓	$Si_2Cl_6 + Fe(CO)_n(PPh_3)_{5-n}$ (n = 3, 4)	(355)
67	$Et_3SiCo(CO)_4/PPh_3$	✓	$Si_2Et_6 + [Co(CO)_3PPh_3]_2$	(227)

Cleavage by silicon, germanium, and tin compounds

68	$Me_3SiMo(CO)_3(Cp)/(Me_3M)OSiMe_3$ (Similarly W; M = Ge, Sn)	✓	$Me_3MMo(CO)_3(Cp) + (Me_3Si)_2O$	(306, 310)
69	$F_3SiFe(CO)_2(Cp)/Me_3P{=}C(SiMe_3)_2$	✓	$Me_3SiFe(CO)_2(Cp) + Me_3P{=}C(SiF_3)(SiMe_3)$	(303)
70	$R_3SiRuH_3L_3/HSiR_3'$ (L = PR_3)	✓	$R'_3SiRuH_3L_3 + HSiR_3$	(226)
71	$R_3SiCo(CO)_4/HSiR_3'$	✓	$R'_3SiCo(CO)_4 + HSiR_3$	(95)
72	$Me_3SiCo(CO)_4/HMMe_3$ (M = Ge, Sn)	✓	$Me_3MCo(CO)_4 + HSiMe_3$	(73)
73	$Me_3SiCo(CO)_4/Me_3GeBr$	✓	$Me_3GeCo(CO)_4 + Me_3SiBr$	(73)
74	$Me_3SiIrH_2(CO)(PPh_3)_2/HMMe_3$ (M = Ge, Sn)	✓	$Me_3MIrH_2(CO)(PPh_3)_2 + HSiMe_3$	(198)
75	$Cl_3SiNi(Cp)(PPh_3)/HGeCl_3$	✓	$Cl_3GeNi(Cp)(PPh_3) + HSiCl_3$	(199)
76	$R_3Si^*PtCl(PPhMe_2)_2{}^d/HSiEt_3$	✓	$Pt–H + HSi^*\ R_3$	(112, 164)
77	$trans\text{-}ClH_2SiPtCl(PEt_3)_2/H_3SiGeCl$	✓	$trans\text{-}ClH_2GePtCl(PEt_3)_2 + H_3SiCl$	(58)
78	$Me_3SiPtX(diphos)^u/HMMe_3$ (M = Ge, Sn)	✓	$Me_3MPtX(diphos) + HSiMe_3$	(112, 113)

Cleavage by hydridic reducing agents

79	$R_3Si^*MnH(CO)_2(Cp')^{c,d}/LiAlH_4$	✓	$HSi^*\ R_3 + H–Mn$	(117, 118)
80	$R_3Si^*Fe(CO)_2(Cp)^d/LiAlH_4$	✓	$HSi^*\ R_3 + ?$	(91, 92)
81	$R_3Si^*Fe(CO)_2(Cp)^d/NaBH_4$	✗	—	(91, 92)
82	$(CH_2{=}CH)Ph_2SiFe(CO)_2(Cp)/NaBH_4$	✗	—	(123)
83	$R_3Si^*Co(CO)_4{}^d/LiAlH_4$	✓	$HSi^*\ R_3 + ?$	(114, 115)
84	$Ph_3SiCo(CO)_4/LiAlH_4$	✓	$HSiPh_3 + ?$	(114, 115)

(table continues)

43

TABLE X (continued)

Entry	Silicon–metal compound/reagent	Success	Products	Ref.
85	$R_3Si^*PtCl(PPhMe_2)_2{}^d/LiAlH_4$	✓	$HSi^*R_3 + Pt + \cdots$	(163, 164)
86	$R_3Si^*PtH(PPh_3)_2{}^d/LiAlH_4$	✓	$HSi^*R_3 + Pt + \cdots$	(163, 164)
	Cleavage by organometallic compounds			
87	$R_3Si^*MnH(CO)_2(Cp')^{c,d}/LiMe$	✓	$HSi^*R_3 + \cdots^o$	(117, 118)
88	$R_3Si^*Co(CO)_4{}^d/LiR$ (excess) (similarly Ph_3Si derivative)	✓	$R_3Si^*Li + ?$	
89	$Ph_3SiCo(CO)_4/RMgBr$	✓	"$Ph_3SiMgBr$" + ?	(114, 115, 120)
	Cleavage by mercury compounds			
90	$R_3SiFe(CO)_2(Cp)/HgBr_2$ (also Ge, Sn analogs)	✓	$R_3SiBr + \cdots$ Kinetic study: Si < Ge ≪ Sn	(102, 103)
91	$Me_nCl_{3-n}SiCoCO)_4/HgX_2$ (n = 1, 2)	✓	$Me_nCl_{3-n}SiX + \cdots$	(215)
92	$R_3Si^*Co(CO)_4{}^d/Hg(CH_3COO)_2$	✓	$R_3Si^*(CH_3COO) + \cdots$	(115)
	Cleavage by alkenes, alkynes, and nitriles			
93	$Me_3SiMn(CO)_5/C_2F_4{}^v/h\nu$ (also Ge, Sn analogs)	✓	$Me_3SiCF_2CF_2Mn(CO)_5$	(106)
94	$Me_3SiMn(CO)_5/CHF{=}CF_2{}^v/h\nu$	✓	cis-$CHF{=}CFMn(CO)_5 + Me_3SiF$	(106)
95	$F_3SiMn(CO)_5/C_2F_4$	×	No clear evidence for insertion (cf. Ref. 397)	(373)
96	$Me_3SiFe(CO)_2(Cp)/CF_3C{\equiv}CR/h\nu$ (also Ge, Sn analogs) (R = H, CF$_3$)	✓	$Me_3SiC(R){=}C(CF_3)Fe(CO)_2(Cp)$	(66)
97	$Me_3SiFe(CO)_2(Cp)/CF_2CF_2CF{=}CF/h\nu$	×	—	(66)
98	cis-$Cl_3SiFeH(CO)_4/C_2F_4$	×	$Cl_3SiFe(CO)_4SiCl_3 + ?$	(259)
99	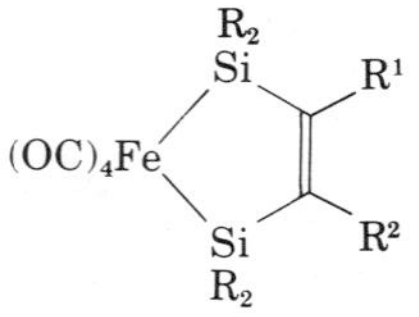 (structure: $(OC)_4Fe$ ring with R_2Si–SiR_2, R^1, R^2) /$PhC{\equiv}CPh$	✓	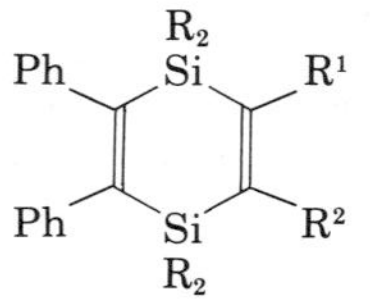 (structure: Ph-substituted ring with R_2Si, R^1, R^2) + ?	(386)

44

	Reactants		Products	Ref.
100	$(OC)_4Fe$ $\overset{Ph_2}{Si}$...$\overset{Ph_2}{Si}$ $Fe(CO)_4$ /$RC{\equiv}CR$	$\checkmark$	$(OC)_4Fe$ $\overset{Ph_2}{Si}$...$\overset{Si}{Ph_2}$ $\overset{R}{\underset{R}{\Vert}}$ $+\cdots q$	(124)
101	$(OC)_4Fe$ $\overset{Ph_2}{Si}$...$\overset{Si}{Ph_2}$ $\overset{R}{\underset{R}{\Vert}}$ /$R'CH_2CN/h\nu$	$\checkmark$	$R'CH{=}CHN$ $\overset{Ph_2}{Si}$...$\overset{Si}{Ph_2}$ $\overset{R}{\underset{R}{\Vert}}$ $+\cdots$	(124)
102	$[Me_3SiRu(CO)_4]_2/C_7H_8$	$\checkmark$	$(Me_3Si)Ru_2(CO)_5(C_7H_6SiMe_3)^r$	(74, 75)
103	$(Me_3Si)_2Ru(CO)_4/C_7H_8$	$\checkmark$	$Me_3SiRu(CO)_2(C_7H_8SiMe_3)$	(75, 245)
104	$(Me_3Si)_2Ru(CO)_4/C_8H_{10}$	$\checkmark$	$Me_3SiRu(CO)_2(\text{tetrahydropentalenyl})^s$	(412)
105	$(Me_3Si)_2Ru(CO)_4/C_8H_{14}$	$\checkmark$	s	(412)
106	$(Me_3Si)_2Ru(CO)_4/C_{12}H_{18}$	$\checkmark$	$Me_3SiRu(CO)_2(\text{tetrahydropentalenyl})^s$	(281)
107	$[Me_3SiRu(CO)_4]_2/C_8H_{12}$	$\checkmark$	$Me_3SiRu(CO)_2(\text{tetrahydropentalenyl})^s$	(281)
108	$[Me_3SiRu(CO)_4]_2/C_8H_{10-n}(SiMe_3)_n$ $(n = 2, 3)$	$\checkmark$	$Ru_3(CO)_8[C_8H_{6-m}(SiMe_3)_m]^{t,u}$ $(m = 2, 3)$	(244, 247, 279)
109	$(OC)_4Ru$ $\overset{SiMe_2}{\underset{SiMe_2}{\big\langle}}$ /C_8H_8	$\checkmark$	$Ru_2(CO)_5(Me_2SiCH_2CH_2SiMe_2C_8H_8)$	(170, 171)
110	$(OC)_4Ru$ $\overset{SiMe_2}{\underset{SiMe_2}{\big\langle}}$ /C_7H_8	$\checkmark$	$Ru(CO)_2(Me_2SiCH_2CH_2CH_2SiMe_2C_7H_8)$ (see text)	(170, 171)
111	$(Me_3Si)_2Ru(CO)_4/C_5H_6$	$\checkmark$	$Me_3SiRu(CO)_2(Cp)$	(409)
112	$(Me_3Si)_2Ru(CO)_4/C_7H_{10}$	$\checkmark$	$Me_3SiRu(CO)_2(C_7H_9)$	(409)
113	$[Me_3SiRu(CO)_4]_2/C_8H_8$	$\checkmark$	$Ru_3(CO)_8(C_8H_6)^t$	(246, 279)

(table continues)

TABLE X (Continued)

Entry	Silicon–metal compound/reagent	Success	Products	Ref.
114	$(Me_3Si)_2Ru(CO)_4/C_8H_8$	✓	$Me_3SiRu(CO)_2(C_8H_8SiMe_3)$ + $(Me_3Si)_2Ru_2(CO)_4(C_8H_6)^f$	(217)
115	$(X_3Si)_2Ni(bipy)^v/PhC{\equiv}CPh$ ($X_3 = Cl_3$, $MeCl_2$)	✓	$(bipy)Ni[Ph(SiX_3)C{=}CPh(SiX_3)]$(**XVII**)	(275, 277)
116	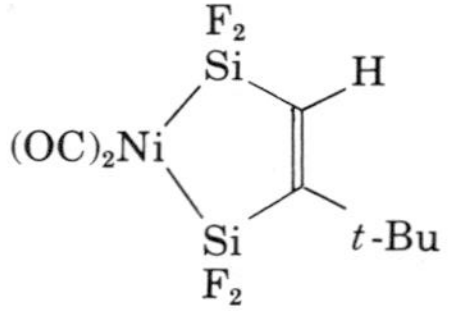 (nickelacycle: $(OC)_2Ni$, $\overset{F_2}{Si}$, H, t-Bu) / $t\text{-}BuC{\equiv}CH$	✓	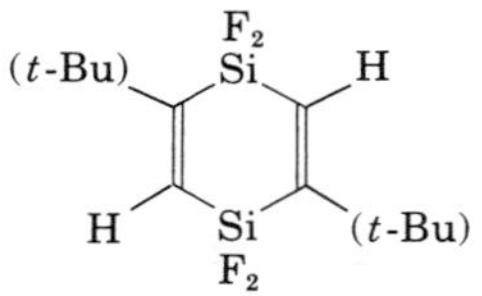 w (ring: $(t\text{-}Bu)$, $\overset{F_2}{Si}$, H, $\overset{F_2}{Si}$, (t-Bu), H)	(100, 297)
117	$Ph_2MeSiPtHL_2/PhC{\equiv}CR$ ($L = PPh_3$; $R = H$, Ph)	✓	$Ph_2MeSiH + L_2Pt(PhC{\equiv}CR)$	(161)
118	$R_3Si^*PtHL_2{}^d/PhC{\equiv}CH$ ($L = PPh_3$)	✓	$R_3Si^*H + L_2Pt(PhC{\equiv}CH)$	(163, 164)
119	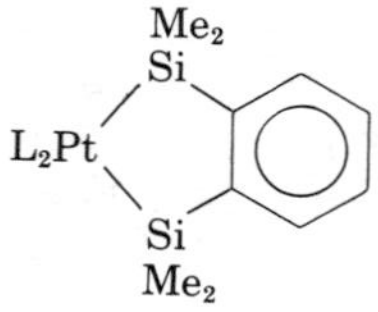 (benzene ring with $\overset{Me_2}{Si}$–L_2Pt–$\overset{Me_2}{Si}$) / $PhC{\equiv}CH$ ($L = PPh_3$)	✓	$L_2Pt(PhC{\equiv}CH) + ?$	(158)
120	$cis\text{-}(HPh_2Si)_2PtL_2/C_2H_2/\Delta$ ($L = PPhMe_2$)	✗	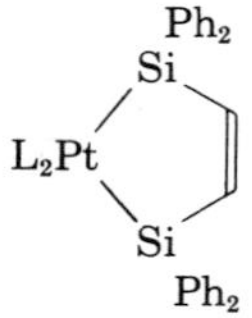 (ring: $\overset{Ph_2}{Si}$, L_2Pt, $\overset{Ph_2}{Si}$) (cf. entry 115)	(159)
121	$(Cl_3Si)_2PtL_2/PPh_3/PhC{\equiv}CPh$ ($L = PPh_3$)	✗	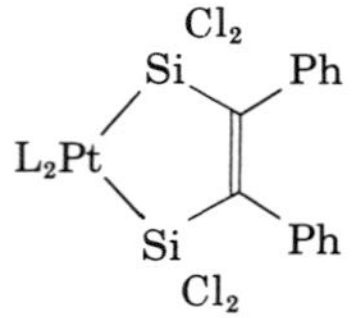 (ring: $\overset{Cl_2}{Si}$, Ph, L_2Pt, Ph, $\overset{Cl_2}{Si}$) + Ph_3PCl_2	(390)

46

122	$(Cl_3Si)_2Pt(PPh_3)_2/PhC\equiv CPh/PPh_3$	No reaction; cf. entries 120 and 121	×	(197)
123	$PhMe_2SiCu \cdot LiCN/$ (i) n-$BuC\equiv CH$ (ii) NH_4Cl	n-$BuCH=CH(SiMe_2Ph)$ (major) $+ n$-$BuC(SiMe_2Ph)=CH_2$ (minor)	✓	(185)
124	$(PhMe_2Si)_2CuLi \cdot LiCN/$ (i) $(n$-$Bu)C\equiv CH$; (ii) NH_4Cl	Only n-$BuCH=CH(SiMe_2Ph)$	✓	(185) [see also Ref. (6)]

Miscellaneous

| 125 | $(Me_3Si)_2Ru(CO)_4/Me_3SiRu(CO)_2(C_8H_8SiMe_3)$ | $Me_3SiRu_2(CO)_5(C_8H_8SiMe_3)^x$ | ✓ | (170, 280) |

[a] diphos = $Ph_2PCH_2CH_2PPh_2$.
[b] Other F_3Si–M derivatives behave similarly, where M = $Fe(CO)_2(Cp)$, $Mo(CO)_3(Cp)$, $W(CO)_3(Cp)$, and $Mn(CO)_4(PPh_3)$.
[c] Cp' = η^5-$CH_3C_5H_4$.
[d] $R_3Si^* = MePh(1$-naphthyl)Si.
[e] Also $[Fe(CO)_2(Cp)]_2$.
[f] And other products.
[g] Only at high pressure and 90°C.
[h] Probably arises from disproportionation of $(SiH_3)_2O$, etc.
[i] Silicic acid.
[j] Also $PhCH=CHPh$, etc.
[k] Adduct formed at 25°C.
[l] $Fe(CO)_2(Cp)$ derivative reacts similarly.
[m] Via adduct as intermediate (see text).
[n] $Mo(CO)_3(Cp)$, $W(CO)_3(Cp)$, and $Co(CO)_4$ derivatives react similarly.
[o] Compare (119) and abstraction of H^+ from analogous species.
[p] Similar insertion or (insertion – Me_3SiF) products from cyclo-C_3F_6, $CF_3C\equiv CH$, $CF_3C\equiv CCF_3$, and $\overline{CF_2CF_2CF}=CF$.
[q] Also forms $Fe_2(CO)_6(R_2C_2)_2$ and $Fe_2(CO)_6(R_2C_2)SiPh_2$.
[r] Also (as major product) $(Me_3Si)_2Ru_2(CO)_5(C_7H_7)$.
[s] Major products are silicon-free compounds.
[t] Pentalene complex.
[u] Via $Me_3SiRu(CO)_2$(tetrahydropentalenyl) intermediates.
[v] bipy = 2,2'-bipyridyl.
[w] Also 2,4-di-t-butyl isomer.
[x] Also open-chain $Me_3SiRu_2(CO)_4(C_8H_8SiMe_3)$ and $Ru_3(CO)_8$-pentalene complexes. Note that $(Me_3Si)_2Ru(CO)_4$ acts as source of "$Ru(CO)_3$."

Sections III,D and III,G,2, respectively. Cleavages initiated thermally or photolytically will also be considered later, in Section V,A.

1. Cleavage by Dihydrogen

Silicon–platinum compounds still afford the only examples of this behavior (entries 1–3). While the reaction is believed to proceed via the formation of a Pt(IV) dihydrido intermediate (134, 235), this species has not yet been isolated.

2. Cleavage by Halogen Compounds

Hydrogen halides usually cleave the silicon–metal bond in simple compounds containing Si–V, Si–Cr, Si–Fe, and Si–Co bonds; cleavage becomes more difficult with heavier transition-metal homologs. Compounds with Si–Mn (30) and especially Si–Re (7, 32, 121) bonds are generally much more resistant, although the special case of $R_3SiMn(H)(CO)_2(\eta^5\text{-}CH_3C_5H_4)$ should be noted (entry 13); this and related derivatives often react readily by a kind of reductive elimination process (compare entries 12, 63, and 87), associated with the close Si· · ·H approach in the molecule (Section IV,A).

The direction of attack is normally that expected for interaction between $H^{\delta+}–X^{\delta-}$ and a bond of polarity $Si^{\delta+}–M^{\delta-}$, and silicon becomes attached to the most electronegative partner. Exceptions may be noted in the case of (1) a compound with a high formal metal oxidation state (entry 5), (2) anionic derivatives (entry 7), (3) the manganese hydrido compound referred to above (entry 13), and (4) in certain vinyl–silicon compounds (entry 22, but not entry 7a), where formal reversal of polarity ("Umpolung") occurs in the Si–M bond. The lack of reactivity shown in entries 20 and 23 is probably due chiefly to steric factors. It is interesting that the silacyclobutane derivative in entry 27 reacts predominantly by ring opening, although some Si–Fe bond cleavage occurs while the ring remains intact.

Halogens themselves almost invariably bring about cleavage, often at low temperatures. An exception, involving $Ph_3SiMn(CO)_5$ and I_2, is shown in entry 11, although the trimethyl analog (entry 9) reacts readily. The interhalogen compound ICl reacts so that the most electronegative halogen atom becomes attached to silicon (entry 17), and the halogen analog CF_3I undergoes loss of CF_2 to achieve the same result (entry 18):

$$Me_3SiFe(CO)_2(Cp) + CF_3I \rightarrow Me_3SiF + IFe(CO)_2(Cp) \tag{49}$$

Halogenated hydrocarbons often bring about substitution at silicon (see Section III,C), but in the case shown in entry 4, Si–Ti (but not Si–Si) bonds are cleaved, and rearrangement to a cyclic Si_5 derivative occurs. A radical process is postulated, similar to that observed in Ti–Ge (*422*) and H–Mn–CO (*72*) systems. Silicon–zirconium bonds are cleaved in an analogous way (entry 5).

3. Cleavage by Oxygen and Sulfur Compounds

Water and alcohols (or alkoxide ions) usually cause cleavage, although some exceptions amongst sterically hindered iron compounds in entries 46–49 should be noted. The direction of attack is such that Si–O compounds seem always to result, even in the case of $R_3SiMnH(CO)_2(Cp)$ derivatives (cf. Section III,B,2).

Insertion reactions with carbon monoxide or sulfur dioxide are well known for compounds with carbon–transition-metal bonds, e.g.,

$$RM(CO)_n \xrightarrow{CO} RCOM(CO)_n \tag{50}$$

$$RM(CO)_n \xrightarrow{SO_2} ROS(O)M(CO)_n \tag{51}$$

They have still not been achieved, however, for silicon–metal systems (entries 34, 35, 43, 44, 54) even in reactions at high pressure (entry 54). Moreover, one example is known of the reverse process, in which CO is extruded from a sila–acyl metal derivative (cf. Section II,G,4,c) (entry 43):

$$fac\text{-}(Ph_3SiCO)Re(CO)_3(diphos) \xrightarrow{>180°C} mer\text{-}Ph_3SiRe(CO)_3(diphos) + CO \tag{52}$$
$$\text{(XIII)} \qquad\qquad\qquad\qquad\qquad\qquad \text{(XV)}$$

These observations are often ascribed to the greater element-to-metal bond strength on passing from carbon to silicon. But other factors must be involved, since insertion *does* occur in some related Ge–metal and Sn–metal systems (e.g., entry 44) (*24*), and it is known that Sn–metal bonds are generally stronger than the corresponding Si–metal bonds (see Section IV,D).

The reaction of benzaldehyde with various silicon–metal compounds has been studied with interesting results (entries 37, 51, 52). Apparently, insertion with Si–O bond formation is the first step, followed by loss of a metal-containing fragment. The following stages are proposed (*264*):

$$Me_3SiMn(CO)_5 + PhCHO \xrightarrow[\text{slow}]{5°C} (Me_3SiO)CHPhMn(CO)_5$$

$$\downarrow 80°C$$

$$[Me_3SiOCHPh]_2 + Mn_2(CO)_{10} \qquad (53)$$

$$(Me_3Si)_2Fe(CO)_4 + PhCHO \xrightarrow{5°C} Me_3SiFe(CO)_4CHPh(OSiMe_3)$$

$$\downarrow 25°C$$

$$[Me_3SiOCHPh]_2 + ? \qquad (54)$$

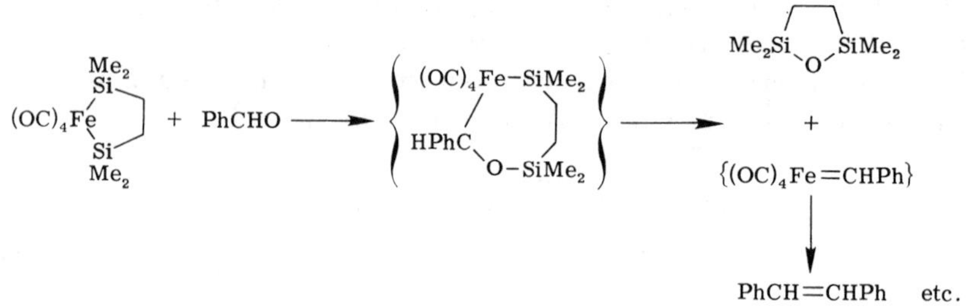

$$(55)$$

The relationship between these processes and those believed to occur in metal-catalyzed hydrosilation of aldehydes and ketones (e.g., *231, 343*) has been emphasized.

Reactions of PhSH with Si–Pt compounds (entries 56 and 57) give the products expected on the basis of soft acid–soft base Pt–S interaction; it is surprising, however, that the reaction in entry 58 yields a product with Si–H rather than Si–O or Si–Cl bonds.

4. Cleavage by Nitrogen and Phosphorus Compounds

While tertiary amines often form quite stable adducts with N→Si bonds (see Section III,D), ammonia and primary or secondary amines normally cleave silicon–metal bonds; intermediate adducts are either not observed or found only at low temperatures (*134, 235*). Typical behavior is shown by (*28, 30, 31*)

$$H_3SiMn(CO)_5 \begin{cases} \xrightarrow{NMe_3} [H_3Si\cdot 2NMe_3]^+Mn(CO)_5^- \\ \xrightarrow{NH_3} (SiH_3)_2NH + HMn(CO)_5 \end{cases} \qquad (56)$$

In the case of entry 59, however, an adduct between $H_3SiMo(CO)_3(Cp)$ and dimethylamine can be isolated at room temperature, although it decomposes in the expected way on warming.

The fine balance between Si–M bond formation, cleavage, and adduct formation is well illustrated by the following system (34).

$$F_3SiNMe_2 + HCo(CO)_4 \rightleftarrows F_3SiCo(CO)_4 + HNMe_2 \rightleftarrows adduct \qquad (57)$$
$$A B C$$

When the components A are mixed, there is rapid formation of adduct C in equilibrium with the other compounds. On fractionation *in vacuo* at low temperatures, the most volatile component, dimethylamine, is readily removed, giving good yields of $F_3SiCo(CO)_4$.

The normal reaction of tertiary phosphines is to replace metal-coordinated carbonyl groups (Section III,E,1) or, less commonly, to form adducts with P→Si bonds (Section III,D). In special cases, however, cleavage of the Si–metal bond may occur directly (entries 63, 66, 66a, and 67); the manganese derivative is discussed further in Section IV,E, while the silylene–iron compound is considered again in Section V,C.

Phosphites often cause cleavage of Si–M bonds (e.g., entries 62 and 64), but not invariably (see Section III,E,1) (93). The reaction shown in entry 64 is an interesting example of *ortho*-hydrogen abstraction; it is suggested, by analogy with similar processes involving $R_3SiMnH(CO)_2(Cp)$ derivatives, that initial production of an Fe–H bonded intermediate is followed by reductive elimination of a hydridosilane.

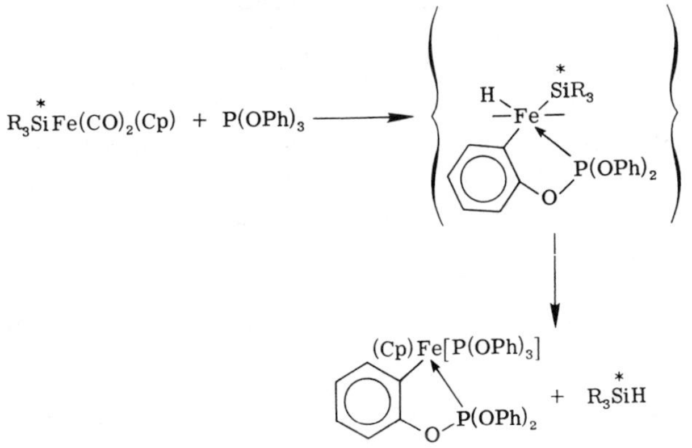

$$R_3Si\overset{*}{F}e(CO)_2(Cp) + P(OPh)_3 \longrightarrow \left\{ \overset{H\diagdown \quad \overset{*}{S}iR_3}{-Fe-} \quad P(OPh)_2 \right\}$$

$$\downarrow$$

$$(Cp)Fe[P(OPh)_3] \quad + \quad \overset{*}{R}_3SiH \qquad (58)$$

The stereochemical implications of this and related reactions are considered in Section IV,E.

Phosphine methylenes, $R_3P{=}CH_2$, are strong Lewis bases and bring about cleavage of Si–M bonds with intermediate formation of an adduct (entries 61 and 65). This is a useful way of synthesizing C-silyl derivatives.

$$FMe_2SiFe(CO)_2(Cp) + Me_3P{=}CH_2 \rightarrow [Me_3PCH_2SiMe_2F]^+[Fe(CO)_2(Cp)]^-$$

$$\Big\downarrow \scriptstyle{Me_3P{=}CH_2}$$

$$Me_3P{=}CH(SiMe_2F) + Me_4P^+[Fe(CO)_2(Cp)]^- \qquad (59)$$

5. Cleavage by Main Group IV Compounds

It has been known for some time (*134, 235*) that R_3Si groups in silicon–metal compounds can be replaced by R_3Ge or R_3Sn groups, the strength of attachment to metal being $Si < Ge < Sn$. This order is consistent with what is known of bond strengths (see Section IV,D). Some further examples of this special kind of cleavage reaction are shown in entries 72–75, 77, and 78. Replacements of one silyl group by another are also known (entries 70 and 71), e.g. (*226*),

$$R_3SiRuH_3L_3 + HSiR_3' \rightarrow R_3'SiRuH_3L_3 + HSiR_3 \qquad (60)$$

where $L = PR_3$, $R_3 = (OEt)_3$ or Cl_2Me, and $R' = F$. This reaction probably proceeds via successive reductive elimination and oxidative addition: the general rule is that the silyl group with the most electronegative substituents becomes attached to the metal. The displacement in entry 76, on the other hand, is probably a result of steric congestion in the initial Si–Pt compound.

Not only hydrido derivatives can enter into this reaction: entries 68 and 69 show that siloxy compounds and C-silyl-substituted phosphine methylenes also undergo exchange (*303, 306, 310*). In the latter case, the silyl group with the most electronegative substituents moves to carbon. Respective examples are

$$Me_3SiM(CO)_3(Cp) + Me_3M'OSiMe_3 \xrightarrow{\Delta} Me_3M'M(CO)_3(Cp) + (Me_3Si)_2O \qquad (61)$$

where $M = Mo$ or W, and $M' = Ge$ or Sn, and

$$F_3SiFe(CO)_2(Cp) + Me_3P{=}C(SiMe_3)_2 \xrightarrow[30 \ h]{60°C}$$

$$Me_3SiFe(CO)_2(Cp) + Me_3P{=}C(SiF_3)(SiMe_3) \qquad (62)$$

6. Cleavage by Hydridic Reducing Agents

Entries 79, 80, and 83–86 show that the potent reducing agent $LiAlH_4$ cleaves a variety of silicon–metal bonds, giving a hydridosilane. Nucleophilic attack by H^- is indicated; a discussion of the associated stereochemistry is deferred until Section IV,E. When the silicon-platinum compounds shown in entries 85 and 86 are reduced, metallic platinum is formed, but the metal-containing products in other cases have not generally been identified. Equation (63) provides an example of a compound with a robust Si–Fe bond that resists reductive cleavage (237) (cf. Section III,C,3).

$$Cl_3SiFe(CO)_2(Cp) \xrightarrow{\text{LiAlH}_4} H_3SiFe(CO)_2(Cp) \qquad (63)$$

Sodium borohydride seems less ready to cleave silicon–metal bonds (entries 81 and 82), and could be effective in selective reductions of the type shown above; no other hydridic reducing agents seem to have been investigated in this context.

7. Cleavage by Organometallic and Mercury Compounds

As shown in entry 87, methyllithium promotes the reductive elimination of R_3Si^*H from $R_3Si^*MnH(CO)_2(\eta^5\text{-}CH_3C_5H_4)$ (cf. entries 11 and 63). Stoicheiometric amounts of organolithium compounds and $R_3SiCo(CO)_4$ derivatives give products in which attack on coordinated carbonyl groups has occurred (120) (see Section III,F,1), but addition of further LiR leads to cleavage and formation of silicon–alkali-metal compounds (114, 115). Grignard reagents also induce cleavage.

Breaking of silicon–transition-metal bonds by mercury compounds is one aspect of the well-known "conversion series" in organosilicon chemistry (22, 24, 154). It was established at an early stage that equilibrium in the system

$$2H_3SiI + Hg[Co(CO)_4]_2 \rightleftarrows 2H_3SiCo(CO)_4 + HgI_2 \qquad (64)$$

lies well to the left (26, 29). Since most other metal carbonyl and metal carbonyl cyclopentadienyl groups are even less electronegative than $Co(CO)_4$ (see Section II,A,1), and to the extent that the conversion series is based on electronegativity values, it appears that all silicon–transition-metal bonds are likely to be cleaved by mercury halides, even the iodide. The more recent examples shown in entries 90–92 support this idea.

8. Cleavage by Alkenes, Alkynes, and Nitriles

The first six entries in this section of Table X (entries 93–98) relate to attempted insertion reactions of fluoroalkenes or fluoroalkynes into Si–Mn or Si–Fe bonds. Even in the successful cases, yields are generally low, and in some instances (e.g., entry 94) Me_3SiF is lost from the insertion product to leave a fluoroalkenyl manganese carbonyl. In the case of the iron hydrido compound in entry 98, there is formal loss of hydrogen to give the binuclear derivative $[Cl_3SiFe(CO)_4]_2$.

With alkynes, the ferradisilacyclopentene and diferradisilacyclobutane derivatives in entries 99 and 100 suffer cleavage of all or some of their Si–Fe bonds to yield cyclic and other products; the former compound also reacts with nitriles, forming a novel heterocycle (**XVI**).

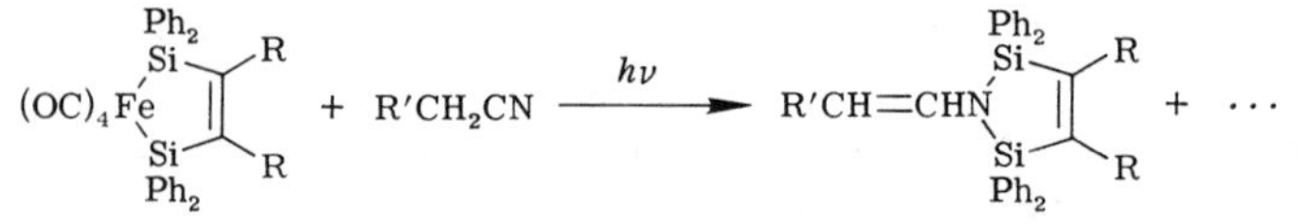

$$(65)$$

An analogous cleavage of a nickeladisilacyclopentene with alkyne is shown in entry 116. It has been pointed out (297) that these reactions show certain similarities to metal-catalyzed cyclotrimerization of alkynes or to cycloaddition of alkynes and substituted disilanes, postulated to involve Si–metal intermediates.

The nickel complex $(X_3Si)_2Ni(bipy)$ ($X_3 = Cl_3$ or $MeCl_2$) reacts with diphenylacetylene to form a violet solid, believed to be the trans isomer (**XVII**) (entry 115).

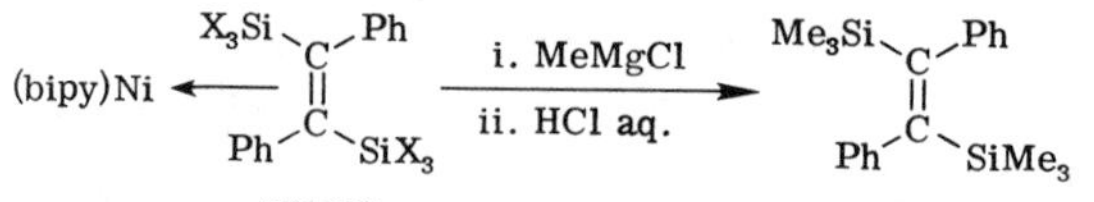

$$(66)$$

Alkylation then yields the trans alkene shown; other alkynes can give rise to cis products. It seems likely that the metal–alkene complex (**XVII**) is similar to intermediates believed to occur in various catalytic processes involving nickel–bipyridyl complexes.

Analogous complexes have not been observed in the case of platinum: instead, a disilylplatinum complex (entry 120) adds ethyne with loss of hydrogen (but no Si–Pt bond cleavage) to give a platinadisilacyclopentene. A similar product (**XIX**) was claimed from the reactions of entry

121, in which a disilylene complex (**XVIII**) was thought to act as an intermediate (L = PPh$_3$):

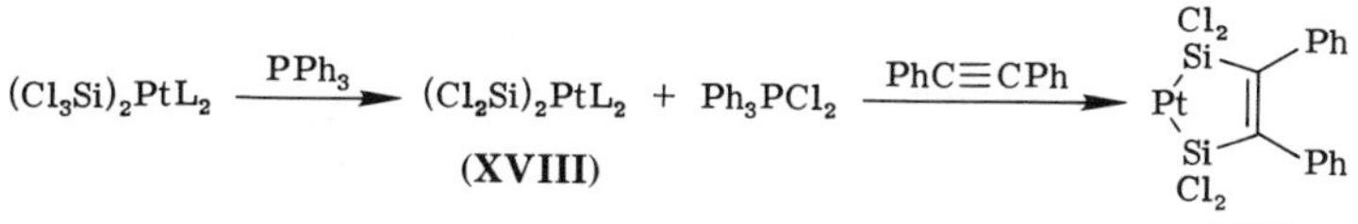

$$(Cl_3Si)_2PtL_2 \xrightarrow{PPh_3} \underset{(XVIII)}{(Cl_2Si)_2PtL_2} + Ph_3PCl_2 \xrightarrow{PhC \equiv CPh}$$

(**XIX**)   (67)

However, a reexamination of the first stage showed that PPh$_3$ did not abstract chlorine from (Cl$_3$Si)$_2$PtL$_2$, but rather that adventious mois-ture produced a cyclodisiloxane (**XX**) with properties similar to those reported for complex (**XVIII**); no addition of alkyne could be effected (entry 122).

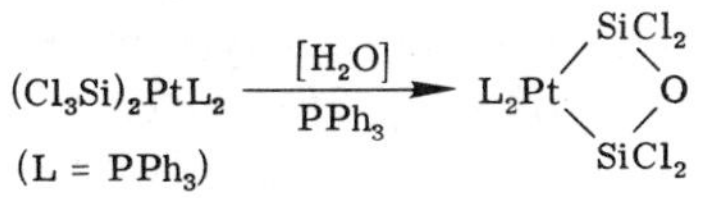

$$\underset{(L\,=\,PPh_3)}{(Cl_3Si)_2PtL_2} \xrightarrow[PPh_3]{[H_2O]} L_2Pt \underset{SiCl_2}{\overset{SiCl_2}{<}} O$$

(**XX**)   (68)

Addition of phenylacetylene to the platinum hydrido complexes R$_3$Si*PtH(PPh$_3$)$_2$ (entries 117 and 118) leads to reductive elimination of HSi*R$_3$ and formation of a Pt(0)–alkyne complex. A similar Pt(0) prod-uct is observed from the cyclic precursor in entry 119.

Representatives of the little-known class of silicon–copper com-pounds are formed *in situ* from organosilyllithium compounds and Cu(I) halides or pseudohalides; they can then undergo cleavage at the Si–Cu bond with alkynes (entries 123 and 124). The reaction, particu-larly with (R$_3$Si)$_2$CuLi, is regiospecific, and a route to useful vinyl de-rivatives.

The remaining entries in Table X summarize the extensive work carried out by Stone and his co-workers at Bristol on cleavage reactions of Si–Ru compounds by cyclic alkenes. Several themes may be dis-cerned: one is the migration of SiMe$_3$ groups from ruthenium to the carbocyclic system. Thus, cycloheptatriene gives rise either to a com-pound in which a trimethylsilylcycloheptatrienyl ring bridges two ruthenium atoms (entry 102) or to one in which a trimethylsilylcy-cloheptadiene ring is attached to a single ruthenium atom (entry 103). Also, cyclooctatetraene can form a derivative with a C$_8$H$_8$SiMe$_3$ ring linked to ruthenium (entry 114): these migrations are believed to be intramolecular. Analogous reactions take place with the cyclic precur-sor (**XXI**) (entries 109 and 110).

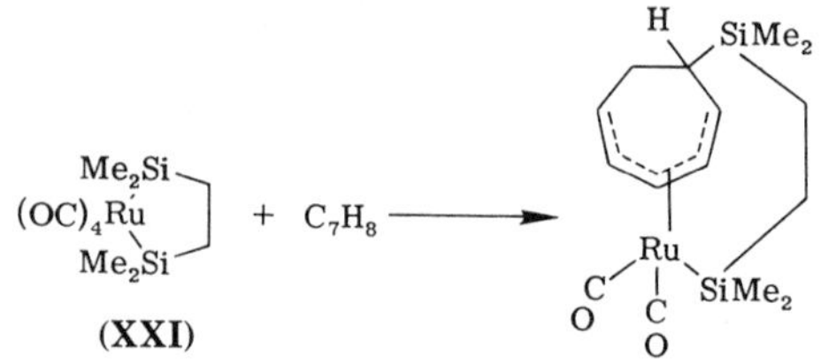

$$
\begin{array}{c}
\text{Me}_2\text{Si} \\
\text{(OC)}_4\overset{\text{Me}_2\text{Si}}{\underset{\text{Me}_2\text{Si}}{\text{Ru}}} \\
\end{array}
\ \Bigg] \ + \ \text{C}_7\text{H}_8 \ \longrightarrow
$$

(XXI)   (69)

Another general theme is that of dehydrogenation. The simplest examples are those of entries 111 and 112, in which cyclopentadienyl and cycloheptadienyl complexes are produced from C_5H_6 and C_7H_{10} respectively. With larger rings as starting materials, however, transannular dehydrogenative ring closure can occur to produce (presumably) $HSiMe_3$ and derivatives of pentalene (C_8H_6) or tetrahydropentalene (C_8H_{10}); in the former case, some ring substitution by Me_3Si groups is possible (entry 108). The bicyclic ring systems may be attached to one (entries 104, 106, 107), two (entry 114), or two out of a cluster of three ruthenium atoms (entries 108, 113) [see also (280) and entry 125]. The structures of some of these interesting compounds are discussed further in Section IV,A.

9. Metal-Exchange Processes

Both cleavage and re-formation of silicon–metal bonds occur in the transmetallation reactions shown in Table XI; some analogous reactions of Ge–metal and Sn–metal derivatives are shown for comparison (entries 4 and 5).

While silyl metal carbonyl derivatives are unreactive towards neutral metal carbonyls (7), metal carbonyl anions give rise to various displacement processes (entries 1–3). The order of displacing ability seems generally similar to that noted earlier for nucleophilicities (Section II,A,1):

$$Fe(CO)_2(Cp)^- \sim Re(CO)_5^- > Mn(CO)_5^- > Co(CO)_4^-$$

$$Mn(CO)_5^- > MH_3Fe(CO)_4^- \quad (M = Si, Ge)$$

$$Fe(CO)_2(Cp)^- > W(CO)_3(Cp)^-$$

Processes of this kind have some synthetic utility, since they permit exchange of silicon-containing groups from the readily prepared Si–Co compounds to other metal centres. The partial exchange of groups on silanediyl derivatives also provides a route to mixed-metal systems:

$$H_2Si(ML_n)_2 + M'L_m^- \rightarrow H_2Si(ML_n)(M'L_m) + ML_n^- \qquad (70)$$

TABLE XI

METAL EXCHANGE PROCESSES

Entry	Reaction	Ref.
1	$H_3SiCo(CO)_4 + Mn(CO)_5^- \rightarrow H_3SiMn(CO)_5 + Co(CO)_4^-$	(7)
	$H_3SiMn(CO)_5 + Co(CO)_4^- \rightarrow$ no reaction	(7)
2	$(H_3Si)_2Fe(CO)_4 + Mn(CO)_5^- \rightarrow H_3SiMn(CO)_5 + SiH_3Fe(CO)_4^-$	(429)
3	$ClMe_2SiSiMe_2W(CO)_3(Cp) + 2Fe(CO)_2(Cp)^- \rightarrow [(Cp)(CO)_2FeSiMe_2]_2 + Cl^- + W(CO)_3(Cp)^-$	(305)
4	$R_3GeCoCo(CO)_4 + Mn(CO)_5^- \rightarrow R_3GeMn(CO)_5 + Co(CO)_4^- \ (R_3 = Me_nH_{3-n}; \ n = 0-2)$	(189)
	$2H_3GeCo(CO)_4 + Fe(CO)_4^{2-} \rightarrow (H_3Ge)_2Fe(CO)_4 + 2Co(CO)_4^-$	(189)
	$(H_3Ge)_2Fe(CO)_4 + Mn(CO)_5^- \rightarrow H_3GeFe(CO)_4^- + H_3GeMn(CO)_5$	(189)
5	$Ph_3SnCo(CO)_4 + ML_n^- \rightarrow Ph_3SnML_n + Co(CO)_4^- \ [ML_n^- = Mn(CO)_5^-, Re(CO)_5^-, Fe(CO)_2(Cp)^-]$	(333)
	$Ph_3SnMn(CO)_5 + Fe(CO)_2(Cp)^- \rightarrow Ph_3SnFe(CO)_2(Cp) + Mn(CO)_5^-$	(333)
	$Ph_3SnRe(CO)_5 + Fe(CO)_2(Cp)^- \rightarrow$ no reaction	(333)

C. LIGAND EXCHANGE AT SILICON

This process, shown as mode 2 in Fig. 2, involves the production and reaction of functionally substituted silyl and silanediyl species. Since this occurs while the silicon–metal bond remains intact, it is not surprising that most of the entries in Table XII relate to compounds with robust Si–Mn and Si–Fe bonds; other examples involve bonds to Cr, Mo, W, Re, Pt, and even Co. The types of exchange reaction observed will be dealt with in turn.

1. Hydrogen–Halogen Exchange (Entries 1, 2, 5–9, 12, 14, 18, 19, 24, 28, 36, 37)

This is typically effected by (1) complex fluorides, such as $AgBF_4$, Ph_3CBF_4, $AgPF_6$, or $AgSbF_6$, or (2) halogenated hydrocarbons such as CCl_4, $CHCl_3$, or CBr_4. In most cases the reactions proceed at a convenient rate at room temperature, but irradiation may be needed when reagents in category (2) are used: radical processes are probably then involved. In these cases it has been noted that reaction becomes easier as the value of the Si–H infrared stretching frequency decreases (307). Two typical reactions are shown below (entries 1 and 6).

$$HMe_2SiCr(CO)_3(Cp) + AgBF_4 \rightarrow FMe_2SiCr(CO)_3(Cp) + Ag + BF_3 + \tfrac{1}{2}H_2 \quad (71)$$

$$H_3SiMn(CO)_5 + n\,CDCl_3 \xrightarrow{h\nu} H_{3-n}Cl_nSiMn(CO)_5 + n\,CHDCl_2 \quad (n = 1\text{–}3) \quad (72)$$

Reaction (72) proceeds sequentially to form mono-, di-, and trisubstituted products, and can conveniently be followed by ^{1}H-NMR spectroscopy (Fig. 3).

Other reagents useful in special cases are PCl_5 (entry 14) and heated HCl (entry 5); in the latter case, hydrogen is evolved.

$$H_3SiMn(CO)_5 + n\,HCl \xrightarrow{75^\circ C} H_{3-n}Cl_nSiMn(CO)_5 + n\,H_2 \quad (n = 1\text{–}3) \quad (73)$$

2. Halogen–Halogen Exchange (Entries 1, 3, 10, 20, 23, 27–29, 38, 40)

Chlorine or bromine attached to silicon can readily be exchanged for fluorine by such reagents as $AgBF_4$, $HF+BF_3$, or $AgPF_6$, as in entry 10:

$$ClR_2SiMn(CO)_5 + AgBF_4 \rightarrow FR_2SiMn(CO)_5 + AgCl + BF_3 \quad (74)$$

This particular reaction was first carried out in the hope that it would lead via Cl⁻ abstraction to a cationic species (**XXII**), as shown in reaction (75); this may be regarded as either a substituted silicenium

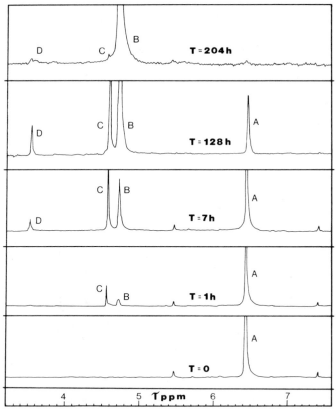

FIG. 3. NMR (^{1}H) spectra of ligand exchange between $H_3SiMn(CO)_5$ and $CDCl_3$. Key: A, $H_3SiMn(CO)_5$; B, $CHDCl_2$; C, $ClH_2SiMn(CO)_5$; D, $Cl_2HSiMn(CO)_5$.

ion (SiR_3^+) or as a metal-stabilized silylene fragment ($M^+\leftarrow SiR_2$), analogous to known metal–carbene complexes.

$$ClR_2SiMn(CO)_5 + AgBF_4 \nrightarrow [R_2SiMn(CO)_5]^+BF_4^- + AgCl \qquad (75)$$
$$\textbf{(XXII)}$$

However, no evidence for the compound (**XXII**) was obtained, even as a transient intermediate. Evidently the cation is a stronger Lewis acid (for F^-) than BF_3, and the reaction proceeds as shown in Eq. (74).

It seems likely that iodine attached to silicon might be replaced by chlorine with the aid of reagents such as AgCl or $PbCl_2$, but no such reactions have been reported. The replacement of chlorine by azide, cyanate, or thiocyanate groups does proceed readily, however (entry 35).

TABLE XII

Ligand Exchange at Silicon

Entry	Reactants	Products	Ref.
1	$X_3SiCr(CO)_3(Cp)/AgBF_4$ (also AgF, $BF_3 \cdot OEt_2$) $(X_3 = Cl_xMe_{3-x}$ $(x = 1–3)$, HMe_2, $HCl_2)$	$X'_3SiCr(CO)_3(Cp)^a$ $[X'_3 = F_xMe_{3-x}, FMe_2, F_3$ (via HF_2)]	(304)
2	$HR_2SiCr(CO)_3(Cp)/CX_4$ (X = Cl, Br)	$XR_2SiCr(CO)_3(Cp)^a$	(307)
3	$(CH_2{=}CH)X_2SiMo(CO)_3(Cp)/AgBF_4$ or $(HF + BF_3)$	$F_3SiMo(CO)_3(Cp)^b$	(308)
4	$(CH_2{=}CH)X_2SiMo(CO)_3(Cp)/HBr$	$(BrCH_2CH_2)X_2SiMo(CO)_3(Cp)^b$	(308)
5	$H_3SiMn(CO)_5/HCl/75°C$	$H_{3-n}Cl_nSiMn(CO)_5$ $(n = 1–3)$	(28, 30, 83)
6	$H_3SiMn(CO)_5/CDCl_3/25°C$	$H_{3-n}Cl_nSiMn(CO)_5$ $(n = 1–3)$	(28, 30, 83)
7	$H_2Si[Mn(CO)_5]_2/Ph_3CBF_4/CH_2Cl_2$	$F_2SiMn(CO)_5]_2 ^c$	(32)
8	$H_2Si[Mn(CO)_5]_2/CDCl_3/25°C$	$Cl_2Si[Mn(CO)_5]_2 ^c$	(32)
9	$H_2Si[Mn(CO)_5]_2/CBr_4/h\nu$	$Br_2Si[Mn(CO)_5]_2$	(32)
10	$XR_2SiMn(CO)_5/AgBF_4$ (X = Cl, Br)	$FR_2SiMn(CO)_5$	(314)
11	$Cl_{3-n}Ph_nSiMn(CO)_5/LiC_6F_5/ether$ $(n = 0–2)$	$(C_6F_5)_{3-n}Ph_nSiMn(CO)_5$	(107)
12	$HX_2SiMn(CO)_5/CX'_4$ $(X_2 = Me_2, MeCl, Cl_2; X' = Cl, Br)$	$X'X_2SiMn(CO)_5$	(307)
13	$ClR_2SiMnH(CO)_2(Cp')^d/LiAlH_4$	$HR_2SiMnH(CO)_2(Cp')$	(119)
14	$HR_2SiMnH(CO)_2(Cp')^d/PCl_5$	$ClR_2SiMnH(CO)_2(Cp')$	(119)
15	$(MeO)R_2SiMnH(CO)_2(Cp')^d/BF_3$	$FR_2SiMnH(CO)_2(Cp')$	(119)
16	$[ClR_2SiMn(CO)_2(Cp')]^-M^+ {}^d/LiR'$	$[R'R_2SiMn(CO)_2(Cp')]^-M^+$	(119)
17	$[ClR_2SiMn(CO)_2(Cp')]^-M^+ {}^d/LiAlH_4$	$[HR_2SiMn(CO)_2(Cp')]^-M^+$	(119)
18	$HMe_2SiFe(CO)_2(Cp)/CCl_4/25°C$	$ClMe_2SiFe(CO)_2(Cp)$	(272)
19	$H_2SiFe(CO)_2(Cp)]_2/CDCl_3$	$Cl_2Si[Fe(CO)_2(Cp)]_2$	(32)
20	$XMe_2SiSiMe_2Fe(CO)_2(Cp)/AgBF_4$ (X = Cl, Br)	$FMe_2SiSiMe_2Fe(CO)_2(Cp)^e$	(305)
21	$Cl_3SiFe(CO)_2(Cp)/ROH$ or ROH/RO^-	$(RO)_{3-n}Cl_nSiFe(CO)_2(Cp)$ $(n = 0–2)$	(236)
22	$(RO)_3SiFe(CO)_2(Cp)/HPF_6$	$(RO)_{3-n}F_nSiFe(CO)_2(Cp)$ $(n = 1–3)$	(316, 317)
23	$Cl_3SiFe(CO)_2(Cp)/AgBF_4 {}^f$	$F_3SiFe(CO)_2(Cp)$	(316, 317)
24	$HX_2SiFe(CO)_2(Cp)/CX'_4$ $(X_2 = Me_2, MeCl, Cl_2; X' = Cl, Br)$	$X'X_2SiFe(CO)_2(Cp)$	(307)
25	$Cl_3SiFe(CO)_2(Cp)/LiC_6F_5$	$(C_6F_5)_3SiFe(CO)_2(Cp)$	(107)

26	$Cl_3SiFe(CO)_2(Cp)/NHR^1R^2$	$(R^1R^2N)_nCl_{3-n}SiFe(CO)_2(Cp)$	(238)
		[$R^1 = R^2 = Me$, $n = 2$; Et, $n = 1$; i-Pr, $n = 0$.	
		$R^1 = H$; $R^2 = n$-Bu or i-Bu, $n = 3$; s-Bu, $n = 2$; t-Bu, $n = 1$.]	
27	$Cl_nMe_{3-n}SiFe(CO)_2(Cp)/AgBF_4$	$F_nMe_{3-n}SiFe(CO)_2(Cp)$	(304)
28	$HCl_2SiFe(CO)_2(Cp)/AgBF_4$	$F_3SiFe(CO)_2(Cp)^{g,h}$	(304)
29	$(CH_2{=}CH)Cl_2SiFe(CO)_2(Cp)/AgBF_4$ or $(HF + BF_3)$	$F_3SiFe(CO)_2(Cp)^i$	(308)
30	$(CH_2{=}CH)Cl_2SiFe(CO)_2(Cp)/HBr$	$(BrCH_2CH_2)Cl_2SiFe(CO)_2(Cp)$	(308)
31	$(MeO)_3SiFe(CO)_2(Cp)/HBr$	$Br_2(MeO)SiFe(CO)_2(Cp)$	(237)
32	$(RHN)_3SiFe(CO)_2(Cp)/HCl$ or HBr	$X_3SiFe(CO)_2(Cp)$ ($X = Cl$, Br)	(237)
33	$(RHN)_3SiFe(CO)_2(Cp)/MeI$	$I(RHN)_2SiFe(CO)_2(Cp)$	(237)
34	$Cl_3SiFe(CO)_2(Cp)/LiAlH_4$	$H_3SiFe(CO)_2(Cp)$	(237)
35	$Cl_3SiFe(CO)_2(Cp)/M^+X^-$ ($X^- = N_3^-$, OCN^-, SCN^-)	$X_3SiFe(CO)_2(Cp)$	(237)

Entry 36:

$$\underset{(Cp)(CO)Fe}{\overset{SiHMe}{\diamond}}{\overset{}{\underset{C}{\overset{}{}}}}Fe(CO)(Cp) \quad / CCl_4 \longrightarrow \underset{(Cp)(CO)Fe}{\overset{SiClMe}{\diamond}}{\underset{C}{}}Fe(CO)(Cp)$$

(309)

37	$HRSi[Fe(CO)_2(Cp)]_2/CCl_4$	$ClRSi[Fe(CO)_2(Cp)]_2$	(309)
38	$ClRSiFe(CO)_2(Cp)]_2/AgBF_4$	$FRSi[Fe(CO)_2(Cp)]_2$	(309)
39	$PhMe_2SiFe(CO)_2(Cp)/Cr(CO)_6/\Delta$	$[(OC)_3Cr-\eta^6-C_6H_5]Me_2SiFe(CO)_2(Cp)^j$	(311)
40	$Cl_3SiCo(CO)_4/AgBF_4$, $AgPF_6$	$F_3SiCo(CO)_4$	(316)

Entry 41:

$(HPh_2Si)_2PtL_2/C_2H_2/60°C$ (L = $PPhMe_2$) $\longrightarrow$

$$L_2Pt\underset{SiPh_2}{\overset{SiPh_2}{\diamond}}$$

(159)

42	$trans$-$ClH_2SiPtClL_2/NHMe_2$ (L = PEt_3)	$trans$-$(Me_2NH_2SiPtClL_2$	(58)

---

a Also Mo, W analogs.

b Also W analogs.

c Also Re analogs.

d $R_2 = Ph(1\text{-naphthyl})$; $Cp' = \eta^5$-$CH_3C_5H_4$.

e Similarly with $M(CO)_3(Cp)$ derivatives (M = Cr, Mo, W).

f Also works with $AgSbF_6$, $AgPF_6$.

g H/F exchange also with AgF, $BF_3 \cdot OEt_2$.

h Via $HF_2SiFe(CO)_2(Cp)$ intermediate.

i Via $(CH_2{=}CH)F_2SiFe(CO)_2(Cp)$ intermediate.

j A similar cobalt derivative is prepared from $[(OC)_3Cr-\eta^6-C_6H_5]Me_2SiH$ and $Co_2(CO)_8$.

3. Other Exchange Reactions

a. *Alkenyl–halogen exchange* (*Entries* 3 *and* 29). The fluorinating agents $AgBF_4$ and $HF+BF_3$ can displace alkenyl groups from silicon attached to fluorine and a transition-metal group.

b. *Halogen–alkyl* (*aryl*) *exchange* (*Entries* 11, 16, *and* 25). Here, organolithium reagents are used to replace chlorine attached to silicon by fluoroaryl or other organo groups. Attempts to extend the reaction in entry 25 to the compounds $Cl_{3-n}Ph_nSiFe(CO)_2(Cp)$ ($n = 1,2$) led to cleavage of the Si–Fe bond.

c. *Halogen–hydrogen exchange* (*Entries* 13, 17, *and* 34). In certain cases where strong Si–metal bonds are present, $LiAlH_4$ can be used successfully to convert Cl–Si–M into H–Si–M units. This exchange is the reverse of that discussed in Section III,C,1.

d. *Alkoxy–halogen exchange* (*Entries* 15, 22, *and* 31). This can be brought about by BF_3, HPF_6, or HBr, and may be incomplete, as in the last example.

e. *Halogen–alkoxy exchange* (*Entry* 21). When alcohols are used, exchange of chlorine is only partial, but addition of alkoxide anions leads to complete replacement.

f. *Amino–halogen exchange* (*Entries* 32 *and* 33). Hydrogen halides and even methyl iodide will cleave Si–N bonds in metal–Si–N systems to give metal–Si–halogen species.

g. *Halogen–amino exchange* (*Entries* 26 *and* 42). Steric effects are very noticeable in the former entry; this is a general characteristic of Si–N bond formation from Si–Cl compounds and amines (*21*). As with the reverse process in (f) above, the reaction is driven toward completion by the formation of amine salts.

A few other miscellaneous reactions may be noted. Vinylsilicon–metal compounds in entries 4 and 30 undergo addition of HBr in an anti-Markownikoff sense to give 2-bromoethyl derivatives; in some other related systems, HBr induces cleavage (see Section III,B,2). Also, the phenyl group in a phenyl–silicon–metal compound can undergo hexahapto interaction with a $Cr(CO)_3$ group (entry 39).

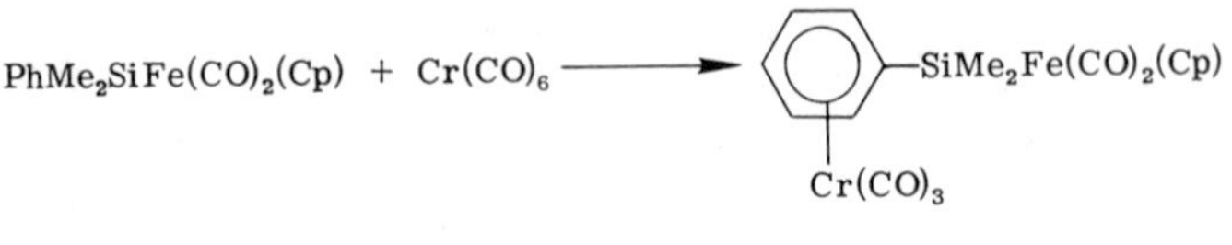

$$PhMe_2SiFe(CO)_2(Cp) + Cr(CO)_6 \longrightarrow \underset{Cr(CO)_3}{\underbrace{\bigcirc}}\!\!-SiMe_2Fe(CO)_2(Cp) \qquad (76)$$

Finally, the effect of the reaction in entry 41 is to replace hydrogen at silicon by an alkenyl group.

In several other cases, compounds have been synthesized that contain potentially reactive groups attached to silicon, although their functionality has not yet been exploited. These include the methylchlorosilyl cobalt derivatives $Cl_nMe_{3-n}SiCo(CO)_4$ (n = 1,2) (215) and platinum complexes such as $trans$-$(Et_3P)_2Pt(X)SiH_2P(SiH_3)_2$ (168) (cf. Table VI, entries 8–10).

D. ADDUCTS WITH LEWIS BASES AT SILICON

There are many instances of the formation of simple adducts between silicon–transition-metal compounds and Lewis bases (Table XIII), corresponding to mode 3 in Fig. 2. Compounds with SiH_3 groups react particularly readily, in the same way that iodosilane has long been known to form adducts such as $H_3SiI \cdot nNMe_3$ (n = 1,2) ($36, 40, 84$).

Other possibilities include compounds with Me_3Si, Me_5Si_2, and F_3Si groups, although in the last case the reaction may be complicated by simultaneous disproportionation leading to formation of $SiF_4 \cdot 2B$, where B is a Lewis base (entry 31). The Cl_3Si derivative in entry 26 does not form adducts with nitrogen or phosphorus bases, although phosphines cause slow elimination of CO (cf. Section III,E,1).

$$Cl_3SiCo(CO)_4 + PMe_3 \xrightarrow{\text{slow}} Cl_3SiCo(CO)_3(PMe_3) + CO \qquad (77)$$

Rhenium derivatives (entries 10–12) are much less ready to form adducts than their manganese analogs (entries 6, 7, 9). Thus, the adducts in entry 10 dissociate completely at room temperature.

$$H_3SiRe(CO)_5 \cdot 2NMe_3 \underset{-23°C}{\overset{25°C}{\rightleftharpoons}} H_3SiRe(CO)_5 + 2NMe_3 \qquad (78)$$

The diiron compound in entry 21 is also unreactive.

Tertiary amines and phosphines have received the most attention, but the use of the strong base $Me_3P{=}CH_2$ should be noted, e.g. (entry 19),

$$R_3SiFe(CO)_2(Cp) + Me_3P{=}CH_2 \rightarrow (Me_3PCH_2SiR_3)[Fe(CO)_2(Cp)] \qquad (79)$$

The methyl(trimethylsilyl)phosphines $PMe_n(SiMe_3)_{3-n}$ decrease in base strength as n decreases, until finally $P(SiMe_3)_3$ does not react at all (entries 28, 29).

TABLE XIII

Adducts of Silicon–Transition-Metal Compounds with Lewis Bases

Entry	Silicon compound	Lewis base, B^a	Combining ratio	Ref.
1	$H_3SiV(CO)_6$	NMe_3, py	1:2	(8, 121)
2	$H_3SiCr(CO)_3(Cp)^b$	NMe_3	1:1	(214)
3	$H_3SiCr(CO)_3(Cp)^b$	$NHMe_2$	1:1.3	(214)
4	$R_3SiMo(CO)_3(Cp)$ ($R_3 = F_3^c$; MeF_2^d)	$Me_3P{=}CH_2$	1:1	(303)
5	$Me_5Si_2Mo(CO)_3(Cp)^c$	$Me_3P{=}CH_2$	1:1	(305)
6	$H_3SiMn(CO)_5$	NMe_3, py	1:2	(27, 31)
7	$H_3SiMn(CO)_5$	bipy	1:1	(31)
8	$Me_3SiMn(CO)_5$	NMe_3	1:1	(62a)
9	$H_2Si[Mn(CO)_5]_2$	py	1:4	(32)
10	$H_3SiRe(CO)_5$	NMe_3, py	1:2 (at $-23°C$)	(7)
11	$H_3SiRe(CO)_5$	bipy	e	(7)
12	$H_2Si[Re(CO)_5]_2$	py	e	(32)
13	$(H_3Si)_2Fe(CO)_4$	NMe_3, py	1:2	(38, 39, 80)
14	$(H_3Si)_2Fe(CO)_4$	bipy	1:1	(38, 39, 80)
15	$H_3SiFeH(CO)_4$	NMe_3, py	1:2	(80)
16	$H_3SiFeH(CO)_4$	bipy	1:1	(80)
17	$R_3SiFeH(CO)_4$ (R = Cl, Ph)	NEt_3	$NHEt_3^+[R_3SiFe(CO)_4]^{-f}$	(261)
18	$(Cl_3Si)_2Fe(CO)_4$	NMe_3, MeCN	$(Cl_3Si \cdot nB)^+[Cl_3SiFe(CO)_4]^-$ (see text)	(261)

19	$R_3SiFe(CO)_2(Cp)$	$Me_3P{=}CH_2$	1:1	(303)
	[$R_3 = Me_xF_{3-x}$ ($x = 0-2$); H_3; $(CH_2{=}CH)F_2$]			
20	$Me_3Si_2Fe(CO)_2(Cp)$	$Me_3P{=}CH_2$	1:1	(305)
21	$H_2Si[Fe(CO)_2(Cp)]_2$	py	e	(32)
22	$H_3SiCo(CO)_4$	NMe_3, py	1:2	(27, 31)
23	$H_3SiCo(CO)_4$	bipy	1:1	(31)
24	$Me_3SiCo(CO)_4$	NMe_3, PMe_3, PEt_3	1:1	(44, 298)
25	$Et_3SiCo(CO)_4$	PEt_3	1:1	(44, 298)
26	$Cl_3SiCo(CO)_4$	NMe_3, PMe_3	g	(44, 298)
27	$Me_3SiCo(CO)_4$	$NHMe_2$	1:1^h	(233)
28	$Me_3SiCo(CO)_4$	$PMe_x(SiMe_3)_{3-x}$ ($x = 1-3$)	1:1	(298, 387)
29	$Me_3SiCo(CO)_4$	$P(SiMe_3)_3$	e	(298, 387)
30	$H_3SiCo(CO)_4$	PMe_3	1:1	(298, 387)
31	$F_3SiCo(CO)_4$	NMe_3, py	1:2 (approx)^i	(34)
32	$trans$-$H_2XSiPtCl(PEt_3)_2$	NMe_3	1:1	(58)

^{a} py = pyridine; bipy = 2,2'-bipyridyl.
^{b} Mo and W similar.
^{c} W similar.
^{d} Cr similar.
^{e} No interaction.
^{f} A similar adduct, $NHEt_3^+[Cl_3SiMn(CO)_2(Cp)]^-$, is formed from $Cl_3SiMnH(CO)_2(Cp)$ and NEt_3.
^{g} No adduct formed; slow CO substitution may occur.
^{h} The same adduct, $Me_3SiNHMe_2^+Co(CO)_4^-$, is also formed from Me_3SiNMe_2 and $HCo(CO)_4$; an excess of $NHMe_2$ leads to Si–Co bond cleavage.
^{i} Some $SiF_4 \cdot 2B$ and $F_2Si[Co(CO)_4]_2$ are also formed.

Infrared spectral studies of a number of these systems have shown that adduct formation is accompanied by a fall in the carbonyl stretching frequency to a value close to that associated with the corresponding metal carbonylate anion (Table XIV). Thus bands assigned to the carbonyl stretching frequencies $\nu(CO)$ in $H_3SiCo(CO)_4$ occur between 2105 and 2025 cm^{-1}; in the adduct $H_3SiCo(CO)_4 \cdot 2NMe_3$ there is only one broad strong band at 1882 cm^{-1}, close to those observed with $Co(CO)_4^-$ at 1883 and 1861 cm^{-1}. The adduct is therefore formulated as $[H_3Si \cdot 2NMe_3]^+[Co(CO)_4]^-$. Indeed, it is believed that in all the cases shown in Table XIII except entry 17, the Lewis base molecules are

TABLE XIV

CARBONYL STRETCHING FREQUENCIES, $\nu(CO)$, OF ADDUCTS

Entry	Compound	$\nu(CO)$, cm^{-1} [a]	Ref.
1	$H_3SiMn(CO)_5(g)$	2107, 2023, 2020	(31)
2	$Mn(CO)_5^-$ (soln)[b]	1898, 1864	(31)
3	$H_3SiMn(CO)_5 \cdot 2NMe_3(s)$	1860	(27, 31)
4	$H_3SiMn(CO)_5 \cdot 2py(s)$	1860	(27, 31)
5	$H_3SiMn(CO)_5 \cdot bipy(s)$	1850	(31)
6	$H_2Si[Mn(CO)_5]_2(soln)^c$	2086, 2012, 1990	(32)
7	$H_2Si[Mn(CO)_5]_2 \cdot 4py(soln)^d$	1900, 1865	(32)
8	$H_3SiRe(CO)_5(g)$	2110, 2028, 2018	(7)
9	$Re(CO)_5^-(soln)^b$	1850	(7)
10	$H_3SiRe(CO)_5 \cdot 2NMe_3(s)^e$	1990, 1970, 1900	(7)
11	$H_3SiRe(CO)_5 \cdot 2py(s)^e$	1990, 1890	(7)
12	$(H_3Si)_2Fe(CO)_4(g)$	2092, 2040, 2021	(39)
13	$Fe(CO)_4^{2-}(soln)^b$	1786	(39)
14	$FeH(CO)_4^-(soln)^b$	1897	(39)
15	$(H_3Si)_2Fe(CO)_4 \cdot 2NMe_3(s)$	1917, 1868	(39)
16	$(H_3Si)_2Fe(CO)_4 \cdot 2py(s)$	1900, 1880	(80)
17	$(H_3Si)_2Fe(CO)_4 \cdot bipy(s)$	1875	(80)
18	$H_3SiFeH(CO)_4(g)$	2107, 2050, 2044, 2036	(39)
19	$H_3SiFeH(CO)_4 \cdot 2NMe_3(s)$	1920, 1863	(80)
20	$H_3SiFeH(CO)_4 \cdot 2py(s)$	1895, 1880	(80)
21	$H_3SiFeH(CO)_4 \cdot bipy(s)$	1890	(80)
22	$H_3SiCo(CO)_4(g)$	2105, 2050, 2025	(31)
23	$Co(CO)_4^-(soln)^b$	1883, 1861	(31)
24	$H_3SiCo(CO)_4 \cdot 2NMe_3(s)$	1870	(27, 31)
25	$H_3SiCo(CO)_4 \cdot 2py(s)$	1882	(27, 31)
26	$H_3SiCo(CO)_4 \cdot bipy(s)$	1870	(31)

[a] Only chief bands given.
[b] Sodium salt in tetrahydrofuran.
[c] In cyclohexane.
[d] In pyridine.
[e] Cooled below $-40°C$.

attached to silicon, with consequent charge separation, i.e.,

$$R_3SiML_m + nB \rightarrow (R_3Si \cdot nB)^+ML_m^- \qquad (n = 1,2) \qquad (80)$$

In the case of silanediyl derivatives, both metal fragments can be lost as anions, e.g. (entry 9),

$$H_2Si[Mn(CO)_5]_2 + 4NC_5H_5 \rightarrow (H_2Si \cdot 4NC_5H_5)^{2+}[Mn(CO)_5]_2^- \qquad (81)$$

The reluctance of derivatives of $Re(CO)_5$ and $Fe(CO)_2(Cp)$ in entries 10–12 and 21 to form adducts can now be readily understood, since these are known to be poor leaving groups as anions. This is, of course, related to their high nucleophilicity, already referred to in Section II,A,1.

This picture of adduct formation is confirmed by an X-ray diffraction study of the adduct $Me_3SiCo(CO)_4 \cdot PMe_2(SiMe_3)$ (387). This consists of clearly separated $[Me_2P(SiMe_3)_2]^+$ cations and tetrahedral $Co(CO)_4^-$ anions. Further support comes from ^1H-NMR measurements, outlined in Table XV, although these are restricted because of the low solubility of the adducts in most nonreacting solvents. It can be seen that on adduct formation (entries $2 \rightarrow 3$ and $4 \rightarrow 5$) there is a marked decrease in chemical shift and increase in ^{29}Si–H coupling constant. The first effect is related to the deshielding associated with formation of singly or doubly charged cations. It may also be noted that the small change in the value of τ on passing from entry 6 to 7 is consistent with the known

TABLE XV

SPECTROSCOPIC DATA FOR ADDUCTS OF Si–Mn AND Si–Re COMPOUNDS

Entry	Compound	τ(Si–H)	$J(^{29}Si–H)$ (Hz)	ν_{as}(Si–H) (cm^{-1})	Ref.
1	SiH$_4$	6.80[a]	202[a]	2180[b]	(22)
2	H$_3$SiMn(CO)$_5$	6.41[a]	196[a]	2156[b]	(30)
3	(H$_3$Si · 2py)$^+$Mn(CO)$_5^-$	4.95[c]	274[c]	2205[d]	(7, 83)
4	H$_2$Si[Mn(CO)$_5$]$_2$	5.96[e]	173[e]	2060[f]	(32, 121)
5	(H$_2$Si · 4py)$^{2+}$[Mn(CO)$_5$]$_2^-$	3.65[c]	336[c]	2230[c]	(32, 121)
6	H$_3$SiRe(CO)$_5$	6.70[a]	186[a]	2144[b]	(7)
7	H$_3$SiRe(CO)$_5$ · 2NMe$_3$	6.48[g]	—	—	(7)

[a] TMS solution.
[b] Gas phase.
[c] Pyridine-d_5 solution.
[d] Nujol mull.
[e] Benzene-d_6 solution.
[f] Cyclohexane solution.
[g] Trimethylamine solution at 0°C, adduct largely dissociated.

weakness of the rhenium adduct. Regarding the second effect, the size of the ^{29}Si–H coupling constant is related to the amount of s character in the Si–H bond (149), and the observed values show that there is a dramatic increase in s character in passing from entry 1 to 3 to 5. This is consistent with the trend expected for the geometries shown in Fig. 4, although this treatment is undoubtedly oversimplified. Adducts of 2,2'-bipyridyl, for example, will probably favor a structure in which axial–equatorial bridging occurs, e.g. (**XXIII**).

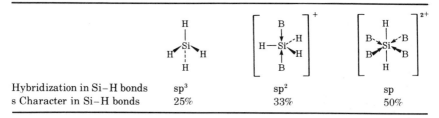

(**XXIII**)

Values of the Si–H asymmetric stretching frequency ν_{as}(Si–H) also increase on passing from entry 1 to 3 to 5, again consistent with the formation of cations having charges of $+1$ and $+2$, and consequent increased ionic character in the Si–H bonds.

The adducts shown in entries 13–16 of Table XIII deserve further attention because there has been some doubt as to their true composition. In the original work (39), the trimethylamine adducts were formulated as: $(H_3Si \cdot NMe_3)_{\frac{1}{2}}^{+}Fe(CO)_4^-$ and $(H_3Si \cdot nNMe_3)^+FeH(CO)_4^-$ (where the value of n was not known); values of ν(CO) in these complexes were more appropriate for a singly charged than a doubly charged anion. Soon afterwards, Graham and his co-workers (261) showed that some related species underwent proton abstraction on reaction with Lewis bases (Table XIII, entry 17).

$$R_3SiFeH(CO)_4 + NEt_3 \rightarrow NHEt_3^+[R_3SiFe(CO)_4]^- \quad (R = Cl,Ph) \quad (82)$$

They therefore suggested that $H_3SiFeH(CO)_4$ was likely to react in the same way. However, further examination of a number of related sys-

Hybridization in Si–H bonds   sp^3 sp^2 sp
s Character in Si–H bonds   25%   33%   50%

FIG. 4. Hybridization and bond type in SiH$_4$, (SiH$_3$ · 2B)$^+$, and (SiH$_2$ · 4B)$^{2+}$.

tems (80) has shown that (1) the stoichiometry of corresponding adducts of $(H_3Si)_2Fe(CO)_4$ and $H_3SiFeH(CO)_4$ is similar, (2) the values of $\nu(CO)$ all correspond to those expected for singly charged anions (see Table XIV), and (3) the catalytic effect of amine salts on the cleavage of Si–Fe bonds by HCl invalidates the original evidence for symmetrical addition of amine molecules to silyl groups in $(H_3Si)_2Fe(CO)_4$ (39). All the evidence is therefore consistent with the formulations shown as (**XXIV**) and (**XXV**).

$$(H_3Si \cdot 2B)^+[H_3SiFe(CO)_4]^- \qquad (H_3Si \cdot 2B)^+[HFe(CO)_4]^-$$
$$\textbf{(XXIV)} \qquad\qquad\qquad\qquad \textbf{(XXV)}$$

Similar anionic species $H_3MFe(CO)_4^-$ (M = Si, Ge) have resulted from exchange reactions (cf. Section III,B,9), while the rather weak unsymmetrical interaction of bases with $(Cl_3Si)_2Fe(CO)_4$ gives rise to a related anion (Table XIII, entry 18).

$$(Cl_3Si)_2Fe(CO)_4 + nB \rightarrow (Cl_3Si \cdot nB)^+[Cl_3SiFe(CO)_4]^- \qquad (B = NMe_3, MeCN) \quad (83)$$

In hydrido species, $R_3SiFeH(CO)_4$, the silicon atom is the most electrophilic site when R is H, but the Fe hydrogen atom takes over that role when more electronegative groups are attached to silicon (R = Cl, Ph).

The chemical reactivity of the adducts can be either greater or less than that of the parent silicon–transition-metal compound. For example, $H_3SiMn(CO)_5$ is unreactive towards anhydrous HCl at room temperature, but its adducts (B = NC_5H_5, NMe_3) react readily at low temperatures (30).

$$H_3SiMn(CO)_5 \cdot 2B + 3HCl \xrightarrow[\text{rapid}]{-80°C} H_3SiCl + HMn(CO)_5 + 2BH^+Cl^- \qquad (84)$$

The rhenium analog behaves similarly, being cleaved quickly by protic reagents in the presence of pyridine, although inert in its absence (7).

On the other hand, some very reactive silyl–metal compounds can be stabilized if they form strong adducts. The silyl iron derivative $H_3SiFeH(CO)_4$ reacts very rapidly with oxygen, giving a bright flash and producing CO, CO_2, volatile siloxanes, and an off-white or brown powder (39). The pyridine adduct $H_3SiFeH(CO)_4 \cdot 2(py)$ gives the same products, but in this case reaction is very slow (80). Moreover, $H_3SiCo(CO)_4$ decomposes slowly *in vacuo* at room temperature, while $H_3SiCo(CO)_4 \cdot 2NMe_3$ is unchanged after a year, and reacts only slowly with air (31). Attempts were made to stabilize the extremely labile vanadium derivative $H_3SiV(CO)_6$ in the same way by complexing, but

without success; the adducts are probably not sufficiently stable towards dissociation (*8, 121*).

E. SUBSTITUTION OF LIGANDS AT THE TRANSITION METAL

This process, represented as mode 4 in Fig. 2, is exemplified by the
reactions in Table XVI. It may be further subdivided according to the
nature of the ligands.

1. Substitution of CO Groups

The most studied reaction involves the replacement of carbonyl
groups by tertiary phosphines. This proceeds most readily in the case of
cobalt derivatives, manganese compounds must usually be heated, and
iron compounds require irradiation (except in the case of entry 13).

The stereochemistry of the final product is normally such that the
tertiary phosphine is trans to the silyl group, both in five-coordinate
(entries 20–25) or six-coordinate (entries 1 and 5) situations. An obvious exception is the cyclic compound in entry 2, which has no option but
to take a cis arrangement. In the case of the six-coordinate complexes
$(Cl_3Si)_2Fe(CO)_3(PPh_3)$ and $Cl_3SiRuH(CO)_3(PPh_3)$ (entries 13 and 15),
the three carbonyl groups adopt a mer arrangement and again the
phosphine is trans to a silyl group.

It had been suggested some years ago (*150*) on the basis of kinetics
measurements that substitution of $Ph_3GeMn(CO)_5$ by phosphines led
initially to the cis isomer; with bulky phosphines, this then rearranged
to the trans species. A similar effect has now been demonstrated for
$Ph_3SiCo(CO)_4$ (*227*) (entry 24) by infrared methods: an equatorially
substituted compound (**XXVI**) is apparently first formed, and this
slowly rearranges to the axially substituted product (**XXVII**).

$$Ph_3SiCo(CO)_4 \xrightarrow{PR_3} Ph_3Si-\underset{\underset{O}{C}\ \underset{O}{C}}{\overset{PR_3}{|}}Co-CO \xrightarrow{O} Ph_3Si-\underset{\underset{O}{C}\ \underset{O}{C}}{\overset{\overset{O}{C}}{|}}Co-PR_3$$

$$\text{(XXVI)} \qquad\qquad \text{(XXVII)} \qquad (85)$$

This is consistent with variable-temperature ^{13}C-NMR studies on a
range of derivatives $R_3ECo(CO)_4$ (E = C, Si, Ge, Sn, Pb; R = F, Cl, Me,
Bu, Bz, Ph). These studies showed that the rate of intramolecular rearrangement, leading to exchange of axial and equatorial carbonyl
groups, is strongly dependent on steric factors (*296*). Thus isomers with
one or two bulky substituents should be quite long-lived.

TABLE XVI

SUBSTITUTION OF TRANSITION-METAL LIGANDS

Entry	Reactants	Products	Ref.
A. Carbonyl substitution			
1	Ph$_3$SiMn(CO)$_5$/PPh$_3$/decalin/130°C	*trans*-Ph$_3$SiMn(CO)$_4$(PPh$_3$)	(376, 377)
2	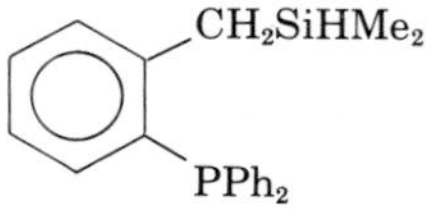 CH$_2$SiHMe$_2$ benzene /PPh$_2$ /Mn$_2$(CO)$_{10}$/50°C	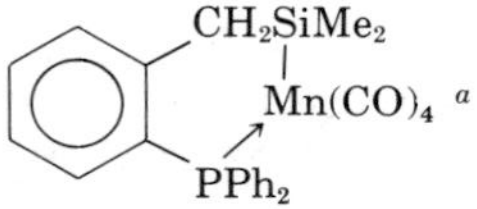 CH$_2$SiMe$_2$ /Mn(CO)$_4$[a] /PPh$_2$	(13)
3	Ph$_3$SiMn(CO)$_5$/diphos[b]	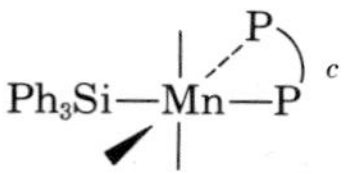 Ph$_3$Si—Mn—P[c] — P like (**XV**)	(377)
4	Ph$_3$SiMn(CO)$_5$/bipy[d]	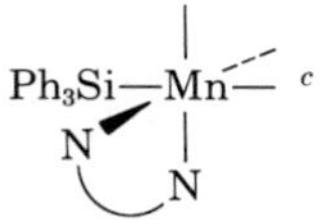 Ph$_3$Si—Mn—[c] —N N	(377)
5	F$_3$SiMn(CO)$_5$/PPh$_3$	*trans*-F$_3$SiMn(CO)$_4$(PPh$_3$)[e]	(373)
6	Me$_3$SiFe(CO)$_2$(Cp)/PR$_3$/$h\nu$	Me$_3$SiFe(CO)(Cp)(PR$_3$)	(270)
7	Ph$_3$SiFe(CO)$_2$(Cp)/PPh$_3$/$h\nu$	Ph$_3$SiFe(CO)(Cp)(PPh$_3$)	(123)
8	R$_3$Si*Fe(CO)$_2$(Cp)/PPh$_3$/$h\nu$	R$_3$Si*Fe(CO)(Cp)(PPh$_3$)	(91, 92)
9	R$_3$Si*Fe(CO)$_2$(Cp)/P(OEt$_3$)/$h\nu$	R$_3$Si*Fe(CO)(Cp)[P(OEt)$_3$]	(93)
10	(Me$_3$Si)$_2$CH$_2$Fe(CO)$_2$(Cp)/PPh$_3$/$h\nu$	Me$_3$SiCH$_2$SiMe$_2$Fe(CO)(Cp)(PPh$_3$)	(353)
11	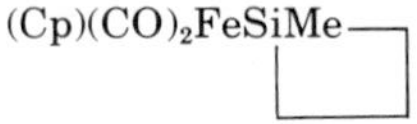 (Cp)(CO)$_2$FeSiMe⎯ /PPh$_3$/$h\nu$	 (Ph$_3$P)(Cp)(CO)FeSiMe⎯	(136)
12	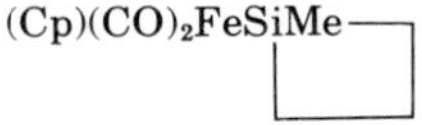 (Cp)(CO)$_2$FeSiMe⎯ /PMePh$_2$/$h\nu$	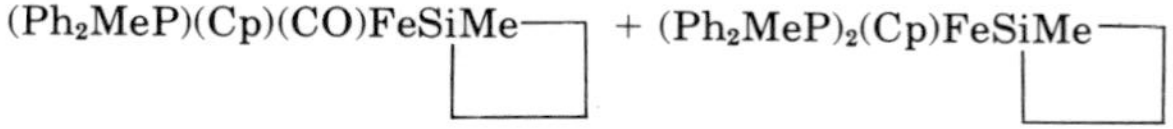 (Ph$_2$MeP)(Cp)(CO)FeSiMe⎯ + (Ph$_2$MeP)$_2$(Cp)FeSiMe⎯	(136)

(*table continues*)

TABLE XVI (Continued)

Entry	Reactants	Products	Ref.
13	$(Cl_3Si)_2Fe(CO)_4/PPh_3$/heat (similarly, Ru,[g] Os)	mer-$(Cl_3Si)_2Fe(CO)_3(PPh_3)$	(361)
14	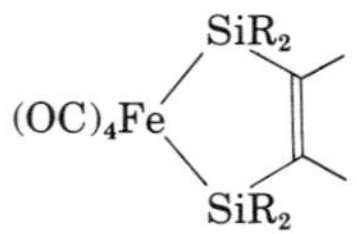 $(OC)_4Fe$ with SiR_2 ... SiR_2 ring /PR_3/$h\nu$	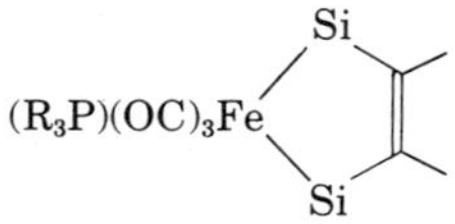 $(R_3P)(OC)_3Fe$ with Si...Si ring	(124)
15	$Me_3SiFe(CO)_2(Cp)/RNC/h\nu$ (R = cyclohexyl)	$Me_3SiFe(CO)(Cp)(CNR) + Me_3SiFe(Cp)(CNR)_2$	(96)
16	cis-$(Cl_3Si)_2Ru(CO)_4/{}^{13}CO/25°C$	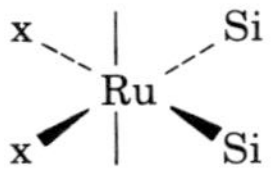 $x\text{---}Ru\text{---}Si$ with Si, x ligands; x = exchange site	(365)
17	$trans$-$(Cl_3Si)_2Ru(CO)_4/{}^{13}CO/h\nu$	cis Isomer as above	(365)
18	cis-$Cl_3SiRuH(CO)_4/PPh_3$	mer-$Cl_3SiRuH(CO)_3(PPh_3)$	(361)
19	cis-$(Cl_3Si)_2Ru(CO)/diphos$,[b] bipy,[d] cyclic polyenes[h]	 $B\text{---}Ru\text{---}Si$ with Si,[i] B ligands	(364)
20	$H_3SiCo(CO)_4/PPh_3$[j]	$trans$-$H_3SiCo(CO)_3(PPh_3)$	(29)
21	$Cl_3SiCo(CO)_4/PR_3$ (R = Me, Et, F)	$trans$-$Cl_3SiCo(CO)_3(PR_3)$	(44)
22	$Me_3SiCo(CO)_4/PPh_3/h\nu$	$trans$-$Me_3SiCo(CO)_3(PPh_3)$	(44)
23	$Cl_3SiCo(CO)_4/P(n\text{-}Bu_3)$	$trans$-$Cl_3SiCo(CO)_3[P(n\text{-}Bu)_3]$ (compare Ge, Sn analogs—see text)	(341)

24	Ph$_3$SiCo(CO)$_4$/PR$_3$	cis-Ph$_3$SiCo(CO)$_3$(PR$_3$) → trans isomer	(227)
25	Ph$_3$SiCo(CO)$_4$/PMePh$_2$	Ph$_3$SiCo(CO)$_3$(PMePh$_2$) + Ph$_3$SiCo(CO)$_2$(PMePh$_2$)$_2$	(227)
B. Hydrogen substitution or reaction			
26	Cl$_3$SiWH(Cp$_2$)/CCl$_4$	CHCl$_3$ + ?	(88)
27	Cl$_3$SiMnH(CO)$_2$(Cp)/CCl$_4$	(?)Cl$_3$SiMnCl(CO)$_2$(Cp) + CHCl$_3$	(259)
28	Cl$_3$SiFeH(CO)$_4$/isoprene	Cl$_3$SiFe(CO)$_3$[η^3-(CH$_2$CHCHCMe$_2$)]^k	(122)
29	R$_3$SiRuH$_3$(PR$_3$)$_3$/D$_2$	R$_3$SiRuH$_{3-x}$D$_x$(PR$_3$)$_3$	(226)
30	R$_3$SiRuH$_3$(PR$_3$)$_n$/CCl$_4$, CDCl$_3$, or I$_2$ (n = 2, 3)	R$_3$SiRuH$_2$X(PR$_3$)$_n$ (X = Cl or I)	(226)
31	F$_3$SiCoH$_2$(PPh$_3$)$_3$/CO	F$_3$SiCo(CO)$_2$(PPh$_3$)$_2$(+ H$_2$ + PPh$_3$)	(17)
C. Substitution of other ligands			
32	[Me$_3$SiRu(CO)$_4$]$_2$/X$_2$ (X = Br,I)	trans-Me$_3$SiRuX(CO)$_4$	(19)
33	trans-Me$_3$SiRuX(CO)$_4$/PPh$_3$/LiAlH$_4$	Me$_3$SiRuH(CO)$_3$(PPh$_3$)	(19)
34	trans-Me$_3$SiRuX(CO)/M(CO)$_5^-$ (M = Mn,Re)	Me$_3$SiRu(CO)$_4$[M(CO)$_5$]	(19)

[a] Re analogue does not eliminate CO (cf. Table IV, entry 7).
[b] diphos = Ph$_2$PCH$_2$CH$_2$PPh$_2$.
[c] Inferred structure from infrared data.
[d] bipy = 2,2'-bipyridyl.
[e] Re similar.
[f] R$_3$Si* = PhMe(1-naphthyl)Si.
[g] Reacts at room temperature.
[h] See also Ref. (279).
[i] Bridged derivatives also possibly formed.
[j] Slow reaction.
[k] Via substituted σ-allyl derivative.

In the case of disubstituted cobalt derivatives, it is believed that representatives of the three isomers (**XXVIII**)–(**XXX**) can be prepared by reaction (86) (entry 25), reaction (87) (entry 31), and reaction 88 (Table IV, entry 33).

$$
\begin{array}{ccc}
\text{L} & \text{L} & \text{SiR}_3 \\
| & | & | \\
\text{R}_3\text{Si}-\text{Co}-\text{L} & \text{R}_3\text{Si}-\text{Co}-\text{CO} & \text{L}-\text{Co}-\text{L} \\
\diagup\;\diagdown & \diagup\;\diagdown & \diagup\;\diagdown \\
\text{C}\quad\text{C} & \text{L}\quad\text{C} & \text{C}\quad\text{C} \\
\text{O}\quad\text{O} & \text{O} & \text{O}\quad\text{O} \\
\textbf{(XXVIII)} & \textbf{(XXIX)} & \textbf{(XXX)}
\end{array}
$$

$$\text{Ph}_3\text{SiCo(CO)}_4 + 2\text{PMePh}_2 \rightarrow \text{Ph}_3\text{SiCo(CO)}_2(\text{PMePh}_2)_2 + 2\text{CO} \qquad (86)$$
$$\textbf{(XXVIII)}$$

$$\text{F}_3\text{SiCoH}_2(\text{PPh}_3)_3 + 2\text{CO} \rightarrow \text{F}_3\text{SiCo(CO)}_2(\text{PPh}_3)_2 + \text{H}_2 + \text{PPh}_3 \qquad (87)$$
$$\textbf{(XXIX)}$$

$$\text{HCo(CO)}_2(\text{PPh}_3)_2 + \text{HSiF}_3 \rightarrow \text{F}_3\text{SiCo(CO)}_2(\text{PPh}_3)_2 + \text{H}_2 \qquad (88)$$
$$\textbf{(XXX)}$$

These compounds seem stable with respect to isomerization, in accord with the ideas discussed in the previous paragraph.

The cis isomer of $(\text{Cl}_3\text{Si})_2\text{Ru(CO)}_4$ undergoes easy exchange of the two CO groups trans to silicon (but not the other two) at room temperature (entry 16); the process is thought to be intramolecular, with a five-coordinate intermediate (**XXXI**) in which the nonexchanging carbonyl groups occupy axial positions of the trigonal bipyramid.

$$
\begin{array}{ccccc}
\text{O} & & \text{O} & & \text{O} \\
\text{C} & & \text{C} & & \text{C} \\
\text{OC}_{\diagdown}\;|\;_{\diagup}\text{SiCl}_3 & \xrightarrow{-\text{CO}} & |\;_{\diagup}\text{SiCl}_3 & \xrightarrow{^{13}\text{CO}} & \text{C}_{\diagdown}\;|\;_{\diagup}\text{SiCl}_3 \\
\text{Ru} & & \text{OC}-\text{Ru}_{\diagdown} & & \text{Ru} \\
\text{OC}^{\diagup}\;|\;^{\diagdown}\text{SiCl}_3 & & |\;^{\diagdown}\text{SiCl}_3 & & \text{O}^{13}\text{C}^{\diagup}\;|\;^{\diagdown}\text{SiCl}_3 \\
\text{C} & & \text{C} & & \text{C} \\
\text{O} & & \text{O} & & \text{O} \\
& & \textbf{(XXXI)} & & \qquad\qquad (89)
\end{array}
$$

The trans isomer will exchange only under photochemical excitation, and then gives rise to the same products as the cis compound (entry 17). Certain bidentate ligands also react with cis-$(\text{Cl}_3\text{Si})_2\text{Ru(CO)}_4$ to give chelate derivatives of the same stereochemistry (entry 19). Apart from this work, there are few reports about the effect of bidentate ligands in general, although it is interesting that nitrogen and phosphorus donors seem to give different types of derivatives with $\text{Ph}_3\text{SiMn(CO)}_5$ (entries 3 and 4).

Five other points may be noted. (1) Triethyl phosphite reacts in the same way as a tertiary phosphine in entry 9, although phosphites often behave differently (see Section III,B,4). (2) The monosubstituted iron derivatives in entries 6–12 and 15 have a chiral center at iron. This point is taken up further in Section IV,E. (3) In no case has carbonyl insertion been observed as a concomitant to carbonyl substitution (compare Section III,B,3), even though this process is common for corresponding organo derivatives, as in

$$RMn(CO)_5 + PPh_3 \rightarrow RCOMn(CO)_4(PPh_3) \tag{90}$$

Indeed, almost the reverse process of C–M bond breaking and Si–M bond formation may be seen in entry 10. (4) Germanium and tin analogs of $Cl_3SiCo(CO)_4$ react with tri-n-butylphosphine to give ionic products (*341*), as in

$$Cl_3MCo(CO)_4 + 2P(n\text{-}Bu)_3 \rightarrow \{Co(CO)_3[P(n\text{-}Bu)_3]_2\}^+MCl_3^- \tag{91}$$

where M = Ge or Sn, but there is no indication that the silicon compound can behave in this way (entry 23) (cf. Section II,F). Nor is there any indication that radical chain processes occur with the silicon compound, although there is evidence for them in the reaction between $Cl_3SnCo(CO)_4$ and group V donors (*3*). (5) Isocyanide substitution appears similar to that of tertiary phosphines, but has been little studied (entry 15).

2. Substitution or Reaction of Hydrogen Ligands

When the silicon–transition-metal bond is reasonably strong, hydrogen attached to the metal may be replaced (mode 4b in Fig. 2) by halogens (entries 26, 27, and 30) or deuterium (entry 29). In the case of the ruthenium example, halogenation can be followed by reductive elimination of R_3SiH (*226*).

$$R_3SiRuH_2Cl(PR_3)_n \rightarrow R_3SiH + RuHCl(PR_3)_3 \qquad (n = 2, 3) \tag{92}$$

In entry 28 is an example of Fe–H addition to conjugated dienes, while the Si–Fe bond remains intact. With isoprene, the initial product from 1,4-addition is a σ-allyl compound, which then rearranges with loss of CO to a π-allyl derivative. This opens a useful route to other functionally substituted derivatives.

$$Cl_3SiFeH(CO)_4 + CH_2{=}C(Me)CH{=}CH_2 \rightarrow Cl_3SiFe(CO)_4(CH_2CH{=}CMe_2)$$

$$\Big\downarrow -CO \tag{93}$$

$$Cl_3SiFe(CO)_3(\eta^3\text{-}CH_2CHCMe_2)$$

Reductive elimination of dihydrogen occurs in entry 31, together with loss of PPh$_3$, to give the usual five-coordination found for Co(I) derivatives.

3. Substitution or Addition of Other Ligands

As yet, there are few examples of this process (mode 4c in Fig. 2), although it offers considerable potential when strong Si–M bonds are present. After the metal–metal bond in [Me$_3$SiRu(CO)$_4$—]$_2$ has been replaced by a metal–halogen bond (entry 32), the resulting trans-Me$_3$SiRuX(CO)$_4$ can undergo either reduction (entry 33) or substitution by metal carbonylate groups (entry 34). Under this heading may also be mentioned an example of oxidative addition at the metal center while the metal–silicon bond remains intact (56, 58).

$$trans\text{-}IH_2SiPt(II)I(PEt_3)_2 \xrightarrow{HI} IH_2SiPt(IV)HI_2(PEt_3)_2 \qquad (93a)$$
$$[4] \qquad\qquad\qquad [6]$$

F. Reactions at the Metal Carbonyl Group

These reactions correspond to mode 5 in Fig. 2, and fall into two categories. The first involves the formation of carbene precursors, while the second concerns the migration of silicon from metal to oxygen.

1. Reactions with Organolithium Derivatives

The reaction of transition-metal carbonyls with organolithium compounds to give metal carbene complexes is now well known; it can also take place when the metal is linked to a group IV atom, as in the following series of reactions (426).

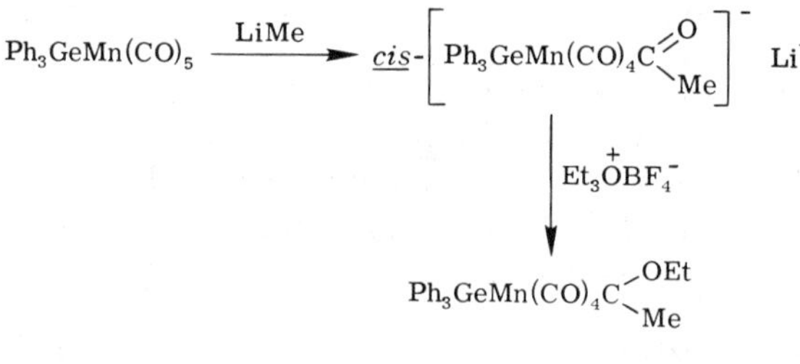

$$Ph_3GeMn(CO)_5 \xrightarrow{LiMe} cis\text{-}\left[Ph_3GeMn(CO)_4C\overset{O}{\underset{Me}{\diagdown}} \right]^- Li^+$$

$$\downarrow Et_3\overset{+}{O}BF_4^-$$

$$Ph_3GeMn(CO)_4C\overset{OEt}{\underset{Me}{\diagdown}} \qquad (94)$$

Analogous germanium–molybdenum, germanium–tungsten (145), germanium–cobalt (89), tin–cobalt, and lead–cobalt (144) carbene derivatives are known.

In the case of silicon, addition of phenyllithium occurs quite analogously, and the resulting ionic derivative (**XXXII**) has been isolated as a salt of the $[(Ph_3P)_2N]^+$ cation. On heating, the lithium salt decomposes to give a benzoylsilane (*120*) (cf. Section III,B,7).

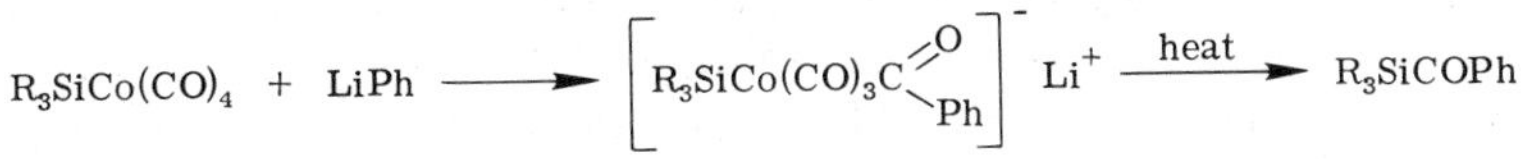

$$R_3SiCo(CO)_4 \ + \ LiPh \ \longrightarrow \ \left[R_3SiCo(CO)_3C{\overset{\nearrow O}{\underset{\searrow Ph}{}}} \right]^- \ Li^+ \ \xrightarrow{heat} \ R_3SiCOPh$$

$[R_3 = Ph_3 \text{ or } MePh(1\text{-naphthyl})]$ (**XXXII**) (95)

There is no evidence as yet for silyl-substituted carbyne derivatives, analogous to the tin–chromium compound $Ph_3SnCr(CO)_4(\equiv CNEt_2)$ (*184*).

2. Migration of Silicon from Metal to Oxygen

It was recognized at an early stage in the development of silicon-transition-metal chemistry that silicon–oxygen compounds often appeared, either as by-products from preparations or as decomposition products on heating or even on storage. At first, adventitious hydrolysis or oxidation was blamed, but it soon became clear that attack on silicon by oxygen of coordinated carbonyl groups was responsible. Since metal carbonyls are known to form adducts with Lewis acids such as compound (**XXXIII**) (*286*),

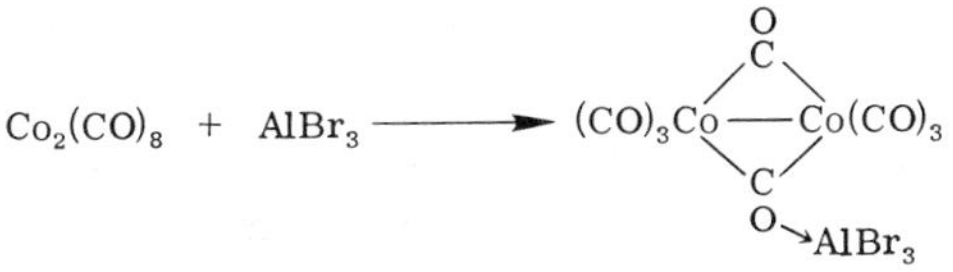

$$Co_2(CO)_8 \ + \ AlBr_3 \ \longrightarrow \ (CO)_3Co\underset{\underset{O\rightsquigarrow AlBr_3}{\overset{|}{C}}}{\overset{\overset{O}{\overset{\|}{C}}}{-\!\!\!-}}Co(CO)_3$$

(**XXXIII**) (96)

while silicon compounds often function as Lewis acids (*23, 24*), this seems very reasonable.

Reference has already been made in Section II,A,3 to the harmful effects of polar solvents, in particular tetrahydrofuran, in the preparation of silicon–transition-metal compounds from a silicon halide and a transition-metal anion (*61, 137, 138, 300, 305, 306, 310, 331, 337*). Frequently, silicon–oxygen rather than silicon–metal compounds result, e.g. (*138, 306, 336*),

$$Mo(CO)_2(Cp)(PPh_3)^- + Ph_3SiCl \xrightarrow[\text{(Ref. 138)}]{THF} (Ph_3Si)_2O + HMo(CO)_2(Cp)(PPh_3) + \cdots \quad (97)$$

$$W(CO)_3(Cp)^- + Me_3SiBr \xrightarrow[\text{(Ref. 306)}]{THF} (Me_3Si)_2O + [W(CO)_3(Cp)]_2 + \cdots \quad (98)$$

$$Co(CO)_4^- + MeSiCl_3 \xrightarrow[\text{(Ref. 336)}]{ether} MeCl_2SiOCCo_3(CO)_9 + \cdots \quad (99)$$

Disiloxanes may be a major product even when the reaction is carried out in nonpolar solvents (136) (see Table I, entry 21).

In all such cases, the reaction is seen as an electrophilic attack by the silicon compound on an oxygen atom of a coordinated carbonyl group, leading initially to a Si–O–C–M linkage (336). This process will be facilitated by a polar solvent of high dielectric constant, since under these conditions the metal carbonylate group will become more like a free anion (i.e., less ion pairing will occur) and consequently the oxygen atom will carry a greater effective negative charge.

A related effect is that silicon–transition-metal compounds, once formed, undergo Si–M bond cleavage when dissolved in polar solvents (252, 262, 300, 305, 306, 310). The products are often complex, but have been shown to include a cluster compound in one case involving cobalt (252).

$$Me_3SiCo(CO)_4 \xrightarrow{\text{THF}} Me_3SiOCCo_3(CO)_9 + Me_3SiOCH{=}CHCH_2CH_3 + \cdots \quad (100)$$

The first product of this reaction, which should be compared with Eq. (18) in Section II,A,3, is a derivative of the well-known —$CCo_3(CO)_9$ cluster, with a tetrahedral CCo_3 grouping; the product in Eq. (99) is similar. The second product in Eq. (100) is clearly derived from attachment of tetrahydrofuran to a trimethylsilyl group, followed by rearrangement.

Also, cis-$(Me_3Si)_2Fe(CO)_4$ dissolves in tetrahydrofuran to give a red solution that contains Si–O derivatives (262). These seem to be different, however, from the product of the reaction between Me_3SiI and $Fe(CO)_4^{2-}$ in the same solvent (331) [referred to in Eq. (13) and Table I, entry 22], which is known to be a tetrasiloxy derivative of a ferracyclopentadiene (54).

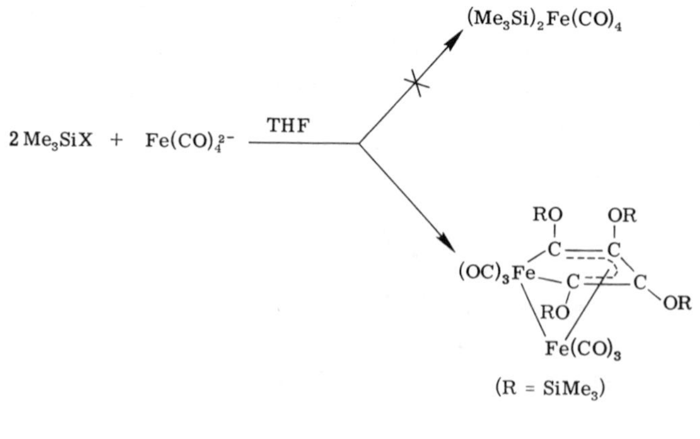

$$(R = SiMe_3) \quad (101)$$

It seems likely that tetrahydrofuran, acting as a Lewis base toward silicon centers, promotes charge separation and incipient formation of carbonylate anions, just as described in Section III,D. The oxygen atoms in coordinated carbonyl groups are thereby made more nucleophilic, leading to formation of Si–O–C–M links, as discussed above. Consistent with this, $Cl_3SiCo(CO)_4$ is unchanged after a solution of it in tetrahydrofuran has been refluxed; it is also known to be unreactive towards Lewis bases such as NMe_3 (Table XIII, entry 26) and $Co(CO)_4^-$ (336).

On heating, a number of silicon–transition-metal compounds give a disiloxane as one of the volatile products [29, Eq. (102); 41; 42; 95, Eq. (103); 215; 306, Eq. (104)]. For example,

$$H_3SiCo(CO)_4 \xrightarrow{\Delta} H_2 + CO + SiH_4 + (SiH_3)_2O + HCo(CO)_4$$
$$+ SiH_2[Co(CO)_4]_2 + \cdots \tag{102}$$

$$Et_3SiCo(CO)_4 \xrightarrow{\Delta} (Et_3Si)_2O + \cdots \tag{103}$$

$$Me_3SiMo(CO)_3(Cp) \xrightarrow{\Delta} (Me_3Si)_2O + \cdots \tag{104}$$

An extreme example of this behavior is afforded by $H_3SiV(CO)_6$, which decomposes almost quantitatively at room temperature according to Eq. (105). (8, 121).

$$2H_3SiV(CO)_6 \xrightarrow{25°C} (SiH_3)_2O + V(CO)_6 + [V(CO)_5C]_n \tag{105}$$

After removal of volatile products, a yellow vanadium carbonyl carbide remains. In all these cases, it seems likely that there is initial electron donation from oxygen to silicon, followed by Si–M bond cleavage. As supporting evidence, two products from the controlled heating of $Me_3SiCo(CO)_4$ are reasonable intermediates in such a process (252).

$$Me_3SiCo(CO)_4 \xrightarrow[50 \text{ h}]{105°C} Me_3SiOCCo_3(CO)_9 + (Me_3SiOC)_4Co_2(CO)_4 + \cdots \tag{106}$$

This migration of silicon from metal to oxygen will be discussed further in connexion with silicide formation in Section V,A.

G. DISPROPORTIONATION, REDUCTIVE ELIMINATION, AND OTHER REACTIONS

1. Disproportionation about Silicon or Metal

The disproportionation equilibrium (107) is common among silicon compounds, especially hydrides and fluorides (22, 24); the right-hand side is generally favored (R = H, halogen, etc.).

$$2R_3SiX \rightleftarrows R_2SiX_2 + R_4Si \qquad (107)$$

Relevant examples include

$$2R_3SiCo(CO)_4 \rightarrow R_4Si + R_2Si[Co(CO)_4]_2 \qquad (108)$$

[R = H (Ref. 29); R = F (Ref. 34)] and

$$2H_3SiMn(CO)_5 \rightarrow SiH_4 + H_2Si[Mn(CO)_5]_2 \qquad (109)$$

(Ref. 30). Also, $Cl_3SiCo(CO)_4$ yields $SiCl_4$ on heating at 150°C, although the nature of the other products is not known (42). These reactions are strongly base-catalyzed (cf. Ref. 35 and Table XIII, entry 31); examples are also known with other group IV elements, e.g. (341),

$$Cl_3SnCo(CO)_4 \xrightarrow{\text{THF}} SnCl_4 + Cl_2Sn[Co(CO)_4]_2 \qquad (110)$$

Disproportionation about the transition metal is much less common: one example involves a silyl iron hydride (259).

$$cis\text{-}Cl_3SiFeH(CO)_4 \xrightarrow{\Delta} (Cl_3Si)_2Fe(CO)_4 + \cdots \qquad (111)$$

2. Reductive Elimination of Silicon Hydrides

It is useful to gather together under this heading a number of reactions of compounds in which an R_3Si group and a hydrogen atom are attached to the same transition-metal center:

$$R_3Si\text{-}M(H)L_m \rightarrow R_3SiH + ML_n \qquad (112)$$

These may be described either as a type of Si–M cleavage reaction or as a migration of hydrogen from metal to silicon. Clearly they represent the reverse process to the well-known oxidative addition reaction discussed in Section II,F.

Compounds of the class $R_3SiMnH(CO)_2(Cp)$ have been mentioned already in this connection (117–119, 225, 259) (Table X, entries 12, 13, 63, and 87). Reactions with tertiary phosphines, chlorine, methyllithium, or an excess of HCl all lead to elimination of R_3SiH ("deinsertion"), as in

$$R_3SiMnH(CO)_2(Cp) \xrightarrow{\text{PPh}_3} R_3SiH + Mn(CO)_2(Cp)(PPh_3) \qquad (113)$$

In a related reaction between $R_3SiFe(CO)_2(Cp)$ and $P(OPh)_3$, migration of *ortho*-hydrogen from a phenyl group first provides an iron–hydrogen bond, and this is followed by reductive elimination of R_3SiH (*93*) [Section III,B,4, Eq. (58)].

The *cis*-hydrido compound $Ph_3SiFeH(CO)_4$ eliminates triphenylsilane when either heated alone (*259*) or treated with triphenylphosphine (*355*).

$$Ph_3SiFeH(CO)_4 \begin{cases} \xrightarrow{\Delta} HSiPh_3 + Fe(CO)_5 + Fe_3(CO)_{12} + \cdots & (114) \\ \\ \xrightarrow{PPh_3} HSiPh_3 + Fe(CO)_n(PPh_3)_{5-n} \quad (n = 3, 4) & (115) \end{cases}$$

Silylhydrido derivatives of ruthenium (*226*) and cobalt (*17*) undergo reactions rather similar to that shown in Eq. (115).

$$R_3SiRuH_3(PR_3)_n \begin{cases} \xrightarrow{CO} HSiR_3 + RuH_2(CO)(PR_3)_3 & (116) \\ \\ \xrightarrow{CCl_4} R_3SiRuH_2Cl(PR_3)_n \end{cases}$$
$$(n = 2, 3)$$
$$\downarrow$$
$$R_3SiH + RuHCl(PR_3)_3 \qquad (117)$$

$$R_3SiCoH_2(PR_3')_3 \xrightarrow{CO} HSiR_3 + CoH(CO)(PR_3')_3 \qquad (118)$$

These eliminations generally proceed more readily as R in R_3Si becomes larger and less electronegative.

Finally, reactions (119) and (120) make an interesting comparison with reactions (115) and (118).

$$(Cl_3Si)_2Fe(CO)_4 \xrightarrow[\text{Ref. 355}]{PPh_3} Si_2Cl_6 + Fe(CO)_n(PPh_3)_{5-n} \quad (n = 3, 4) \qquad (119)$$

$$Et_3SiCo(CO)_4 \xrightarrow[\text{Ref. 227}]{PPh_3} Si_2Me_6 + [Co(CO)_3(PPh_3)]_2 \qquad (120)$$

3. Miscellaneous Reactions

The preparation of the interesting silaallyl species (**XIV**)* has already been described (Section II,G,4,c). On photolysis, it gives trimethyl-

---

* See footnote, p. 37, and Appendix.

vinylsilane as one product; the mechanism is as yet not certain, but it is interesting that a very similar yield of the same product, $CH_2{=}CHSiMe_3$, is obtained by irradiation of $CH_2{=}CHSi_2Me_5$ and $Fe(CO)_5$, and it may well be that the same intermediate is involved (385).

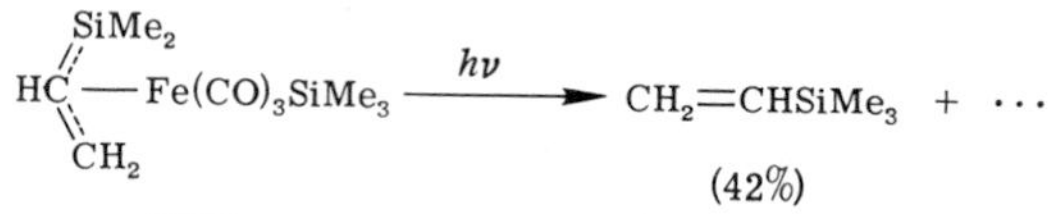

$$
\begin{array}{c}
\overset{\displaystyle SiMe_2}{\underset{\displaystyle CH_2}{HC{-}Fe(CO)_3SiMe_3}} \xrightarrow{\;h\nu\;} CH_2{=}CHSiMe_3 \;+\; \cdots
\end{array} \qquad (121)
$$

(42%)

(XIV)

IV. Information from Physical Methods

A. ELECTRON AND X-RAY DIFFRACTION STUDIES

Table XVII lists compounds studied by these techniques and the silicon–metal bond lengths obtained.

The volatile compounds in the first four entries were suitable for gas-phase electron diffraction studies. Their octahedral (entries 1–3) and trigonal bipyramidal (entry 4) geometries are shown in formulas (XXXIV) and (XXXV). In each case, it was found that the carbonyl

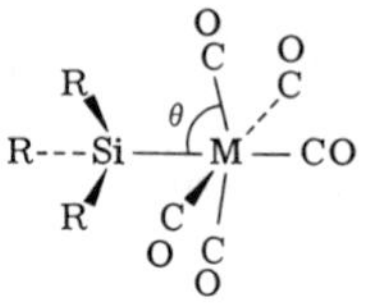

(XXXIV; M = Mn or Re) 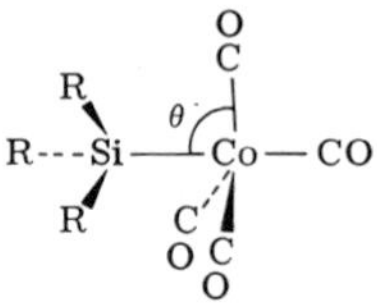　　　　　(XXXV)

groups cis to the silyl group were tilted towards it, such that values of θ (SiMC$_{cis}$) were (XXXIV, M = Mn, R = H), $85\frac{1}{2}°$; (XXXIV, M = Mn, R = F), 87°; (XXXIV, M = Re, R = H), 85°; and (XXXV, R = H), 82°. Such distortions are frequently seen in the structures of solid metal-carbonyl derivatives, but it is important to establish that they also occur in molecules in the gas phase, uncomplicated by intermolecular interactions and crystal packing effects. It has been suggested that these distortions arise as a result of (1) unequal competition between a silyl group and the carbonyl group trans to it for π-bonding interactions with the transition-metal atom, and/or (2) some bonding interaction between silicon and the carbonyl groups cis to it in the molecule (i.e., equatorial CO) (63, 374, 380). No clear preference between these proposals can be established at present.

TABLE XVII

STRUCTURAL DETERMINATIONS OF SILICON–METAL COMPOUNDS

Entry	Compound	d(Si–M) (pm)^a	Ref.
A. Electron diffraction determinations			
1	$H_3SiMn(CO)_5$ (**XXXIV**)	241	(369)
2	$F_3SiMn(CO)_5$ (**XXXIV**)	236	(371)
3	$H_3SiRe(CO)_5$ (**XXXIV**)	256	(370)
4	$H_3SiCo(CO)_4$ (**XXXV**)	238	(28, 374)
B. X-Ray diffraction determinations			
5	$[H_2SiTi(Cp)_2]_2{}^b$ (**XXXVII**)	216	(232)
6	$Ph_3SiZrCl(Cp)_2{}^b$ (**XXXVI**)	281	(326)
7	$(Et_2Si)_2W_2(CO)_8H_2$ (**XLII**)	259; 270	(52)
8	$Cl_2PhSiMnH(CO)_2(Cp)^b$ (**XLIII**)	231	(403)
9	$[Ph_2SiMn(CO)_4]_2$ (**LXI**)	240	(402)
10	$Ph_3SiMnH(CO)_2(Cp)^b$ (**XLIII**)	242	(248, 251)
11	trans-$Me_3SiMn(CO)_4(PPh_3)$ (**XXXIV**; axial CO replaced)	245	(125)
12	$Me_3SiMn(CO)_5$ (**XXXIV**)	250	(216)
13	$(Me_3Si)_3SiMn(CO)_5$ (**XXXIV**)	256	(338)
14	$Ph_3SiReH(CO)_2(Cp)^b$ (**XLIII**)	249	(404)
15	$Cl_2PhSiRe_2(CO)_9H$	251	(404)
16	$(Et_2Si)_2Re_2(CO)_6H_4$ (**XLI**)	253	(127)
17	$(Ph_2Si)_2Re_2(CO)_8$ (**XXXIX**)	254	(53, 127)
18	$(Ph_2Si)Re_2(CO)_8H_2$ (**XXXVIII**)	254	(173)
19	$(Et_2Si)_2Re_2(CO)_7H_2$ (**XL**)	255	(128)
20	$Me_3SiRe(CO)_5$ (**XXXIV**)	260	(126)
21	$(Me_3Si)_3SiRe(CO)_5$ (**XXXIV**)	267	(126)
22	$Cl_3SiFe(CO)_4^- NEt_4^+$ (analogous to **XXXV**)	222	(256)
23	$(MeF_2Si)_2FeH(CO)(Cp)^b$ (**XLIV**)	225	(405)
24	$(Cl_3Si)_2FeH(CO)(Cp)^b$ (**XLIV**)	225	(313)
25	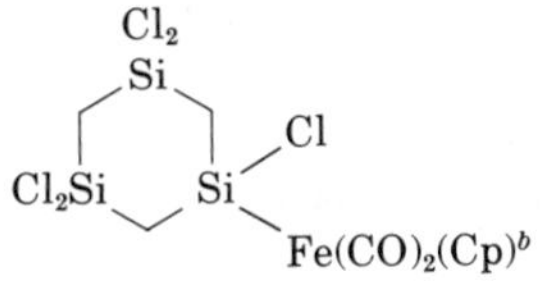	226	(240)
26	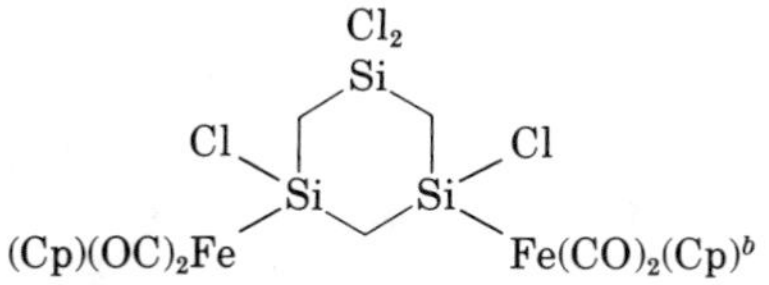	228	(240)
27	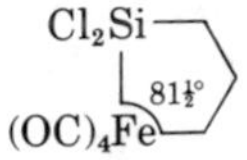	231	(399)

(table continues)

TABLE XVII (*Continued*)

Entry	Compound	d(Si–M) (pm)^a	Ref.
28	*trans*-(Cl$_3$Si)$_2$Fe(CO)$_4$ (**XLV**)	233	(*419*)
29	(Me$_2$PhSi)$_2$FeH(CO)(Cp)^b (**XLIV**)	234	(*403*), [cited in Ref. (*419*)]
30	Me$_2$PhSiFe(Cp)(PMe$_2$Ph)$_2$ ^{b}	234	(*419*)
31	*cis*-Ph$_3$SiFeH(CO)$_4$, analogous to (**XLVI**)	242	(*403*), [cited in Ref. (*128*)]
32	*cis*-(Me$_3$Si)$_2$Fe(CO)$_4$ (**XLVI**)	246	(*419*)
33	(Me$_3$Si)$_2$(μ-SiMe$_2$)$_2$Ru$_2$(CO)$_6$ (**XLVII**)	239, 249, 251	(*130*)
34	Me$_3$SiRu(CO)$_2$[C$_7$H$_7$(C$_6$F$_5$)(SiMe$_3$)] (**XLVIII**)	243	(*245*)
35	Me$_3$SiRu$_2$(CO)$_4$(C$_8$H$_8$SiMe$_3$) (**IL**)	244	(*170, 204*)
36	(C$_8$H$_8$SiMe$_2$CH$_2$CH$_2$)Me$_2$SiRu$_2$(CO)$_5$ (**L**)	246	(*170, 204*)
37	Me$_3$SiRu$_2$(CO)$_5$(C$_7$H$_6$SiMe$_3$) (**LI**)	246	(*74, 242*)
38	(OC)$_4$CoSiCo$_3$(CO)$_9$ (**LII**)	222, 229	(*396*)
39	F$_3$SiCo(CO)$_4$ (**XXXV**)	223	(*177*)
40	Cl$_3$SiCo(CO)$_4$ (**XXXV**)	225	(*375*)
41	Me$_2$P(SiMe$_3$)$_2^+$Co(CO)$_4^-$ (**LIII**)	see text	(*387*)
42	Cl$_3$SiRhH(Cl)(PPh$_3$)$_2 \cdot x$SiHCl$_3$ ($x \approx 0.4$) (**LIV**)	220	(*327*)
43	$\overline{\text{SiMe}_2\text{OMe}_2\text{Si}}$IrH(CO)(PPh$_3$)$_2$ (**LXII**)	241	(*140, 207*)
44	(μ-SiMe$_2$)Pt$_2$[P(cyclohexyl)$_3$]$_2$(C≡CPh)$_2$ (**LIX**)	226; 244	(*105*)
45	N(C$_2$H$_4$O)$_3$SiPtCl(PMe$_2$Ph)$_2$ (**LVI**)	229	(*160*)
46	*trans*-Ph$_2$MeSiPtCl(PMe$_2$Ph)$_2$ (**LV**)	229	(*321*), [cited in Ref. (*134*)]
47	[Ph$_2$MeSiPt(t-BuNC)(μ-t-BuN=CH)]$_2$ (**LX**)	231	(*104*)
48	(*S*)-(+)-*trans*-R$_3$Si*PtCl(PMe$_2$Ph)$_2$ ^{c} (**LV**)	232	(*156*)
49	{Et$_3$SiPt[P(cyclohexyl)$_3$](μ-H)}$_2$ (**LVIII**)	233	(*205*)
50	*trans*-H$_3$SiPtH[P(cyclohexyl)$_3$]$_2$ (**LV**)	238	(*169*)

^{a} To nearest pm; precision normally such that s.d. $\leq$ 0.5 pm.
^{b} Cp = (η^5-C$_5$H$_5$).
^{c} R$_3$Si* = MePh(1-naphthyl).

Regarding silicon–metal bond lengths, it can be seen by comparison of entries 1–4, 12, 13, 20, 21, and 39 that d(Si–M) decreases as R changes in (**XXXIV**) and (**XXXV**) in the order R = Me$_3$Si > Me > H > F. This implies that both steric and electronic factors are significant; the longest bonds are observed when bulky ligands of low electronegativity are attached to silicon. Whether, as has been proposed (*126, 338*), the long bond lengths in entries 13 and 21 correspond to an essentially single covalent bond, all others being more or less contracted by d_π–d_π interaction (or some other effect), remains an open question. The difficulties of estimating such "covalent single-bond lengths" are emphasised in Ref. (*134*) and in many of the papers cited in Table XVII. Moreover, now that experimental bond data are available for a wider range of compounds, it is clear that there are wide variations in the length of any given bond, amounting to about 25 pm in the case of Si–Mn and Si–Fe compounds. It still remains true, however, that by far the longest silicon–metal bond recorded is that in the zirconium derivative of entry 6 (**XXXVI**), a compound in which the metal has a formal d^0 configuration and in which, consequently, no d_π–d_π interaction is expected.

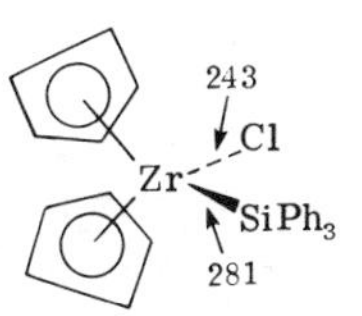

(**XXXVI**)* 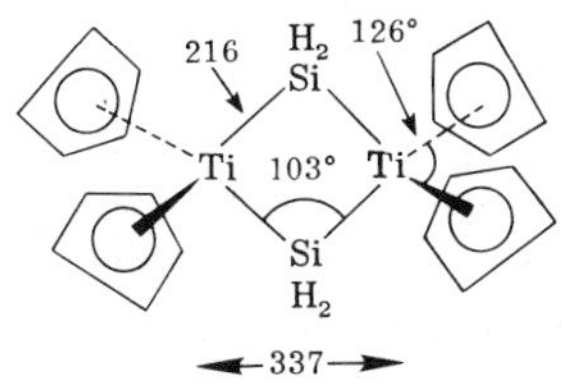   (**XXXVII**)

Strangely, the shortest Si–M distance reported (entry 5) is also in a formally d^0 situation. The compound (**XXXVII**) is prepared by the reaction of $KSiH_3$ and $(\eta^5\text{-}C_5H_5)_2TiCl_2$ in glyme: the reason for its anomalously short Si–Ti bond is not clear.

There has been much interest in the possibility of Si$\cdots$H$\cdots$M bridging, first suggested in connection with the compound $(Ph_2Si)Re_2(CO)_8H_2$ (entry 18) (*173, 250*). Although the hydrogen atoms were not located, infrared spectral evidence supported a structure (**XXXVIII**), with some H$\cdots$Si interaction. However, a series of related compounds (**XXXIX**)–(**XLI**) (entries 16, 17, and 19) have now

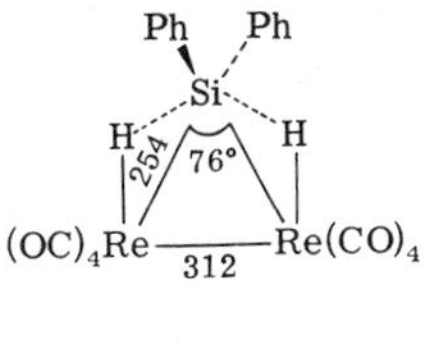

(**XXXVIII**) 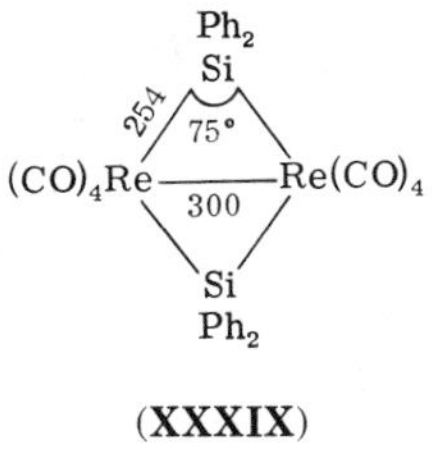   (**XXXIX**)

been studied, and since their structural parameters are very similar to each other and to those of compound (**XXXVIII**), it seems most reasonable to suppose that only terminal Re–H bonds are present. In particu-

* All distances in this and following structures in picometers.

lar, the fact that all Si–Re distances in the unsymmetrical compound (**XL**) are the same argues strongly against any hydrogen bridging. A Raman study of compound (**XXXVIII**) and its deuterated version was interpreted by the authors (*12*) in terms of a bridged structure, but the data are equally consistent with terminal Re–H (Re–D) bonds. It appears that M—H···Si interaction will be significant only when steric constraints force the hydrogen atom close to silicon. This seems to

Et$_2$
Si
255 74° 255 H
(OC)$_4$Re —305— Re(CO)$_3$
H
Si
Et$_2$

(**XL**)

Et$_2$
Si
H 253 75° H
(OC)$_3$Re —308— Re(CO)$_3$
H H
Si
Et$_2$

(**XLI**)

occur in the tungsten derivative (**XLII**, entry 7), in which the Si–W distances are markedly different; the W$_2$Si$_2$ ring and four of the carbonyl groups lie approximately in a plane.

Et Et
Si
259 270 H
(OC)$_4$W —318— W(CO)$_4$
74°
H
Si
Et Et

(**XLII**)

M
O–C
C–O   H···SiR$_3$

(**XLIII**)

Similarly, the two cis-dicarbonyl manganese compounds (**XLIII**, M = Mn, R$_3$ = PhCl$_2$ or Cl$_3$, entries 8 and 10) are sterically constricted, and location of the positions of the hydrogen atoms has given definite evidence for bridging. The Mn–H distances are 149 and 155 pm respectively, while the Si···H distances are 179 and 176 pm, only about 20% greater than the normal length for a Si–H bond (148 pm). Significant Si···H interaction is consistent with the known tendency of these compounds to lose R$_3$SiH in reductive elimination processes (see Section III,G,2). On the other hand, the analogous rhenium derivative (**XLIII**; M = Re, R = Ph) (entry 14) is less hindered because of the greater size of the rhenium atom; the Si···H distance is correspondingly longer, with no significant interaction. A similar situation obtains in the cis-octahedral compound Ph$_3$SiFeH(CO)$_4$ (entry 31): the hydrogen atom is 164 pm from iron but 273 pm from silicon—much too far for effective interaction. The compound Cl$_2$PhSiRe$_2$(CO)$_9$H (entry

15), made by irradiating a mixture of $HSiPhCl_2$ and $Re_2(CO)_{10}$ (*248, 251*), appears to involve a $Re \cdots H \cdots Re$ bridge, but no details of the structure apart from the Si–Re bond length have yet been published.

There is only one example as yet of the effect of substituting carbonyl by tertiary phosphine: *trans*-$Me_3SiMn(CO)_4(PPh_3)$ (entry 11) adopts a structure like that of formula (**XXXIV**), except that the axial carbonyl group is replaced. The Si–Mn bond length decreases somewhat, but the main change concerns the tilting of equatorial carbonyl groups towards silicon. This is now irregular, with values for θ of about 87° for two cis carbonyls and about 81° for the remaining two. The authors attribute this to a minimizing of steric effects. There is also only a single example of an anionic species: $Cl_3SiFe(CO)_4^-$ (entry 22) has a trigonal bipyramidal arrangement analogous to structure (**XXXV**), and a very short Si–Fe bond.

Among the iron compounds, several of the type $(R_3Si)_2FeH(CO)(Cp)$ (entries 23, 24, and 29) possess the structure shown in (**XLIV**), with trans R_3Si groups. In the case when $R_3 = MeF_2$, $d(Fe–H)$ is about 150 pm, while the $Si \cdots H$ distance (206 pm) is not thought to correspond to significant interaction.

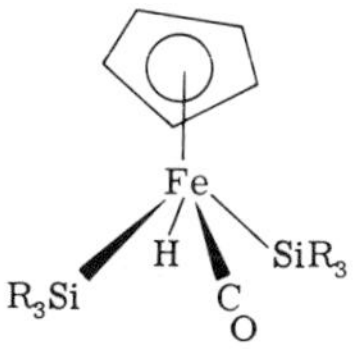

(**XLIV**) 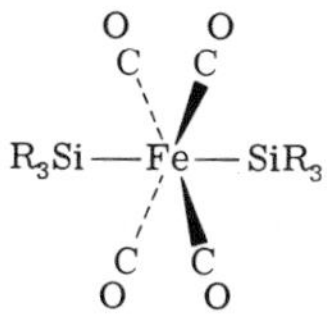   (**XLV**)

Examples of both *trans*- and *cis*-$(R_3Si)_2Fe(CO)_4$ structures are known, (**XLV**) and (**XLVI**), shown in entries 28 and 32 respectively. The *cis*-bis(trimethylsilyl) compound is considerably distorted, to the extent that the $C_{ax}FeC_{ax}$ angle is only 141°. The authors attribute this both to the bulk and the excellent σ-donor characteristics of the Me_3Si groups, and they point out that calculations have suggested that very good σ-donors can stabilize a bicapped tetrahedral geometry (*234*). It is significant that a related compound with less steric congestion and poorly σ-donating ligands, *cis*-$(Cl_3Ge)_2Ru(CO)_4$, adopts an almost regular octahedral arrangement (*45*).

An interesting diruthenium compound (entry 33) has the structure shown in (**XLVII**); the ring is planar, bisected by a metal–metal bond, and the Si–Ru distances vary noticeably according to their bridging or terminal position and, in the former case, whether they are trans to carbonyl or silicon. The remaining ruthenium derivatives in entries

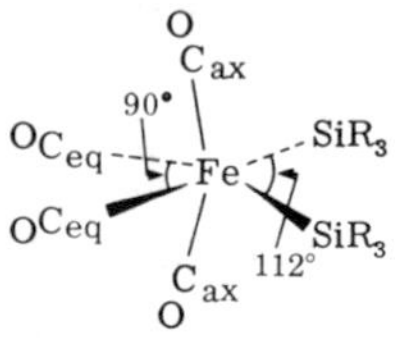

(XLVI) 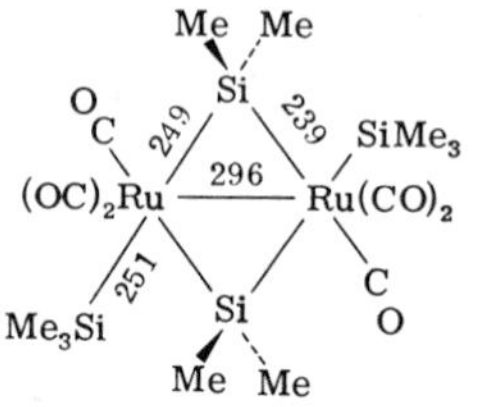   (XLVII)

34–37 will not be discussed in detail, since their main interest lies in the arrangement of the organic portion, but they are depicted in formulas (XLVIII)–(LI).

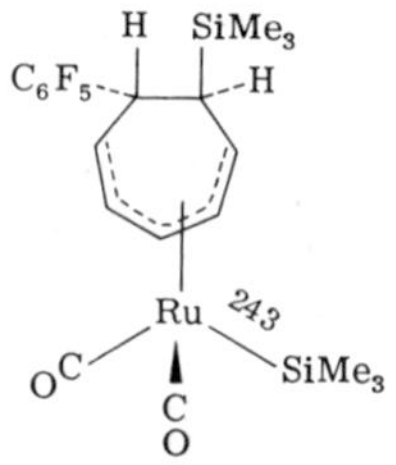

(XLVIII) 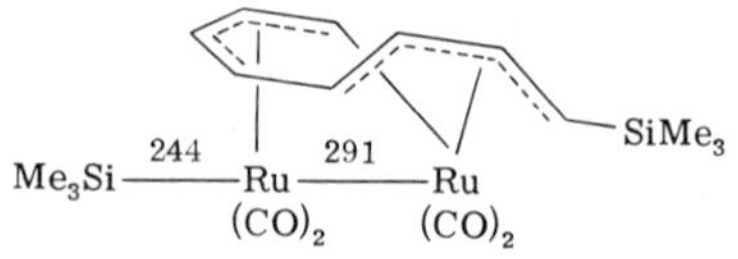   (IL)

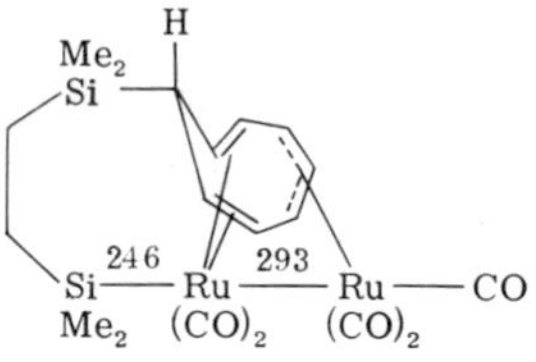

(L) 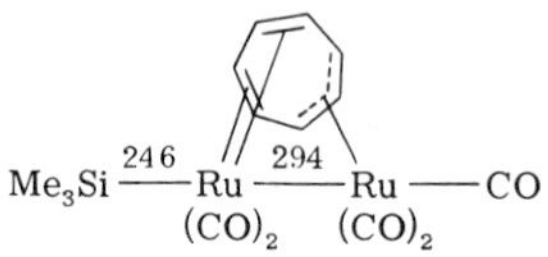   (LI)

It will be seen that the Si–Ru and (where appropriate) Ru–Ru bond lengths are remarkably constant.

When SiI_4 and $NaCo(CO)_4$ are warmed and irradiated together in n-hexane, the cluster compound $(OC)_4CoSiCo_3(CO)_9$ (entry 38) is produced, probably through the intermediacy of $ISiCo_3(CO)_9$. Its structure is shown in formula (LII) and the differing Si–Co(cluster) and Si–Co(exo) bond lengths should be noted.

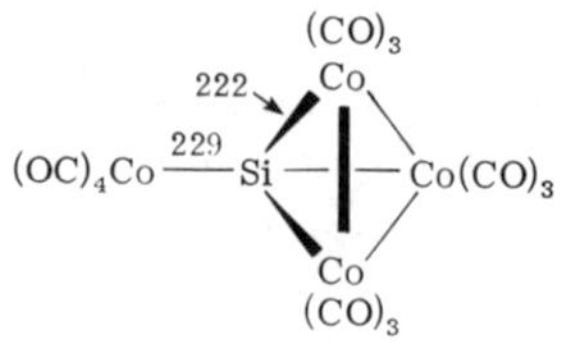

(LII) 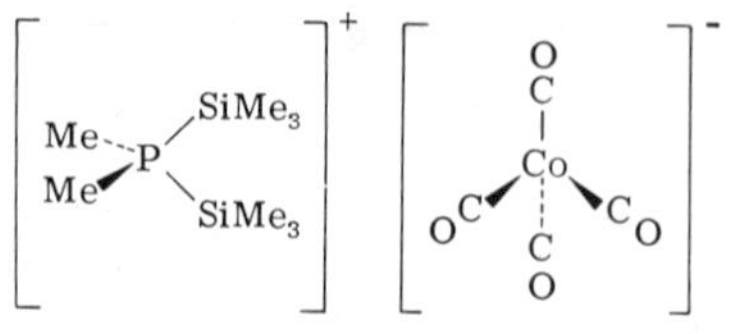   (LIII)

Compound (**LIII**) (entry 41), made from $Me_3SiCo(CO)_4$ and $Me_2P(SiMe_3)$, consists of separated ions, the anion being tetrahedral and the cation approximately so; there is no Si–Co interaction (cf. Section III,D). The coordination around Rh(III) in compound (**LIV**) (entry 42) is approximately trigonal bipyramidal, with phosphines in axial positions. However, an *ortho* hydrogen of one triphenylphosphine ligand molecule is sufficiently close to the metal to raise the coordination number arguably to six; the angle $P\widehat{Rh}P$ is only 162°.

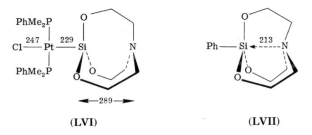

(LIV)   (LV)

Entries 46 and 48 are examples of trans square-planar Pt(II) derivatives (**LV**; X = Cl, R'_3 = $PhMe_2$); the latter, with R_3 = MePh(1-naphthyl), is chiral at silicon, and its structure is particularly significant because it establishes the absolute stereochemistry of the (+)-enantiomer as (*S*). The long Pt–Cl bond (245 pm in each case) should be noted: it is attributed to the large trans influence of silicon (*97, 98, 229*). The simple silyl derivative in entry 50 [**LV**; X = H, R'_3 = (cyclohexyl)$_3$] is prepared from monosilane and $PtH_2[P(cyclo-hexyl)_3]_2$; the Si–Pt distance is appreciably greater than that in other members of this class.

In compound (**LVI**) (entry 45), the silicon is part of a silatrane cage (*24*): in contrast to simple silatranes such as (**LVII**) (*356*), in which there is substantial Si· · ·N interaction, the platinum-substituted version has essentially none, and the coordination about nitrogen is trigonal planar. It is suggested that these effects are a result of electron release from platinum, which provides sufficient negative charge on silicon to discourage electron donation. Again, the Pt–Cl bond trans to silicon is very long.

(LVI)   (LVII)

Another compound containing the tricyclohexylphosphine ligand is the novel hydrogen-bridged species (**LVIII**, cy = cyclohexyl) (entry 49) made from $Pt(C_2H_4)_2[P(cyclohexyl)_3]$ and $HSiEt_3$ at room temperature. The starred atoms in the diagram are nonplanar; they lie in two planes intersecting along the H· · ·H axis, with a dihedral angle of 21°. This compound acts as a potent catalyst for hydrosilation reactions (*205, 206, 415*).

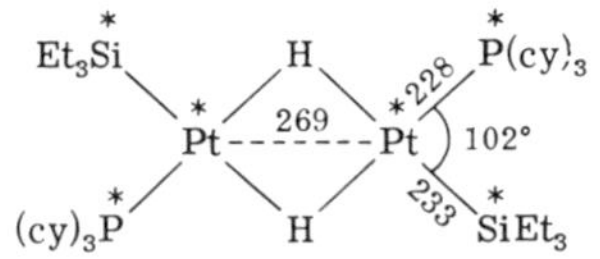

(LVIII)

(cy = cyclohexyl)

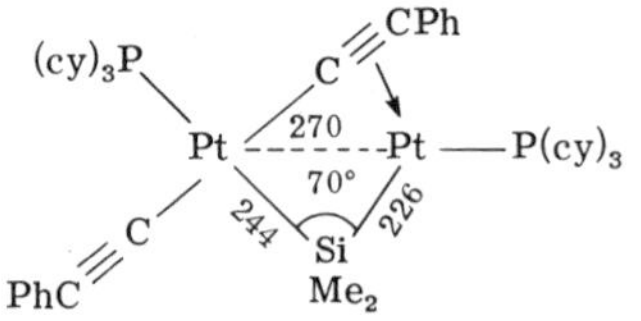

(LIX)

A more complicated structure (**LIX**) (entry 44) results from the reaction of $Pt(C_2H_4)[P(cyclohexyl)_3]_2$ with $Me_2Si(C{\equiv}CPh)_2$; the $SiMe_2$ bridging group is markedly asymmetric. Entry 47 (**LX**) is an example of an isocyanide complex with bridging substituted methyleneimine groups; the central diplatina six-membered ring is in a boat conformation, with no Pt· · ·Pt interaction, and the Pt–N bonds trans to silicon are unusually long.

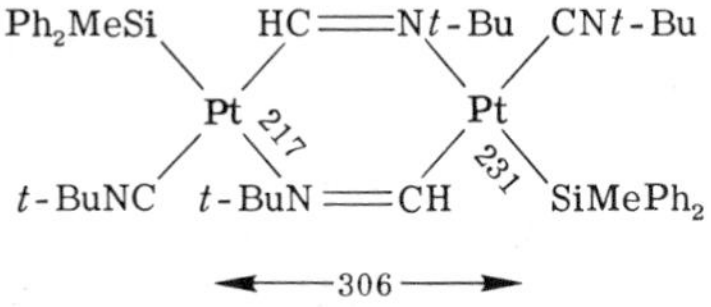

(LX)

Two interesting cyclic compounds are listed as entries 9 and 43. The first (**LXI**) contains a metal–metal bond and a planar Mn_2Si_2 ring, and should be compared with structure (**XXXIX**). The second, (**LXII**), has a four-membered $IrSi_2O$ ring, bent about the Si· · ·Si axis, while the iridium atom is some 26 pm above the P_2Si_2 plane. The SiÔSi bond angle is much less than the normal range (130–180°), and it is clear that the ring is highly strained; this fact may well be related to the compound's ability to catalyze the disproportionation reaction (*139, 207*) of Eq. (122).

$$(Me_2SiH)_2O \rightarrow Me_2SiH_2 + \frac{1}{x}(Me_2SiO)_x \qquad (122)$$

Three other cyclic compounds are depicted in Table XVII as entries 25–27: the first two compounds show a skew-boat conformation, while the last approximates to a chair form.

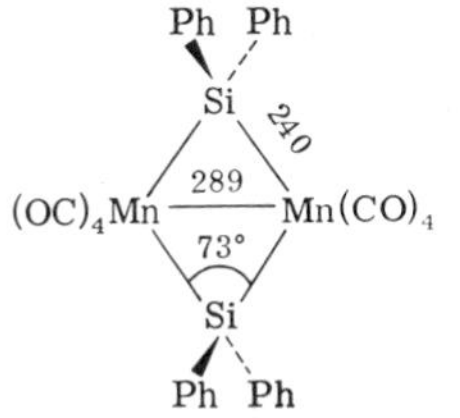

(LXI) 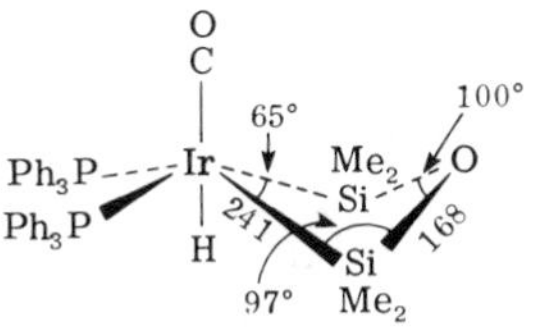   (LXII)

B. Vibrational Spectroscopy

Although routine reports of infrared spectra have often been published, there have been few recent accounts of more complete studies involving Raman measurements, isotopic substitution, or force-constant calculations. Table XVIII also shows that relatively few different types of compounds have received this degree of attention.

It is now clear that some coupling may occur in these molecules between $\nu(Si–M)$, $\nu(CO)$, and Si–X modes. Thus, intense polarized Raman bands assigned to $\nu(Si–M)$ occur at the following frequencies (cm^{-1}): $Me_3SiRe(CO)_5$, 297 (82); $H_3SiRe(CO)_5$, 297; $D_3SiRe(CO)_5$, 284 (121). In $H_2Si[Re(CO)_5]_2$, a similar band occurs at 250 cm^{-1} (32). Also, carbonyl stretching frequencies in the infrared spectra of $H_2Si[M(CO)_5]_2$ (M = Mn, Re) shift by 1–6 cm^{-1} on deuteration (32). It may be noted that there is strong vibrational coupling between the two $M(CO)_5$ units in these compounds. This suggests that a correlation with the established vibrational modes of the parent carbonyls $(OC)_5M—M(CO)_5$ may be made, and indeed bands of the silanediyl compounds have been satisfactorily assigned on this basis.

Very low values of the Si–H stretching frequency are found in some H_2SiM_2 derivatives; the value found for $H_2Si[Fe(CO)_2(Cp)]_2$ of 2042 cm^{-1} (32) compares quite closely with the value of 2037 cm^{-1} predicted from simple additivity considerations (235). These low values are consistent with very low effective electronegativities for metal carbonylate groups (32, 235).

Two quantitative applications may be noted: in the first (entries 12 and 14), infrared intensity measurements are related to the angle of tilt, θ, between equatorial CO groups and the metal–silicon bond [see formulas (XXXIV) and (XXXV)]. Thus in the series $Me_3MCo(CO)_4$ the

TABLE XVIII

Vibrational Spectroscopy of Silicon–Metal Compounds[a]

Entry	Compound	Information[b]	Ref.
1	$H_3SiMn(CO)_5$ [c]	IR/R; AFCC	(7, 30, 299)
2	$Me_3SiMn(CO)_5$ [d]	IR/R; AFCC	(81)
3	$R_3SiMn(CO)_5$ (R = H, Me)	NCA	(349)
4	$Cl_3SiMn(CO)_5$ [e]	IR; NCA	(347, 348)
5	$H_2Si[Mn(CO)_5]_2$ [c]	IR/R; D	(32)
6	$(Ph_2Si)Re_2(CO)_8H_2$ [f]	R; D	(12)
7	$Me_3SiRe(CO)_5$ [d]	IR/R; AFCC	(82)
8	$Cl_3SiFe(CO)_4^-NEt_4^+$	IR/R	(61)
9	$Cl_2MeSiFe(CO)_2(Cp)$ [g]	VT/IR; AFCC; isomers	(141, 142)
10	$R_3SiFe(CO)_2(Cp)$ [g] (R = alkyl, aryl, etc.)	IR	(354)
11	$Cl_3SiCo(CO)_4$ [d]	IR/R; FFC; NCA	(60–62, 148, 424)
12	$X_3SiCo(CO)_4$ (X = Cl, Ph)	IR; intensity/angles	(143)
13	$X_3SiCo(CO)_4$ (X = H, F)	IR/R; FFC	(61)
14	$Me_3SiCo(CO)_4$ [d]	IR/R; intensity/angles	(73, 152)

[a] Routine reports of infrared spectra are generally excluded.

[b] IR = infrared; R = Raman; AFCC = approximate force-constant calculations; NCA = normal coordinate analysis; D = deuterium substitution; VT = variable temperature; FFC = force-field calculations; intensity/angles = infrared intensity measurements used to calculate angles between CO groups.

[c] Also Re.

[d] Also Ge and Sn analogs.

[e] Also X_3Ge (X = Cl, Br); X_3Sn (X = Cl, Br, I).

[f] See Section IV,A for a discussion of these results in connection with possible Re–H · · · Si bridging.

[g] Cp = $(\eta^5\text{-}C_5H_5)$.

angles are estimated to be: M=Si, $82\frac{1}{2}°$; Ge, 94°; Sn, 98° (73). In the second (entry 9), substitution of carbonyl groups by ^{13}CO led to the calculation of force constants for the two rotational conformers of $Cl_2MeSiFe(CO)_2(Cp)$, which arise because of hindered rotation about the Si–Fe bond. Variable-temperature infrared studies on the same compound also led to rough estimates of the small enthalpy and entropy differences between the two forms. In this connection, it has also been reported that more than two carbonyl stretching bands are present in the infrared spectra of $ClMe_2SiFe(CO)_2(Cp)$ and $ClMe(n\text{-}Pr)SiFe(CO)_2(Cp)$, implying that these too exist as different conformers (136).

Continuing attempts have been made to relate vibrational spectra to the bonding in these compounds. In some cases [e.g., entries 11 and 13

(*60, 61*)], the role of d_π–d_π Si–M interactions has been emphasized, while in others (e.g., entry 10), observed effects have been related chiefly to the high σ-donor ability of R_3Si groups. No decision can yet be made as to which is the more realistic of these approaches.

C. NMR SPECTROSCOPY

Developments in instrumental techniques have made possible the direct observation of ^{29}Si-NMR spectra (e.g., *90, 218–220*), and the first results for silicon–transition-metal compounds have now been reported (Table XIXA). The sign convention is such that positive values of δ denote shifts to lower fields (or higher frequencies) with respect to the standard, so that the silicon becomes less shielded. The wide range of values of δ should be noted: the order of decreasing shielding seen in entries 2, 1, 3, and 14, respectively, parallels that observed for tin nuclei in analogous tin compounds, but the factors determining the size of δ are not yet clear, and it certainly cannot be correlated directly with, for example, the extent of d_π–d_π interaction. Entries 11 and 12 show that increasing ring strain is associated with a dramatic increase in δ values.

Measurements of ^{13}C-NMR spectra have provided much useful information concerning the number and proportions of isomers in solution, and their rates of interconversion (Table XIXB). It is generally true that in octahedral derivatives $R_3SiM(CO)_5$ of the type shown in formula (**XXXIV**), the carbon atom of the axial carbonyl group (trans to silicon) is more shielded (i.e., resonates at higher field) than carbon atoms in the equatorial CO groups (entry 7). For disubstituted octahedral compounds $(R_3M)_2M'(CO)_4$ (entry 9), the carbon atoms trans to M [e.g., C_{eq} in (**XLVI**)] are more shielded than the remaining two, except in the solitary case of $(Me_3Sn)_2Fe(CO)_4$ (*420*).

Graham and his colleagues have examined a wide range of compounds (entry 9) to determine the preferred stereochemistry and rates of interconversion of isomers. Some of their results relating to silicon derivatives are shown in Table XX. In certain cases, samples enriched with ^{13}CO were used, while sometimes the presence of the minor isomer was established by infrared methods. For *cis*-$(Me_3Si)_2Fe(CO)_4$, the ^{13}C-NMR spectrum at very low temperatures shows separate resonances for axial and equatorial carbonyl groups (**XLVI**), but at higher temperatures the molecule is fluxional, with a coalescence temperature of $-55°$. Because $^1J(^{13}CO$–$^{57}Fe)$ and $^2J(^{13}CO$–$^{29}Si)$ spin–spin coupling is preserved in the high-temperature limiting spectrum, carbonyl ex-

TABLE XIXA

NMR SPECTROSCOPY OF SILICON–TRANSITION-METAL COMPOUNDS: ^{29}Si MEASUREMENTS

Entry	Compound	δ Value[a]	Ref.
1	$Me_3SiMn(CO)_5$	+17.6	(295)
2	$Me_3SiRe(CO)_5$	−14.4	(295)
3	$Me_3SiFe(CO)_2(Cp)$[b]	+41.0	(295)
		+40.8	(283)
		+41.9	(451)
4	$Me_3Si^2Me_2Si^1Fe(CO)_2(Cp)$[b]	+17.6 (Si^1)	(451)
		−11.1 (Si^2)	(451)
5	$MeClHSiFe(CO)_2(Cp)$[b]	+65.9	(302)
6	$MeF_2SiFe(CO)_2(Cp)$[b]	+68.3	(302)
7	$MeHSi[Fe(CO)_2(Cp)]_2$[b]	+62.5	(302)
8	$ClHSi[Fe(CO)_2(Cp)]_2$[b]	+110.7	(302)
9	$MeClSi[Fe(CO)_2(Cp)]_2$[b]	+142.0	(302)
10	$Cl_2Si[Fe(CO)_2(Cp)]_2$[b]	+146.7	(302)
11	ClMeSi⌐⌐⌐ \ $(OC)_4Fe$—┘	+96	(290)
12	Me_2Si ⟨Fe(CO)_4 / Fe(CO)_4⟩	+173	(68)
13	R^1 Me$_2$Si^1 ⟩C=C⟨ M(CO)_4 / R^2 Si^{2}Me$_2$		(451)
	M = Fe; R^1 = R^2 = Ph	+48.2	
	M = Fe; R^1 = Ph, R^2 = Me$_3$Si^3	+48.8, +45.4 (Si^1, Si^2) −7.6 (Si^3)	
	M = Ru; R^1 = Ph, R^2 = Me$_3$Si^3	+34.7, +32.4 (Si^1, Si^2) −7.9 (Si^3)	
14	$Me_3SiCo(CO)_4$	+44.0	(295)
15	$F_3SiCo(CO)_4$	−29.1^c	(263a)

[a] Referred to Me_4Si standard.
[b] $Cp = (\eta^5\text{-}C_5H_5)$.
[c] Value obtained from ^{19}F double resonance experiments (263a).

change is considered to be a nondissociative, intramolecular process. One possible route would involve rapid cis ⇌ trans ⇌ cis′ equilibria, but other pathways such as a trigonal twist or a tetrahedral jump are not excluded (419). The activation parameters for the positional inter-

TABLE XIXB

NMR Spectroscopy of Silicon–Transition-Metal Compounds: ^{13}C
Measurements and Other Nuclei

Entry	Compound	Comments	Ref.
^{13}C Measurements			
7	X$_3$SiRe(CO)$_5$ [a] (X = Cl, Me)	*trans*-CO more shielded than *cis*-CO	(425)
8	R$_3$SiFe(CO)$_2$(Cp)[b] (R = Me, Ph)	d$_\pi$–d$_\pi$ Si–Fe interaction assumed	(188)
9	(Cl$_n$Me$_{3-n}$Si)$_2$Fe(CO)$_4$ [a,c,d] (n = 0–3)	VT[e]; fluxional (see text)	(361, 362 419, 420)
10	Me$_2$Si $\diagdown$ Fe(CO)$_4$ [c,f] $\diagup$ Me$_2$Si	VT; nonfluxional	(418)
11	(Cl$_3$Si)$_2$Ru(CO)[η^6- C$_6$H$_4$(t-Bu)$_2$-p]	VT; restricted rotation	(363)
12	X$_3$SiCo(CO)$_4$ [f] (X = F, Cl, Ph)	VT; fluxional	(296)
Other nuclei			
13	X$_3$SiMn(CO)$_5$[g] (X = Cl, Ph, C$_6$F$_5$)	^{55}Mn	(47)
14	(μ-SiMe$_2$)$_2$Fe$_2$(CO)$_7$	^{1}H; fluxional	(289)
15	(μ-SiHMe)Fe$_2$(CO)$_3$(Cp)$_2$[b]	^{1}H; isomers	(309)
16	Ph$_3$SiPt(C$_6$H$_4$F)(PMe$_2$Ph)$_2$ and related compounds	^{19}F	(155)
17	*trans*-R$_3$SiPtX(PR$_3'$)$_2$	^{1}H, ^{31}P, ^{195}Pt	(11)

[a] Also Ge, Sn analogs.
[b] Cp = (η^5-C$_5$H$_5$).
[c] Also Ru, Os analogs.
[d] Also ^{1}H study of some Si–Ru, Si–Os derivatives.
[e] VT = Variable-temperature study.
[f] Also ^{1}H study.
[g] Also Ge, Sn, Pb analogs.

change of carbonyl groups in this compound are $\Delta H^{\ddagger}$, 43.5 ± 2.5 kJ mol^{-1}, and $\Delta S^{\ddagger}$, −9.5 ± 11.0 J K^{-1} mol^{-1}, and it appears that the barrier heights increase in the orders Fe < Ru < Os and Me$_3$Si < ClMe$_2$Si < Cl$_2$MeSi < Cl$_3$Si.

Consistent with the idea outlined above, the cyclic compounds in entry 10 are anchored in the cis position, and are shown by ^{13}C- and ^{1}H-NMR measurements to be stereochemically rigid, even at 90°.

The ruthenium compound in entry 11 has a piano-stool structure (**LXIII**), and hindered rotation about the Ru–aromatic ring bond may

TABLE XX

Position of Cis–Trans Equilibria for Compounds $(R_3Si)_2M(CO)_4$ (M = Fe, Ru, Os)[a]

Compound	% Trans isomer	T (K)	Solvent
$(Me_3Si)_2Fe(CO)_4$	0	183	CD_2Cl_2
$(ClMe_2Si)_2Fe(CO)_4$	0	253	Toluene-d_8
$(Cl_2MeSi)_2Fe(CO)_4$	8	253	Toluene-d_8
$(Cl_3Si)_2Fe(CO)_4$	19	300	Toluene-d_8
$(Me_3Si)_2Ru(CO)_4$	0	253	Toluene-d_8
$(ClMe_2Si)_2Ru(CO)_4$	0	303	Toluene-d_8
$(Cl_2MeSi)_2Ru(CO)_4$	12	305	Toluene-d_8
$(Cl_3Si)_2Ru(CO)_4$	[b,c]	—	CD_2Cl_2
$(Me_3Si)_2Os(CO)_4$	63[d]	303	Toluene-d_8
$(ClMe_2Si)_2Os(CO)_4$	60	303	Toluene-d_8
$(Cl_2MeSi)_2Os(CO)_4$	[b]	—	CD_2Cl_2
$(Cl_3Si)_2Os(CO)_4$	100[e]	303	CD_2Cl_2

[a] Data from Refs. (419) and (420).

[b] Cis and trans isomers both isolable; isomerization is negligible at room temperature.

[c] Above 70°C, isomerization proceeds to give equilibrium mixture with 70% trans form.

[d] For cis $\rightleftharpoons$ trans equilibrium, $\Delta H° = 3.5$ kJ mol^{-1}; $\Delta S° = 16$ J K^{-1} mol^{-1}.

[e] Above 120°C, cis isomer isomerizes to 100% trans.

be demonstrated by variable-temperature NMR methods. The least hindered position is that in which the line joining the two bulky t-butyl groups bisects the angle between the two $SiCl_3$ groups.

(LXIII)   (LXIV)

Another example of a fluxional species is $R_3SiCo(CO)_4$, given in entry 12; exchange of axial and equatorial carbonyl groups is easy even at low temperatures, as shown by the low observed coalescence temperatures for R = F, -144°C; Cl, -85°C; and Ph, -61°C. Again the process is seen as intramolecular, with exchange rates chiefly determined by steric factors: the derived free energy of activation when R = F is 29; Cl, 35; and Ph, 40 kJ mol^{-1}. Either Berry pseudorotation or "turnstile" rotation could provide a reasonable pathway for the exchange.

The compound (**LXIV**) (entry 14) provided an early example of a stereochemically nonrigid bridged system. Variable-temperature ^{1}H-NMR spectra showed it to be fluxional at room temperature. A number of related systems have now been studied, including $(OC)_3Co(\mu\text{-}GeMe_2)_2Co(CO)_3$ (5), $(Cp)(CO)Fe(\mu\text{-}CO)(\mu\text{-}GeMe_2)Fe(CO)$ (Cp) (4), and $(OC)_3Fe(\mu\text{-}CO)(\mu\text{-}SnR_2)_2Fe(CO)_3$ (213). All are fluxional, although the second becomes so only above about 90°C.

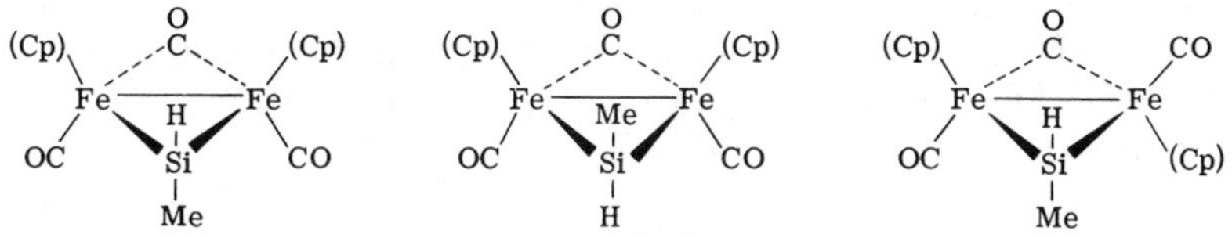

(**LXVa**)   (**LXVb**)   (**LXVc**)

When compound (**LXV**) is prepared by the reaction

$$MeHSi[Fe(CO)_2(Cp)]_2 \xrightarrow{h\nu} (Cp)(CO)Fe(\mu\text{-}CO)(\mu\text{-}SiHMe)Fe(CO)(Cp) + CO$$
$$\text{(\textbf{LXV})} \qquad\qquad\qquad\qquad (123)$$

its solution initially contains the two cis isomers (**LXVa**) and (**LXVb**); however, after a time all three isomers can be detected by ^{1}H-NMR spectroscopy (entry 15).

It was inferred from ^{55}Mn-NMR chemical shift data for the compounds in entry 13 that the ability of the X_3M groups to function as σ-donors increases in the order $M = Ge < Sn < Si$. It is interesting that this order, presumably also one of increasing M–Mn bond strength, agrees with the orders of force constants for group IV–transition-metal bonds derived from vibrational spectra of $Me_3MMn(CO)_5$ (81) and $Cl_3MCo(CO)_4$ (424). Orders of bond dissociation energy derived from appearance potentials are sometimes different, however (see Section IV,D).

The same conclusion, that the Ph_3Si group is a good σ-donor, was drawn from ^{19}F-NMR experiments on fluorophenyl derivatives of Si–Pt compounds (entry 16); additionally, it was concluded that Ph_3Si attached to Pt was a good π-acceptor. A wide range of chemical shifts and coupling constants involving ^{1}H, ^{31}P, and ^{195}Pt nuclei were determined by double-resonance experiments on a large number of Si–Pt derivatives (entry 17).

Useful information can sometimes be obtained from $^1J(^{29}Si\text{-}^1H)$ coupling constants. It has been shown that their magnitude depends chiefly on the Fermi contact term, and hence on the degree of s character in the Si–H bond (149). Exceptionally low values of the $(^{29}Si\text{-}H)$

coupling constant are observed for compounds of the type $H_2Si(ML_n)_2$: for $ML_n = Re(CO)_5$, it is only 165 Hz (32). This implies that the s character of the Si–H bonds is very small and hence, from Bent's rule (55), that the (ML_n) groups have very low effective electronegativities. Similar conclusions have been reached from measurements of Si–H stretching frequencies in these compounds (see Section IV,B) and of $^2J(^{119}Sn–C–H)$ coupling constants in compounds such as $Me_2Sn[Re(CO)_5]_2$ (413). In pictorial terms, this effect is equivalent to a structure (**LXVI**); the bulky (ML_n) groups are as far apart as possible, and the Si–M bonds have essentially sp character. Diffraction experiments are needed to confirm this.

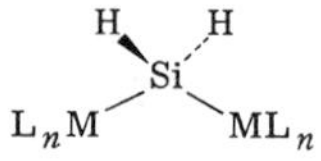

(**LXVI**)

D. MASS SPECTROMETRY AND APPEARANCE POTENTIALS

In Table XXI are listed those compounds which have been studied in some detail by the use of this technique.

The normal fragmentation pattern for a compound $R_3SiM(CO)_5$ involves loss of CO and concomitant loss of R (especially when R = H); although Si–M cleavage may occur, SiM^+ is always an abundant species. For the compounds $H_2Si[M(CO)_5]_2$ (M = Mn, Re; entry 6), the following pattern has been established, where $x = 0-2$ and $y = 0-10$:

$$SiH_2M_2(CO)_{10}^+ \rightarrow SiH_xM_2(CO)_y^+ \rightarrow SiM_2^+ \rightarrow SiM^+ \rightarrow Si^+$$
$$\downarrow \qquad \downarrow$$
$$M_2^+ \quad \rightarrow M^+ \qquad\qquad (124)$$

Other processes (all established by the observation of enhanced metastable peaks) include elimination of R–R, SiR_2, and $M(CO)_5$, e.g. (32),

$$SiR_2Re_2(CO)_9^+ \rightarrow SiRe_2(CO)_9^+ + R_2 \qquad (R = H, D) \qquad (125)$$

$$SiF_2Mn_2(CO)_5^+ \rightarrow Mn_2(CO)_5^+ + SiF_2 \qquad\qquad (126)$$

$$SiF_2M_2(CO)_{10}^+ \rightarrow SiF_2M(CO)_5^+ + M(CO)_5 \qquad (M = Mn, Re) \qquad (127)$$

The cationic product in Eq. (127) is a substituted silicenium ion; attempts to produce a corresponding species chemically by abstraction of R^- from $R_3SiM(CO)_n$ or $R_2Si[M(CO)_n]_2$ have all been unsuccessful [see Section III,C and Ref. (24)]. Thus addition of the chloride-ion acceptor $SbCl_5$ to $Cl_3SiRe(CO)_5$ results in Si–Re bond cleavage (32).

TABLE XXI

Mass Spectroscopic Studies of Silicon–Transition-Metal Compounds

Entry	Compound	Ref.
1	$R_3SiCr(CO)_3(Cp)$ ($R_3 = F_xMe_{3-x}$; $x = 0–3$)	(304)
2	$R_3SiMn(CO)_5$ ($R = H$)	(30)
3	$R_3SiMn(CO)_5$ ($R = Me$)[a] AP[b]	(81, 410)
4	$R_3SiMn(CO)_5$ [$R_3 = (C_6F_5)_xPh_{3-x}$; $x = 0–3$][a] AP[b]	(108)
5	$Me_3SiMn(CO)_n(PF_3)_{5-n}$ ($n = 2–5$)	(378)
6	$R_2Si[Mn(CO)_5]_2$ ($R = H$)[c]	(2, 32)
7	$R_2Si[Mn(CO)_5]_2$ ($R = Cl$)	(2)
8	$Me_3SiRe(CO)_5^g$ AP[b]	(82)
9	$R_3SiFe(CO)_2(Cp)$ ($R_3 = H_3, Br_3, I_3, Br_2OMe, \ldots$)	(237)
10	$R_3SiFe(CO)_2(Cp)$ ($R = Me$)[d] AP[b]	(253, 410)
11	$R_3SiFe(CO)_2(Cp)$ [$R_3 = Cl_x(OR)_{3-x}$; $x = 0–3$]	(236)
12	$R_3SiFe(CO)_2(Cp)$ ($R_3 = F_xMe_{3-x}$; $x = 0–3$)	(304)
13	$Me_3SiFe(CO)(PPh_3)(Cp)$ AP[b]	(410)
14	$R_2Si[Fe(CO)_2(Cp)]_2$ ($R = H$)	(32)
15	$R_2Si[Fe(CO)_2(Cp)]_2$ ($R = HCl, HMe, Cl_2, ClMe, F_2$)	(309)
16	$R_3SiCo(CO)_4$ ($R = H$)	(29)
17	$R_3SiCo(CO)_4$ ($R_3 = Cl_3, F_3, Me_3, MeF_2$) AP[b]	(379, 380)
18	$R_3SiCo(CO)_4$ ($R = Me$)[a] AP[b]	(82)
19	$R_3SiCo(CO)_4$ ($R = C_6F_5$)	(398)
20	$Cl_3SiCo(CO)_n(PF_3)_{4-n}$ ($n = 2, 3$)	(381)
21	$H_2Si[Co(CO)_4]_2$	(2)
22	$Cl_2SiCo_2(CO)_7$	(2)

[a] Also Ge, Sn analogs.
[b] AP = appearance potential measurements.
[c] Also Re analog.
[d] Also Ge, Sn, Pb analogs.

$$Cl_3SiRe(CO)_5 + 2SbCl_5 \rightarrow Re(CO)_5Cl \cdot SbCl_5 + SiCl_4 + SbCl_3 \qquad (128)$$

Fragmentation of compounds with metal–cyclopentadienyl bonds has been observed to give rise to very abundant $C_5H_5Si^+$ ions (e.g., entries 9, 14, and 15). In Eq. (129), they correspond to the base peak (32).

$$X_2Si[Fe(CO)_2(Cp)]_2 \rightarrow C_5H_5Si^+ + \cdots \qquad (X = H, Cl) \qquad (129)$$

It was suggested that this ion might have the penta-*hapto* arrangement (**LXVII**): calculations have since shown that this arrangement is favored energetically over alternative likely structures (287), while X-ray studies of the solid compound $C_5Me_5Sn^+BF_4^-$ reveal that $C_5Me_5Sn^+$ adopts a similar geometry (267).

(LXVII)

Other processes established include the migration of halogen from silicon to metal (e.g., entries 1, 9, and 12) and of fluorine from carbon to metal (entry 4). Ions containing silicon and oxygen are sometimes found (e.g., entries 2 and 16); while they may arise from adventitious hydrolysis, it seems probable that they are also the result of migration of silicon from metal to carbonyl oxygen (see Section III,F,2).

Measurements of appearance potentials can, in favorable circumstances, provide estimates of silicon–transition-metal bond dissociation energies. Thus D(Si–M) in $Me_3SiM(CO)_5$ is 240 kJ mol^{-1} (M = Mn) (*81, 108*) and 300 kJ mol^{-1} (M = Re) (*82*). This illustrates the general point that the heavier transition metals in a group form the strongest bonds with silicon; other group IV elements do the same (*87*). Confirmation of these high values for D(Si–M) comes from measurements of the dissociation of the diatomic molecule AuSi at high temperatures in the gas phase, from which a value of 318 ± 17 kJ mol^{-1} for D(Au–Si) was inferred (*192*).

For a series of compounds $Me_3MMn(CO)_5$, the dissociation energy D(M–Mn) remains similar or falls slightly on passing from M = Si to Ge to Sn, while estimates of force constants from vibrational spectra also suggest that the Si–Mn bond is a little stronger than the Ge–Mn linkage (*129, 410*). In the case of $Me_3MRe(CO)_5$, however, the bond dissociation energies increase in the order M = Si < Ge < Sn (*82*). Although an exact value for D(Si–Co) in $Me_3SiCo(CO)_4$ could not be obtained, it appears to be greater than 230 kJ mol^{-1} and less than the corresponding value for D(Sn–Co) (*82*). It is interesting that the value of D(Si–Fe) rises from 190 kJ mol^{-1} in $Me_3SiFe(CO)_2(Cp)$ to 215 kJ mol^{-1} in the phosphine-substituted derivative $Me_3SiFe(CO)(PPh_3)(Cp)$. In corresponding pairs of compounds, D(Si–Fe) is again less than D(Sn–Fe) (*410*), and bonds to Cr, Mo, and W probably become stronger on passing from Si to Ge to Sn (*87*).

E. Optical Activity in Silicon–Transition-Metal Compounds

The synthesis of a number of compounds of the type $R^1R^2R^3SiM(L)_n$, with a chiral silicon atom, has made it possible to examine the stereochemistry of reactions of silicon in some detail. Table XXII shows that compounds with silicon linked to Mn, Fe, Co, and Pt have been studied.

Compound (**LXVIII**) in entries 1–6 is readily made by the oxidative addition–elimination reaction:

$$(\eta^5\text{-MeC}_5\text{H}_4)\text{Mn(CO)}_3 \;+\; (+)\text{-R}_3\overset{*}{\text{Si}}\text{H} \xrightarrow[-\text{CO}]{h\nu} (-)\text{-}(\eta^5\text{-MeC}_5\text{H}_4)(\text{CO})_2(\text{H})\text{Mn}-\overset{*}{\text{Si}}\overset{\text{Ph}}{\underset{\text{Np}}{\text{Me}}}$$

$$(\textbf{LXVIII})$$

$$(130)$$

It is assumed, in common with other reactions involving metal insertion into Si–H bonds (see below), that this reaction proceeds with retention of configuration at silicon.

Its reactions fall into two chief categories. In the first (entries 1–3), nucleophiles attack the silicon atom with inversion at silicon: this is consistent with the manganese-containing unit behaving as a good leaving group (*24, 406*). In the second type, either nucleophilic (entries 4 and 5) or electrophilic (entry 6) attack leads to reductive elimination of the original silane, with retention of configuration; this process is termed "de-insertion" by the original authors, and is related to the M—H· · ·Si interaction already present in the molecule of (**LXVIII**), evidenced by a Si· · ·H distance of only 176 pm in Ph₃SiMnH(CO)(Cp) (*248, 251*) (see Sections III,G,2 and IV,A).

Compound (**LXIX**) results from the reaction of the appropriate chlorosilane with [Fe(CO)₂(η⁵-C₅H₅)]⁻ anions (*91, 92, 123*); such reactions proceed with inversion of configuration (*406*):

$$(-)\text{-R}_3\overset{*}{\text{Si}}\text{Cl} \;+\; \text{Na}^+\text{Fe(CO)}_2(\text{Cp})^- \xrightarrow[-78°]{\text{THF}} (+)\text{-}(\text{Cp})(\text{CO})_2\text{Fe}-\overset{*}{\text{Si}}\overset{\text{Ph}}{\underset{\text{Me}}{\text{Np}}}$$

$$(\textbf{LXIX}) \qquad (131)$$

Reaction with nucleophiles is difficult (entries 8 and 9), and proceeds with retention in each case; this is consistent with the known behavior of the Fe(CO)₂(Cp) unit as a good nucleophile and a poor leaving group (*147*). Electrophilic attack by chlorine also involves retention (entry 10).

In contrast to the behavior of the manganese derivative (**LXVIII**), the iron compound undergoes carbonyl substitution without cleavage of the Si–M bond (entry 7).

$$(+)\text{-R}_3\text{Si}^*\text{Fe(CO)}_2(\text{Cp}) + \text{PPh}_3 \xrightarrow[-\text{CO}]{h\nu} (\pm)\text{-R}_3\text{Si}^*\text{Fe}^*(\text{CO})(\text{PPh}_3)(\text{Cp}) \qquad (132)$$
$$(\textbf{LXX})$$

The product (**LXX**) contains an additional chiral center at iron (with four different attached groups), and is consequently formed as a pair of

TABLE XXII

Optically Active Silicon–Metal Compounds and Their Reactions

Entry	Compound	Reagent	Silicon product[a]	Reference
1	(−)-R$_3$Si* MnH(CO)$_2$(Cp')[b,c] (**LXVIII**)	H$_2$O	(−)-R$_3$Si*OH[d] (I)	(117, 118)
2		MeOH/NEt$_3$	(−)-R$_3$Si*OMe[d] (I)	(117, 118)
3		LiAlH$_4$	(−)-R$_3$Si*H (I)	(117, 118)
4		PR$_3$ (R = Ph, OPh)	(+)-R$_3$Si*H[e] (R)	(117, 118)
5		LiMe	(+)-R$_3$Si*H (R)	(117, 118)
6		Cl$_2$ or CCl$_4$	(+)-R$_3$Si*H (R)	(117, 118)
7	(R)-(+)-R$_3$Si* Fe(CO)$_2$(Cp)[b,f] (**LXIX**)	PPh$_3$/hv	(±)-R$_3$Si*Fe*(CO)(PPh$_3$)(Cp)[g]	(91, 92)
8		H$_2$O	(R$_3$Si*)$_2$O[d] (R)	(91, 92)
9		LiAlH$_4$	(−)-R$_3$Si*H (R)	(91, 92)
10		Cl$_2$	R$_3$Si*Cl[h] (R)	(91, 92)
11		P(OPh)$_3$	(−)-R$_3$Si*H (R)	(93)
12	(−)-R$_3$Si* Fe*(CO)(PPh$_3$)(Cp)[b,f] (**LXX**)	LiAlH$_4$	(−)-R$_3$Si*H (R)	(91, 92)
13		Cl$_2$	R$_3$Si*Cl[h] (R)	(91, 92)
14		Cl$_2$/PPh$_3$	R$_3$Si*Cl[h] (I)	(91, 92)
15	(S)-(+)-R$_3$Si*Co(CO)$_4^b$ (**LXXI**)	Et$_3$SiH	(+)-R$_3$Si*H (R)	(407, 408)
16		R'OH (R = H, Me)[i]	(−)-R$_3$Si*OR' (I)	(114, 115, 407, 408)
17		LiAlH$_4$	(−)-R$_3$Si*H (I)	(114, 115)
18		Hg(OAc)$_2$	(−)-R$_3$Si*OAc (I)	(114, 115)
19		NMe$_3$(CH$_2$Ph)$^+$F$^-$	(±)-R$_3$Si*F (rac)	(114, 115)
20		X$_2$ (X = Cl, Br)	(−)-R$_3$Si*X[h] (R)	(114, 115)

21	Cl_2/PPh_3	$(\pm)$-R_3Si^*Cl (rac)	(114, 115)
22	LiMe (excess)[j]	$R_3Si^*Li^k$ (R)	(114, 115)
23	LiPh (1 mole)	$trans$-$[R_3Si^*Co(CO)_3COPh]^-Li^{+l}$	(120)
24	$LiAlH_4$	$(+)$-R_3Si^*H (R)	(157, 163, 164)

$trans$-(S)-$(+)$-$R_3Si^*PtCl(L)_2$ **(LXXII)** (L = PMe$_2$Ph)

25	PhSH	$(+)$-R_3Si^*H (R)	(157, 163, 164)
26	Et_3SiH	$(+)$-R_3Si^*H (R)	(157, 163, 164)
27	X_2 (X = Br, I)	R_3Si^*Cl (R)	(157, 163, 164)
28	$LiAlH_4$	$(-)$-R_3Si^*H (R)	(163, 164)

$trans$-$(-)$-$R_3Si^*PtH(L)_2$ **(LXXIII)** (L = PPh$_3$)

29	PhSH	$(-)$-R_3Si^*H (R)	(163, 164)
30	PhCOCl	$(-)$-R_3Si^*H (R)	(163, 164)
31	$PhC{\equiv}CH$	$(-)$-R_3Si^*H (R)	(163, 164)

[a] R = retention, I = inversion, rac = racemization.
[b] R_3Si^* = MePh(1-naphthyl)Si.
[c] Cp′ = η^5-MeC$_5$H$_4$.
[d] Converted into R_3Si^*H by LiAlH$_4$, with retention, to estimate optical purity.
[e] Also (Cp′)Mn(CO)$_2$(PR$_3$).
[f] Cp = η^5-C$_5$H$_5$.
[g] Similar mixtures of diastereoisomers result using P(cyclohexyl)$_3$ and P(OEt)$_3$.
[h] Converted into R_3Si^*H by LiAlH$_4$, with inversion, to estimate optical purity.
[i] With NEt$_3$.
[j] MeMgBr behaves similarly but with lower stereospecificity.
[k] Converted by water into R_3Si^*H, with retention, to estimate optical purity.
[l] On heating, gives R_3Si^*COPh, with retention.

diastereoisomers. These can be separated by fractional crystallization, and their reactions have been studied. The (−) isomer behaves similarly to the parent compound (**LXIX**) in its reactions with LiAlH$_4$ or Cl$_2$ (entries 12 and 13), but predominant inversion of configuration occurs with Cl$_2$ and PPh$_3$ together (entry 14). The reasons for this behavior are not yet clearly understood.

An interesting reaction occurs between compound (**LXIX**) and P(OPh)$_3$ (entry 11); R$_3$Si*H is eliminated in a process reminiscent of reductive elimination, with retention of configuration; the proposed mechanism, involving *ortho*-hydrogen elimination from a phenyl group of the ligand, has already been outlined in Section III,B,4, and the observed retention underlines the similarity of this reaction to that of R$_3$Si*MnH(CO)$_2$(η^5-MeC$_5$H$_4$) in entry 4.

The cobalt compound R$_3$Si*Co(CO)$_4$ (**LXXI**) was historically the first optically active silicon–transition-metal derivative to be prepared (*407*), using the route of Eq. (133).

$$2(+)\text{-}\overset{*}{R_3}SiH \ + \ Co_2(CO)_8 \ \xrightarrow{-H_2} \ 2(+)\text{-} \ \underset{Np}{\overset{Ph}{\underset{\underset{Me}{|}}{\diagdown}}} \overset{*}{Si} - Co(CO)_4 \qquad (133)$$

(**LXXI**)

An exactly similar reaction is observed in the case of (+)-R$_3$Ge*H, and the product, (+)-R$_3$Ge*Co(CO)$_4$, has been shown by X-ray diffraction techniques to have the absolute configuration (S) (*115*). In the same way, diffraction studies show the absolute configuration of (+)-R$_3$Si*H to be (R), and (+)-R$_3$Ge*H is known to be the same (*116*). It may therefore be deduced that the product of reaction (133) is (S)-(+)-R$_3$Si*Co(CO)$_4$, and that the reaction follows the general rule of retention at silicon for reactions involving substitution or metal insertion at a Si–H bond (*164, 406*).

Attack of nucleophiles on R$_3$SiCo(CO)$_4$ generally proceeds readily with Si–Co bond cleavage and inversion at silicon (entries 16–18); this is consistent with Co(CO)$_4$ being a good leaving group. Entry 15 shows a substitution reaction of the Si–H bond, which goes with retention. This is clearly related to the fact that R$_3$Si*Co(CO)$_4$ acts as a catalyst for Si*–H/Si*–D exchange with complete retention of configuration (*407*).

$$(+)\text{-}R_3Si^*H + PhEtMeSiD \xrightarrow{\text{catalyst}} (+)\text{-}R_3Si^*D + PhEtMeSiH \qquad (134)$$

Electrophilic attack by Cl$_2$ causes cleavage with retention (entry 20); a number of other compounds either do not react (e.g., alkyl, aryl, and acyl halides, BF_3) or cause racemization (entries 19 and 21) (115).

Organolithium compounds can yield more than one product with $R_3Si^*Co(CO)_4$; equimolar amounts give an adduct resulting from nucleophilic attack by $R^{\delta-}$ on oxygen of a *trans*-carbonyl group (entry 23; compare Sections III,B,7 and III,F,1).

$$(+)\text{-}R_3Si^*Co(CO)_4 + LiPh \rightarrow trans\text{-}[R_3Si^*Co(CO)_3(COPh)]^-Li^+$$

$$(-)\text{-}R_3Si^*COPh \qquad (+)\text{-}[(Ph_3P)_2N]^+\{trans\text{-}[R_3SiCo(CO)_3(COPh)]\}^- \quad (135)$$

with the arrow labeled *heat* on the left and $(Ph_3P)_2N^+Cl^-$ on the right downward.

The anion can be isolated as its salt with the large cation, $[(Ph_3P)_2N]^+$; on heating, the lithium salt decomposes to give a benzoylsilane with retention. With an excess of organolithium reagent, the optically active silylanion derivative R_3Si^*Li is produced (entry 22), possibly via an adduct of the type shown above. It may be noted that the analogous germyl anion reagents, R_3Ge^*Li, have been used to prepare optically active Ge–Ni, Ge–Mo, and Ge–W derivatives (116), and the silicon analog could also prove to be a useful synthetic agent.

Two methods have been used to make optically active silicon–platinum compounds (157, 163, 164):

$$(+)\text{-}R_3Si^*H + cis\text{-}PtCl_2(L)_2 \xrightarrow{NEt_3} trans\text{-}(+)\text{-}R_3SiPtCl(L)_2 \qquad (136)$$
$$(L = PMe_2Ph) \qquad \textbf{(LXXII)}$$

$$(+)\text{-}R_3Si^*H + Pt(C_2H_4)(L)_2 \longrightarrow (-)\text{-}R_3SiPtH(L)_2 \qquad (137)$$
$$(L = PPh_3) \qquad \textbf{(LXXIII)}$$

Both reactions are thought to proceed with retention of configuration. This has been confirmed in the first case by an X-ray diffraction study of compound (**LXXII**) (156), which has an absolute configuration of (S), corresponding to that of its precursor (R)-(+)-R_3Si^*H [compare Section IV,A, formula (**LV**)]. It will be seen that all the reactions in entries 24–31 proceed with retention, and it has been suggested (163, 164) that this is the norm for Si–Pt compounds, and that the reactions all involve six-coordinate intermediates, with successive steps of oxidative addition and reductive elimination. It is possible to demonstrate stereochemical cycles with a high degree of retention of optical activity [93% in reaction (138)].

$$(+)\text{-}R_3Si^*H \xrightarrow{cis\text{-}PtCl_2(L)_2} trans\text{-}(+)\text{-}R_3SiPtCl(L)_2 \qquad (138)$$

with $LiAlH_4$ closing the cycle back to the starting material.

F. Other Techniques

1. Kinetic Measurements

Several kinetic determinations have been performed with silicon–transition-metal compounds. These include the rates of cleavage of $R_3SiFe(CO)_2(Cp)$ and $Me_3SiMn(CO)_5$ by $HgBr_2$ or iodine (102, 103). In both cases the rates for corresponding group IV compounds increased in the order Si < Ge ≪ Sn, and it was suggested that the reactions proceed via an S_E2 open and a cyclic transition state respectively. Investigations of the kinetics of oxidative addition to Ir(I) compounds by R_3SiH have already been discussed in Section II,F (179, 221, 222, 224), while the kinetics of reductive elimination of R_3SiH have been investigated for systems such as

$$Ph_3SiMnH(CO)_2(Cp) + PPh_3 \rightarrow Ph_3SiH + Mn(CO)_2(PPh_3)(Cp) \qquad (139)$$

It was concluded that $HSiPh_3$ is lost in a concerted unimolecular process (225).

A recent investigation of the hydrosilation of alkynes by Et_3SiH in the presence of the bridged derivative $\{(PhCH_2)Me_2SiPt[P(cyclo-hexyl)_3](\mu\text{-H})\}_2$, analogous to (LVIII), has shown that the rate increases as the π-acceptor power of the alkyne increases, consistent with the formation of Si–Pt–alkyne complexes as intermediates (416). This complements earlier kinetic studies on the hydrosilation of alkenes catalyzed by Co, Rh, and Pt compounds, in which similar Si–metal-alkene intermediates were postulated (85, 134).

2. Photoelectron and Mössbauer Spectroscopy

Photoelectron spectra using ultraviolet excitation have been reported for $H_3SiMn(CO)_5$ (129, 369), $F_3SiMn(CO)_5$ (371), $Me_3SiMn(CO)_5$ (319), $H_3SiRe(CO)_5$ (129), and $H_3SiCo(CO)_4$ (129). In a number of cases, comparisons have been made with similar carbon and germanium derivatives, and attempts have been made to relate the inferred ordering of molecular-orbital energy levels to observed electronic spectra (319). Although it is hard to draw firm conclusions from the results, it seems likely that the σ-acceptor power of the MR_3 groups increases in the order M = C < Ge < Si. There is no clear evidence for d_π–d_π interaction between silicon and the transition metal, and in particular a decrease in Si–M bond length (and hence an increase in bond energy) on passing from $H_3SiMn(CO)_5$ to $F_3SiMn(CO)_5$ is attributed to stronger σ-interaction in the latter compound (371).

The X-ray–excited O(1s) photoelectron spectra of various compounds of the type RMn(CO)$_5$ have been correlated with infrared carbonyl stretching frequencies, and conclusions have been drawn about the extent of π-donation from metal to ligands; in particular, it is inferred that the SiCl$_3$ group in Cl$_3$SiMn(CO)$_5$ is a rather poor π-acceptor (265).

Measurements of Mössbauer spectra have been confined to a number of silicon–iron derivatives (^{57}Fe spectra). Compounds studied include (H$_3$Si)$_2$Fe(CO)$_4$ and H$_3$SiFeH(CO)$_4$ (48); Ph$_n$X$_{3-n}$SiFe(CO)$_2$(Cp) (X = Cl, C$_6$F$_5$) (107); Cl$_3$SiFeH(diphos)$_2$ and (R$_3$Si)$_2$Fe(diphos)$_2$ (R = Cl, OEt; diphos = Ph$_2$PCH$_2$CH$_2$PPh$_2$) (355); [ClMeSiFe(CO)$_4$]$_2$ (395); and a range of derivatives RFe(CO)$_2$(Cp) in which R = Ph$_n$Me$_{3-n}$Si (n = 0–3), Et$_3$Si, Me$_5$Si$_2$, Me$_7$Si$_3$, or various organo groups (354). In the last case, it was shown that the silyl compounds had greater s-electron density at iron than their alkyl analogs, and the authors suggested that this was related to the superior σ-donor ability of silyl groups rather than to significant silicon–metal π-interaction (354). The dimeric character of the compound [ClMeSiFe(CO)$_4$]$_2$, with a four-membered FeSiFeSi ring, was confirmed by its Mössbauer spectrum, typical of that expected for six-coordinate iron (395). There appear to have been no instances of NQR spectra since the early work on a range of group IV compounds of general type X$_3$MCo(CO)$_4$ (M = Si, Ge, Sn, Pb) (78, 341); in the case of Cl$_3$SiCo(CO)$_4$, the ^{59}Co, ^{35}Cl, and ^{37}Cl resonances were observed, and some evidence was adduced for Si–Cl π-bonding.

V. Topics of Special Interest

A. Pyrolysis and Chemical Vapor Deposition: Metal Silicides

Silicon–transition-metal compounds vary widely in their resistance to thermal degradation; in the case of the hydrido derivatives, some approximate decomposition temperatures are: H$_3$SiV(CO)$_6$, 25°C; H$_3$SiFeH(CO)$_4$, 70°C; H$_3$SiCo(CO)$_4$, 100°C; H$_3$SiMn(CO)$_5$, 130°C; (H$_3$Si)$_2$Fe(CO)$_4$, 150°C; H$_2$Si[Re(CO)$_5$]$_2$, 180°C (25). Where comparisons are possible, they are all more stable thermally than their carbon analogs. Although reports of corresponding temperatures for organo- and halogenosilyl derivatives are sparse (42, 213a, 215, 235, 306), it appears that most of these compounds begin to decompose below 200°C.

Early studies on H$_3$SiCo(CO)$_4$ and H$_3$SiMn(CO)$_5$ showed that H$_2$ and CO were important volatile decomposition products, and it was at first supposed that the dark residual solid consisted of a highly cross-linked network of SiH$_x$ and M(CO)$_y$ units, linked by silicon–metal bonds (29, 30), like (**LXXIV**).

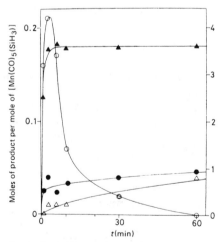

(**LXXIV**)

It soon became clear, however, that other processes, such as Si–O bond formation and disproportionation, were occurring, as already discussed in Sections III,F,2 and III,G,1.

A more detailed study of the pyrolysis of H$_3$SiMn(CO)$_5$ in a sealed tube for various periods of time at 450°C showed that the volatile products were H$_2$, CO, SiH$_4$, and CH$_4$; a metallic-looking brown film covered the walls of the tube, and an apparently amorphous grey powder was also present (33). Figure 5 shows that hydrogen and methane are produced in increasing amounts as the reaction proceeds, while CO reaches a steady concentration after about 10 min. Silane reaches a maximum concentration after about 5 min, then decreases to zero after 60 min.; silane is known to decompose to silicon and hydrogen above about 425°C (368), and its presence is readily accounted for by the disproportionation reaction (109).

Any disiloxane, (SiH$_3$)$_2$O, produced as a result of silicon migration from metal to oxygen (Section III,F,2) will rapidly decompose at 450°C

FIG. 5. Volatile products from the static pyrolysis of H$_3$SiMn(CO)$_5$. Key: left-hand scale, SiH$_4$ ○, CH$_4$ △; right-hand scale, H$_2$ ●, CO ▲.

to yield SiH_4 and polymeric solids containing silicon, oxygen, and possibly hydrogen (22). Carbon monoxide is an inevitable breakdown product of metal carbonyl derivatives, and its slow reaction with hydrogen, possibly catalyzed by the solid metal-containing products, accounts for the formation of methane.

$$CO + 3H_2 \rightarrow CH_4 + H_2O \tag{140}$$

Water will be removed by reaction with Si–H bonds.

After 60 min at 450°C, the inferred composition of the residual solid is $Si_{1.00}Mn_{1.00}C_{1.36}O_{1.40}H_{0.97}$; it will be seen that nearly 30% of the original oxygen and carbon is retained. Bearing in mind the vigorous conditions, this suggests that these elements are present not as residual carbonyl groups but combined with silicon and manganese respectively. It is significant that the decomposition of $Cr(CO)_6$ at 630°C leads to as little as 40% of the chromium as free metal, the remainder being present as carbide and oxide (351). As a route to pure metal silicides, pyrolysis under these conditions offers little hope of success; uv photolysis gives similar unpromising products (80, 83).

Much more promising results have been obtained with a flow system shown diagrammatically in Fig. 6; this allows the vapor of a silyl metal-carbonyl derivative to be rapidly pyrolyzed in a stream of helium

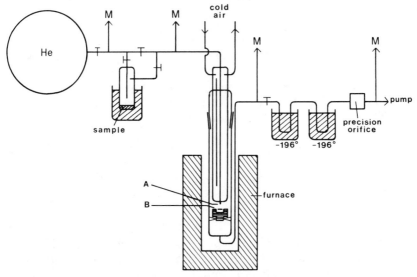

FIG. 6. Chemical vapor deposition apparatus for deposition of metal silicides. Key: A, vapor exit nozzle from cooled inlet tube; B, substrate plate (e.g., Si, SiO_2, metals), with metal support.

as carrier gas at low pressure. Chemical vapor deposition (c.v.d.) processes of this kind are being used increasingly for the production of pure materials needed for electronics applications (366), although their use to prepare metal silicides has until now been limited to the co-reduction of $SiCl_4$ and a volatile transition metal halide, heated together in the vapor phase in a stream of hydrogen (339).

In the present case, conditions are adjusted so that helium and silyl metal-carbonyl vapor (typically with a mole ratio of 30 : 1) have a total pressure of about 1 mmHg. Decomposition is very rapid on entering the furnace: crystalline solids may form around the exit nozzle, while a matt grey or mirror-like thin film is deposited on a small plate about 5 mm from the nozzle. These thin films have been analyzed by X-ray powder and electron microprobe methods, and directly by atomic absorption spectroscopy (33, 121, 187).

Silyl cobalt tetracarbonyl decomposes in a straightforward way to give a single phase:

$$H_3SiCo(CO)_4 \xrightarrow{\text{500°C}} CoSi \ (+ \ H_2 + CO) \tag{141}$$

Similarly, disilyl iron tetracarbonyl gives as a single product the low-temperature (β) form of iron disilicide, a semiconductor; on annealing at 1000°C for 4 days, the β form decomposes to give a mixture of the iron-deficient high-temperature (α) form, together with FeSi.

$$(H_3Si)_2Fe(CO)_4 \xrightarrow{\text{500°C}} \beta\text{-}FeSi_2 \xrightarrow{\text{1000°C}} \alpha\text{-}FeSi_2 + FeSi \tag{142}$$

Silyl manganese pentacarbonyl decomposes in a more complex way to give a mixture of two phases, "MnSi" (actually a silicon-rich nonstoichiometric phase of approximate composition $MnSi_{1.25}$) and stoichiometric Mn_5Si_3, such that the overall metal : silicon ratio is 1 : 1. The analogous rhenium system behaves similarly.

$$H_3SiMn(CO)_5 \xrightarrow{\text{500°C}} \text{"MnSi"} + Mn_5Si_3 \tag{143}$$

These examples illustrate the important general point that the metal : silicon ratio of the silicide film is always the same as that in the volatile precursor.

These reactions appear to involve a rapid stripping process, reminiscent of that occurring in a mass spectrometer (Section IV,D). Hydrogen and CO are lost before the system is able to reach equilibrium: the products are therefore different from those formed in sealed-tube reactions and favored thermodynamically. Provided that it is possible to synthesize sufficiently volatile precursors with the desired metal : silicon ratio, this c.v.d. process affords a route to thin films of a range of pure metal silicides, using relatively low temperatures. In contrast, other methods of preparation usually involve heating silicon–metal mixtures to high temperatures, and often produce a number of different phases (*18, 59, 187*).

B. SILICON–TRANSITION-METAL CLUSTER COMPOUNDS

Following the many recent developments in the chemistry of metal carbonyl clusters and other metal–metal bonded species, interest has developed in silicon–transition-metal clusters. These will be defined for the present purpose as compounds containing (1) at least one metal–metal bond and (2) at least one silicon atom directly linked to two or more metal atoms. Thus compounds such as $Me_3SiOs(CO)_4Os(CO)_4$-$SiMe_3$ and $R_3SiOCCo_3(CO)_9$ are excluded; examples of compounds that meet these criteria are shown in Table XXIII.

Structural features of the compounds in entries 1–6 have already been discussed in Section IV,A. The manganese compound in entry 2 (and its rhenium analog) were prepared by heating Ph_2SiH_2 with the metal carbonyl $M_2(CO)_{10}$; similar germanium– and tin–manganese derivatives have been made via Eq. (144) (*263, 414*).

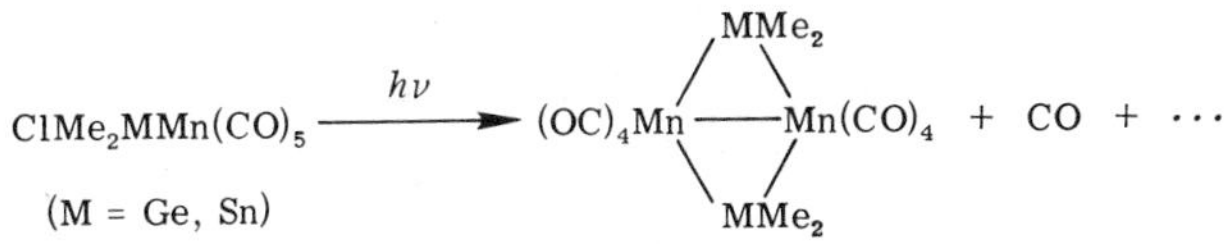

$$ClMe_2MMn(CO)_5 \xrightarrow{h\nu} (OC)_4Mn \overset{\displaystyle MMe_2}{\underset{\displaystyle MMe_2}{\diagup\diagdown}} Mn(CO)_4 + CO + \cdots$$

$$(M = Ge, Sn)$$

$$\tag{144}$$

Related compounds with exocyclic group IV–metal bonds can also be made, e.g. (*282, 367*),

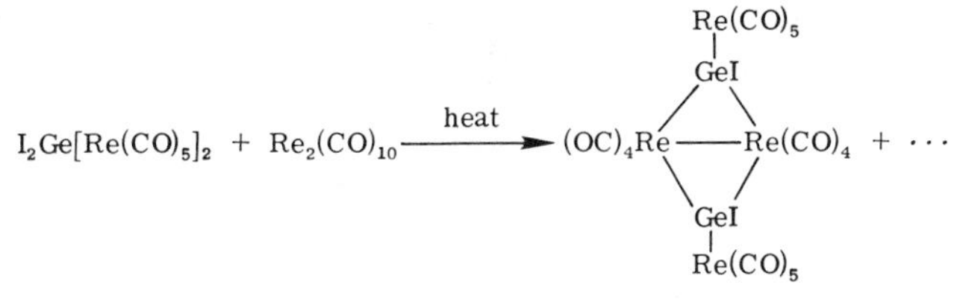

$$I_2Ge[Re(CO)_5]_2 + Re_2(CO)_{10} \xrightarrow{heat} (OC)_4Re \overset{\displaystyle \overset{Re(CO)_5}{|} GeI}{\underset{\displaystyle \underset{Re(CO)_5}{|} GeI}{\diagup\diagdown}} Re(CO)_4 + \cdots$$

$$\tag{145}$$

$$2\, BrSn[Mn(CO)_5]_3 \xrightarrow{heat} (OC)_4Mn \overset{\displaystyle \overset{Mn(CO)_5}{|} SnBr}{\underset{\displaystyle \underset{Mn(CO)_5}{|} SnBr}{\diagup\diagdown}} Mn(CO)_4 + \cdots$$

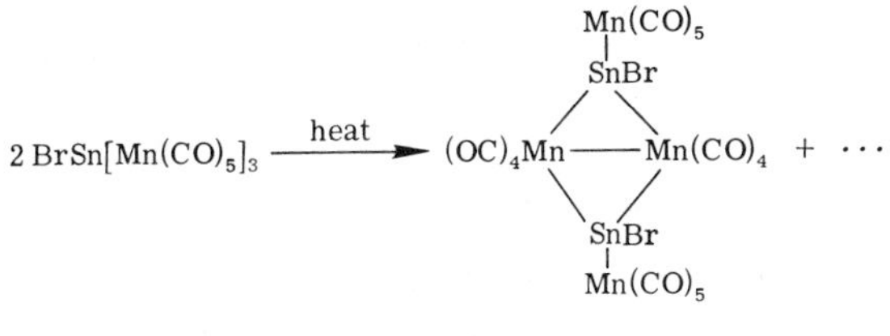

$$\tag{146}$$

TABLE XXIII

Silicon–Transition-Metal Cluster Compounds

Entry	Compound	M–M bridging groups	Ref.
1	$(Et_2Si)_2W_2(CO)_8H_2$ (**XLII**)	$2(Si \cdots H)$	(52)
2	$[Ph_2SiMn(CO)_4]_2$ (**LXI**)	2Si	(402)
3	$[Ph_2SiRe(CO)_4]_2$ (**XXXIX**)	2Si	(53, 127)
4	$(Ph_2Si)Re_2(CO)_8H_2$ (**XXXVIII**)	Si	(173)
5	$(Et_2Si)_2Re_2(CO)_7H_2$ (**XL**)	2Si	(128)
6	$(Et_2Si)_2Re_2(CO)_6H_4$ (**XLI**)	2Si	(127)
7	$(Me_2Si)Fe_2(CO)_8$ (**LXXVIII**)	Si	(68)
8	$(Ph_2Si)Fe_2(CO)_3(Cp)_2$ [a] (**LXXVI**)	Si,CO	(110)
9	$(MeRSi)Fe_2(CO)_3(Cp)_2$ [a,b] (**LXXVI**)	Si,CO	(309)
10	$(Me_2Si)_2Fe_2(CO)_7$ (**LXXVII**)	2Si,CO	(268, 289, 395)
11	$(ClMeSi)_2Fe_2(CO)_7$ (**LXXVII**)	2Si,CO	(395)
12	$(Cl_2Si)_2Fe_2(CO)_7$ (**LXXVII**)	2Si,CO	(391)
13	$(Me_3Si)_2(Me_2Si)_2Ru_2(CO)_6$ [c] (**XLVII**)	2Si	(76, 130)
14	$(Me_2Si)_3Ru_2(CO)_6$ like (**LXXIX**)	3Si	(76)
15	$(Me_2Si)_3Os_3(CO)_9$ (**LXXXI**)	Si	(76)
16	$(R_2Si)Co_2(CO)_7$ [d] (**LXXXII**)	Si,CO	(181, 268)
17	$(Cl_2Si)Co_2(CO)_7$ (**LXXXII**)	Si,CO	(2)
18	$[(OC)_4Co]PhSiCo_2(CO)_7$ (**LXXXII**)	Si,CO	(46)
19	$[(OC)_5Mn]ClSiCo_2(CO)_7$ (**LXXXII**)	Si,CO	(2)
20	H_2Si ⌐⌐ $Si[Co_2(CO)_7]$ / Si / H_2 (**LXXXII**)	Si,CO	(186)
21	$(Me_2Si)_2Co_2(CO)_6$ (**LXXXIII**)	2Si	(268)
22	$[(OC)_4Co]SiCo_3(CO)_9$ (**LII**)	Si	(389, 396)
23	$(OC)_9Co_3SiSiCo_3(CO)_9$	Si	(269, 278)
24	$(Me_2Si)Pt_2[P(cyclohexyl)_3]_2(C{\equiv}CPh)_2$ (**LIX**)	Si,C≡C	(105)

[a] $Cp = \eta^5\text{-}C_5H_5$.
[b] $R = H, Cl$.
[c] Also Os analog.
[d] $R = Me, Et, Ph$.

On the other hand, the photolysis of $Me_2Ge[Mn(CO)_5]_2$ leads to a compound first formulated as (**LXXV**) with no bridging CO groups (263), but now known to be $(OC)_4Mn(\mu\text{-}CO)(\mu\text{-}GeMe_2)Mn(CO)_4$ (413a).

$$Me_2Ge[Mn(CO)_5]_2 \xrightarrow{h\nu} (OC)_4Mn \longleftarrow Mn(CO)_5 + CO$$

(**LXXV**)

(147)

The iron derivatives in Table XXIII fall into three categories based on the structures (**LXXVI**) (entries 8 and 9), (**LXXVII**) (entries 10–12), and (**LXXVIII**) (entry 17).

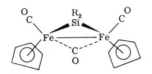

(**LXXVI**; and corresponding trans form)

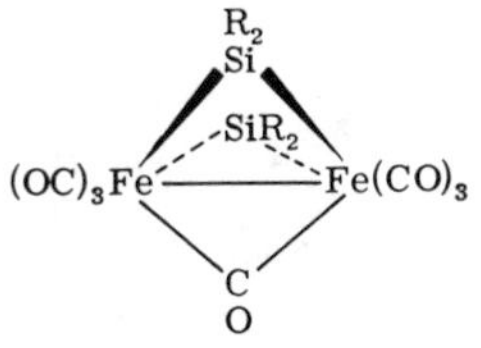

$(OC)_3Fe$ ⎯ $Fe(CO)_3$

(**LXXVII**)

$(OC)_4Fe$ ⎯ $Fe(CO)_4$

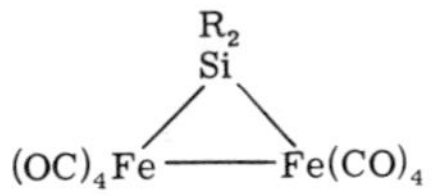

(**LXXVIII**)

They have been prepared by the following reactions:

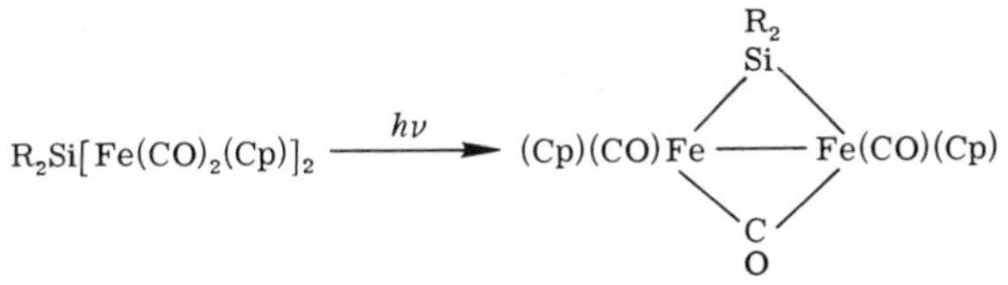

$$R_2Si[Fe(CO)_2(Cp)]_2 \xrightarrow{h\nu} (Cp)(CO)Fe \longrightarrow Fe(CO)(Cp) \tag{148}$$

$HMe_2SiSiMe_2H + Fe(CO)_5 \xrightarrow{h\nu\ or\ heat}$

$HMe_2SiSiMe_2H + Fe_2(CO)_9 \xrightarrow{heat}$ $(OC)_3Fe \longrightarrow Fe(CO)_3 + \cdots$

$[Cl_2SiFe(CO)_4]_3 \xrightarrow{h\nu}$

$[R_2SiFe(CO)_4]_2 \xrightarrow{h\nu}$

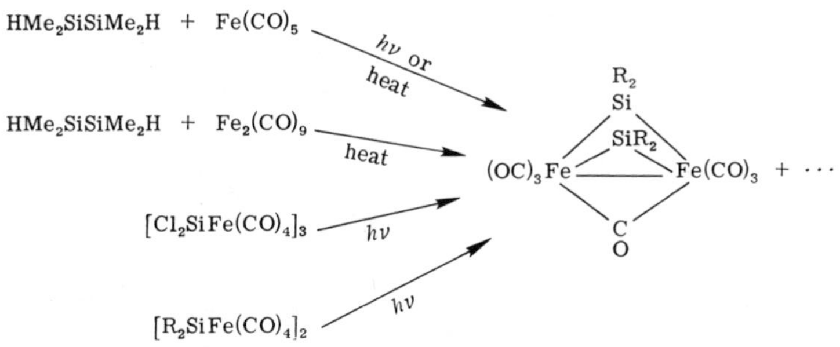

$R_2 = Me_2, HMe$

R = Me; refs. (*268, 289*)
R = Cl; ref. (*391*)
$R_2 = Me_2$, HMe; ref. (*395*)   (149)

where R = Me (Refs. *268, 289*), R = Cl (Ref. *391*), or $R_2 = Me_2$ or HMe (Ref. *395*).

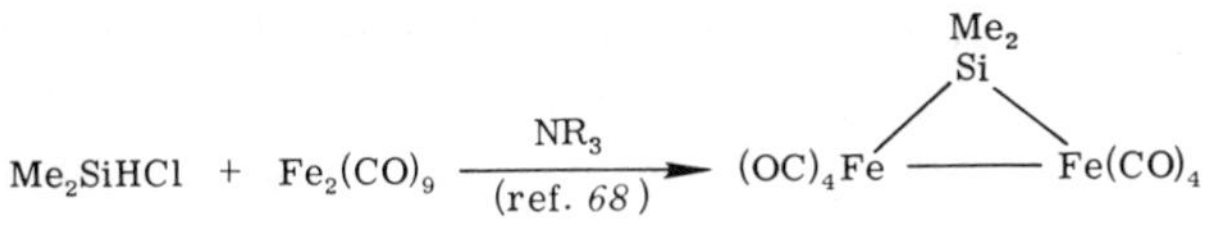

$$Me_2SiHCl + Fe_2(CO)_9 \xrightarrow[\text{(ref. 68)}]{NR_3} (OC)_4Fe \longrightarrow Fe(CO)_4 \tag{150}$$

NMR studies of the cis–trans isomerism of (**LXXVI**; R_2 = HMe) (*309*), the fluxionality of (**LXXVII**; R = Me) (*289*), and the ^{29}Si chemical shift of (**LXXVIII**; R = Me) (*68*) have already been discussed in Section IV,C.

Germanium and tin analogs of structures (**LXXVI**) and (**LXXVII**) are known (*172, 213, 263, 414*), and some of them are fluxional (*4, 213*); a germanium analog of compound (**LXXVIII**) has also been reported (*77*; see also Ref. *324*). No silicon–iron example of the structural types represented by (**LXXIX**) (*77*) or (**LXXX**) (*320*) has yet been prepared,

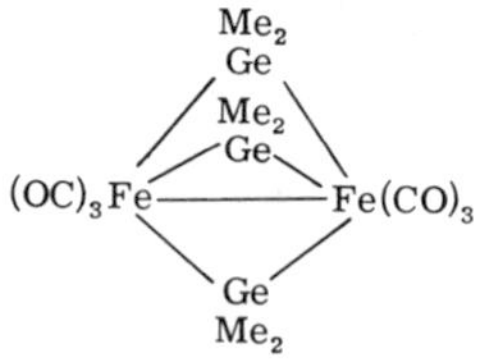

Me$_2$
Ge
Me$_2$
Ge
(OC)$_3$Fe
Fe(CO)$_3$
Ge
Me$_2$

(**LXXIX**)

Fe(CO)$_2$(Cp)
Sn
Fe(CO)$_3$
(OC)$_3$Fe
Fe(CO)$_3$
Sn
Fe(CO)$_2$(Cp)

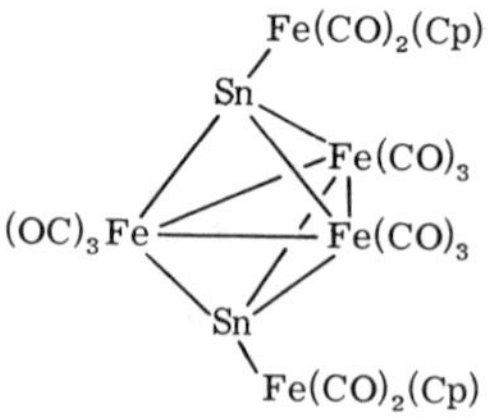

(**LXXX**)

although the Si–Ru analog of (**LXXIX**) is known (entry 14); the disilicon bridged Ru and Os derivatives (entry 13) also have no Si–Fe counterpart. They and the interesting trinuclear osmium cluster (**LXXXI**) (entry 15) are prepared by the extrusion of silylene units from disilanes in the presence of metal carbonyl compounds (*76*) (cf. reactions 149 and 157).

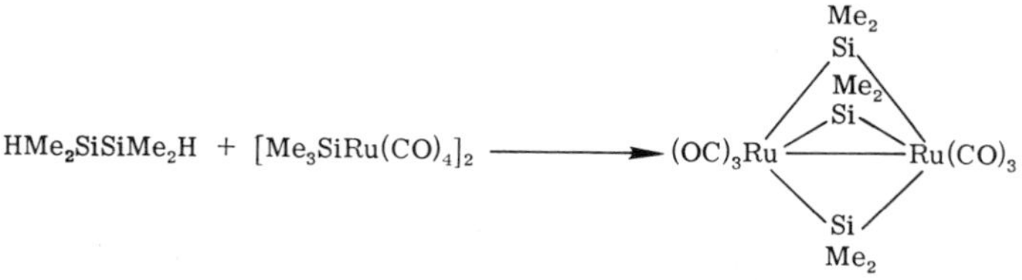

HMe$_2$SiSiMe$_2$H + [Me$_3$SiRu(CO)$_4$]$_2$
Me$_2$
Si
Me$_2$
Si
(OC)$_3$Ru
Ru(CO)$_3$
Si
Me$_2$

(151)

Me$_3$SiSiMe$_2$H + H$_2$Os(CO)$_4$
Δ
(CO)$_3$
Os
Me$_2$Si
SiMe$_2$
(OC)$_3$Os
Os(CO)$_3$
Si
Me$_2$

(**LXXXI**) (152)

The chemistry of silicon–cobalt clusters is more complicated because of the ready elimination of CO from proximate $Co(CO)_4$ groups (*208*) to give $Co_2(CO)_7$ or $Co_3(CO)_9$ units. This elimination is generally easier for silicon–cobalt compounds than for corresponding germanium and tin

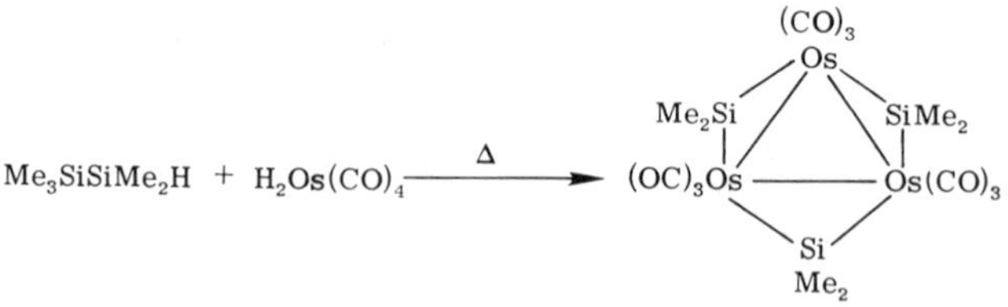

derivatives, so that, for example, no stable silicon analogs of $Sn[Co(CO)_4]_4$ (178, 393) or $RM[Co(CO)_4]_3$ (M = Ge, Sn) (178, 190, 334) are known.

Structures of compounds in entries 16–20 are all based on the type (**LXXXII**), in which R^1 and R^2 can be alkyl, aryl, halogen, a metal carbonyl group, or part of a cyclic system.

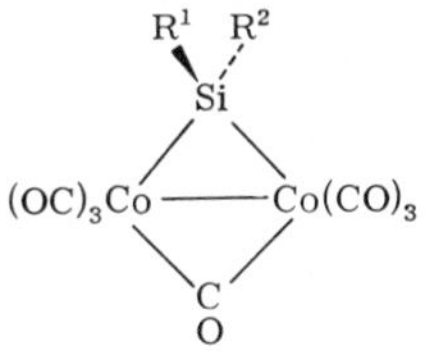

(**LXXXII**) 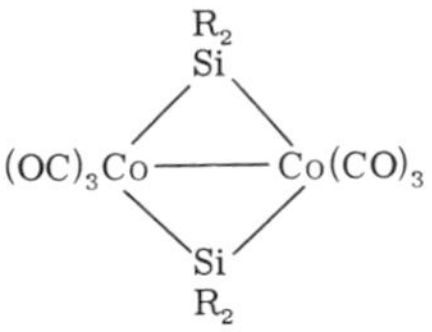   (**LXXXIII**)

They are usually readily prepared by reactions such as Eq. (153) (2, 181), where R = Ph, Et, Cl, or Eq. (154) (46).

$$R_2SiH_2 + Co_2(CO)_8 \rightarrow R_2SiCo_2(CO)_7 + CO + H_2 \tag{153}$$

$$2PhSiH_3 + 3Co_2(CO)_8 \rightarrow 2[(OC)_4Co]PhSiCo_2(CO)_7 + 2CO + 3H_2 \tag{154}$$

There is no evidence that they will react with CO to form an open silanediyl derivative, although germanium analogs undergo the following reversible change (49, 263):

$$R_2GeCo_2(CO)_7 \underset{h\nu}{\overset{CO}{\rightleftharpoons}} R_2Ge[Co(CO)_4]_2 \tag{155}$$

Again, the germanium analog of the product from reaction (154) can gain or lose CO to form an open germanetriyl or a closed tricobalt cluster derivative, but the silicon compound apparently undergoes neither of these reactions (46, 178).

$$RGeCo_3(CO)_9 \overset{\Delta}{\leftarrow} [(OC)_4Co]RGeCo_2(CO)_7 \xrightarrow[300\ atm]{CO} RGe[Co(CO)_4]_3 \tag{156}$$
$$(R = Me, Ph) \qquad\qquad\qquad\qquad\qquad\qquad (R = Ph)$$

Structure (**LXXXIII**) is represented by the compound in entry 21; it is one product from the following reaction (268):

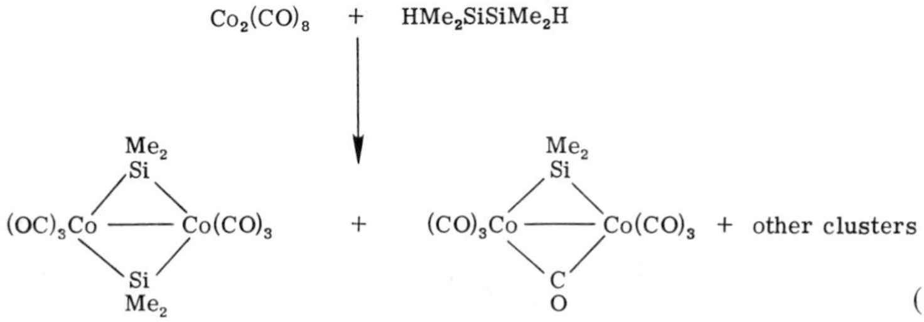

$$\tag{157}$$

Both germanium (5) and tin (414) analogs are known, and their ready fluxionality (down to below $-50°C$) has been studied in detail (5).

The compound in entry 22, $(OC)_4CoSiCo_3(CO)_9$ (**LII**), results when a mixture of SiI_4 and $Na^+Co(CO)_4^-$ in hexane is warmed and irradiated; probably $ISiCo_3(CO)_9$ is formed as an intermediate (396). The tetrahedral $SiCo_3$ unit parallels that found in many methylidyne tricobalt clusters of the type $RCCo_3(CO)_9$ (358, 401); a germanium analog of compound (**LII**) is known, but apparently not one with tin (71, 389). One of the first reported $SiCo_3$ cluster derivatives was the silicon–silicon bonded species in entry 23 (269).

$$Co_2(CO)_8 + SiPh_4 \xrightarrow[\text{pet. ether}]{\text{reflux}} (OC)_9Co_3Si{-}SiCo_3(CO)_9 \tag{158}$$

From subsequent work, however, it appears that the product isolated was $ClCCo_3(CO)_9$ (278). Silicon compounds of the type $RSiCo_3(CO)_9$, corresponding to one of the germanium–cobalt products of reactions (156), have not yet been made; neither has an analog of the spiroderivative $Ge[Co_2(CO)_7]_2$ (191). Synthetic efforts directed towards these systems would be valuable.

The sole example of a silicon–platinum cluster is the compound in entry 24: its structure has been noted in Section IV,A. It seems very likely that many further cluster systems await discovery, particularly with iridium, platinum, and gold, and that this represents an important future area of research. One obvious application is as precursors to metal silicides with high metal:silicon ratios using c.v.d. techniques (compare Section V,A).

C. METAL-COMPLEXED SILYLENE DERIVATIVES

There are a number of successful examples of the stabilization of a short-lived species by coordination to a metal center (388), and to these can now be added the silylene (SiR_2) group, which normally exists only as a transient intermediate (332).

Attempts to produce compounds of the type $R_2SiM(CO)_n$, in which the silylene is acting as a two-electron donor, normally lead to isolation of oligomers (258, 391, 392, 395).

$$Fe(CO)_5 + HSiCl_3 \xrightarrow{h\nu} (OC)_4Fe\underset{\underset{\underset{Cl_2}{Si}}{\diagdown}}{\overset{\overset{Cl_2}{Si}}{\diagup}}Fe(CO)_4 \tag{159}$$

$$Et_4N^+[R^1R^2ClSiFe(CO)_4]^- \xrightarrow{AlCl_3} [R^1R^2SiFe(CO)_4]_n + AlCl_4^- \tag{160}$$

where $n = 3$ for $R^1 = R^2 = Cl$; or $n = 2$ for $R^1R^2 = Me_2$, MeCl, (cyclo-hexyl)Cl.

$$W(CO)_6 + Si_2I_6 \xrightarrow{h\nu} [I_2SiW(CO)_5]_2 \qquad (161)$$

The original authors suggested that the products of reactions (160) and (161) might be halogen-bridged [e.g., (**LXXXIVa**)], but a directly bonded ring such as (**LXXXIVb**) and the product of reaction (159) seem

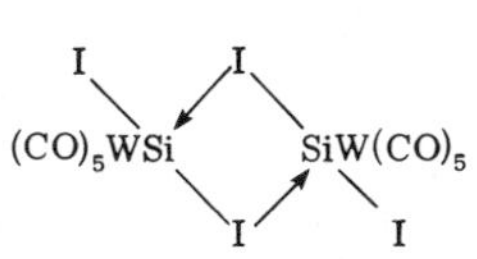

(**LXXXIVa**) 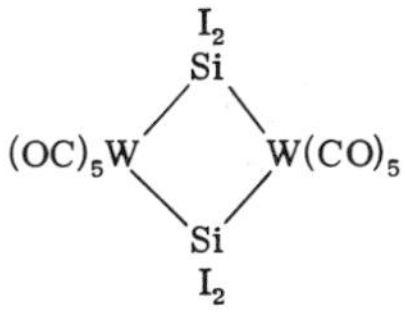   (**LXXXIVb**)

at least as likely.

In the case of germanium and tin, reactions such as (162) lead to the production of $GeMe_2-$ or $SnMe_2-$metal complexes, stabilized by electron donation from the Lewis base, tetrahydrofuran (*315*).

$$Me_2MX_2 + Na_2Cr_2(CO)_{10} \xrightarrow{THF} Me_2MCr(CO)_5$$
$$(M = Ge, Sn) \qquad \uparrow \atop THF \qquad (162)$$

Furthermore, if the groups attached to the group IV atom are large, unsolvated derivatives such as $R_2GeCr(CO)_5$ [R = $CH(SiMe_3)_2$, $N(SiMe_3)_2$, or mesityl] can be isolated (*266, 292*).

With silicon, no exactly analogous products have been reported. A base-stabilized Me_2Si–iron derivative has been produced, however, by the following ingenious reaction (*394*).

$$Me_2SiH(NEt_2) + Fe(CO)_5 \xrightarrow{h\nu} (Et_2N)Me_2SiFeH(CO)_4$$

$$Me_2SiFe(CO)_4$$
$$\uparrow$$
$$NHEt_2 \qquad (163)$$

It is stable only below $-20°C$. Attempts to produce an unsupported silylene–iron complex by the removal of HCl from compound (**LXXXV**) were unsuccessful (*394*).

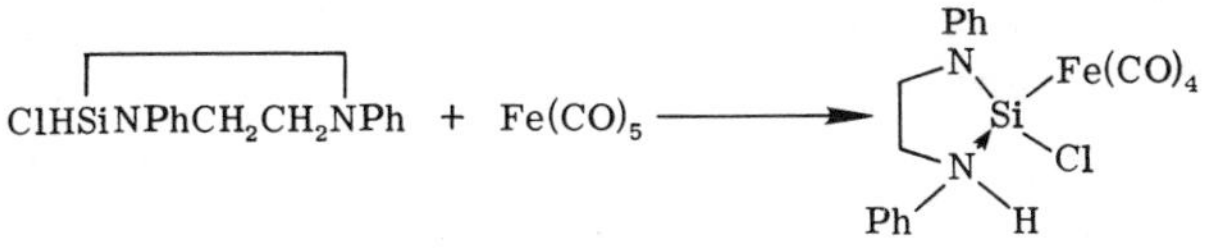

(**LXXXV**)   (164)

The first silylene–metal complex with no base stabilization was prepared by extrusion of SiMe$_2$ from a substituted disilane (384), as in Eq. (165) where R is H or Me.

$$\text{Fe}_2(\text{CO})_9 + \text{HMe}_2\text{SiSiMe}_2\text{R} \xrightarrow[\text{25°C; 15 h}]{\text{benzene}} \text{Me}_2\text{SiFeH(CO)}_3(\text{SiMe}_2\text{R}) \qquad (165)$$

The product is a barely volatile liquid at room temperature; with PPh$_3$, both silicon atoms are displaced from iron, leaving Fe(CO)$_3$(PPh$_3$)$_2$.

Silylene–metal complexes, R$_2$SiM(L)$_n$, have often been invoked as possible intermediates in metal-catalyzed reactions of silicon compounds. Some examples of M include Ni (reaction of organodisilanes with alkynes) (345); Pt (disproportionation of methyldisilanes) (431); Rh (formation of organodisilanes and trisilanes from substituted hydridosilanes) (342); and Pd (reaction of substituted silacyclopropenes with alkynes) (383). Iron–silylene intermediates may also be involved in the decomposition of η^3-silaallyl derivatives* such as (CH$_2$CHSiMe$_2$)Fe(CO)$_3$SiR$_3$ (385) or of (η^4-disilahexadiene)–Fe(CO)$_3$ complexes (328); in both cases, SiR$_2$ species appear as products. It may finally be noted that the proposed di(silylene)–platinum complex (Ph$_3$P)$_2$Pt(SiCl$_2$)$_2$ (390) should now be re-formulated as a cyclic platinadisiloxane (Ph$_3$P)$_2$PtSiCl$_2$OSiCl$_2$ (197) (see Section III,B,8).

Appendix

Since this review was written, a number of significant new papers have appeared; they are briefly noted here in relation to the most relevant section of the text.

Section II,A,3.   The harmful effect of cyclic ethers as solvents is emphasised by a recent study of their reaction with Me$_3$SiMn(CO)$_5$ (cf. Section III,B,3) (437):

$$\text{Me}_3\text{SiMn(CO)}_5 + \overline{\text{CH}_2[\text{CH}_2]_n\text{O}} \rightarrow (\text{OC})_5\text{Mn}[\text{CH}_2]_{n+1}\text{OSiMe}_3$$
$$(n = 2\text{–}4) \qquad\qquad\qquad\qquad\qquad \downarrow \text{ CO}$$
$$(\text{OC})_5\text{MnC(O)}[\text{CH}_2]_{n+1}\text{OSiMe}_3$$

Acyclic ethers R$_2$O also attack (Me$_3$Si)$_2$Fe(CO)$_4$, giving Me$_3$SiOR and Me$_3$SiFe(R)(CO)$_4$ as initial products (435, 436).

Section II,D.   Reaction between Hg(SiMe$_3$)$_2$ and (Cp)$_2$ZrCl$_2$ gives Me$_3$SiZr(Cp)$_2$Cl and (Me$_3$Si)$_2$Zr(Cp)$_2$ (434). Also, the novel reagent

---

* But see footnote, p. 37, and Appendix.

$(Me_3Si)_3Al$ reacts with $(Cp)_2TiCl_2$ to give $Me_3SiTi(Cp)_2Cl$ (*449*). It has previously been reported that Me_3SiLi yielded no Si–Ti product with $(Cp)_2TiCl_2$, although with $TiCl_4$ some evidence for $(Me_3Si)_4Ti$ and $(Me_3Si)_2Ti \cdot nTHF$ was obtained (*447*).

Sections II,E and II,F. The first examples of silylation of metal clusters are reported, using both anionic clusters with H_2 evolution (*454*):

$$HRu_3(CO)_{11}^- + 2Et_3SiH \rightarrow HRu_3(CO)_{10}(SiEt_3)_2^- + H_2 + CO$$

and neutral clusters with oxidative addition (*456*):

$$M_3(CO)_{12} + 3HSiCl_2X \rightarrow (XCl_2Si)_3M_3H_3(CO)_9 + 3CO \qquad (M = Ru, Os; X = Me, Cl)$$

Oxidative addition has also been used to prepare the bicyclic cis derivative $Pt(\leftarrow PPh_2[CH_2]_2SiMe_2)_2$ from $Pt(cycloocta-1,5-diene)_2$ and $Ph_2P[CH_2]_2SiHMe_2$ (*443*), and a range of cyclic compounds $SiF_2C(t-Bu)=CHSiF_2ML_n$ [ML_n = $Mo(CO)_5$, $Mn(CO)_2(Cp)$, and $Fe(CO)_4$] from $SiF_2C(t-Bu)=CHSiF_2$ and appropriate metal carbonyls (compare Table V, entry 39, probably an $Ni(CO)_3$ adduct) (*458*). NMR studies (*440*) of the oxidative addition of SiH_3Q (Q = H, Me, halogen, etc.) to $IrH(CO)-(PPh_3)_2$, yielding $QH_2SiIrH_2(CO)(PPh_3)_2$, have shown that the stereochemistry is more complex than that postulated earlier (*179, 180*). For related studies with $IrI(CO)(PEt_3)_2$, see (*439*).

Section II,G,4,c. It is now clear (*446, 450, 451*) that the products from the reaction of $Fe_2(CO)_9$ with vinyl disilanes are not (η^3-silaallyl)-$Fe(CO)_3$ species, as previously proposed (*382, 385*), but rather η^2-vinyl adducts of $Fe(CO)_4$[e.g., (η^2-$CH_2=CHSiR_2X)Fe(CO)_4$]. All attempts to remove X as X^- have proved fruitless (cf. Section III,G,2).

Section III,E,3. All metal ligands except SiR_3 are substituted when *trans*-$Ph_3SiPtCl(PMe_2Ph)_2$ reacts with KCN and 18-crown-6 in CH_2Cl_2 solution, giving $[K(18\text{-crown-}6)]_2^+[Ph_3SiPt(CN)_3]^{2-}$ (*432*).

Section III,G,2. Sequential oxidative addition, halogenation at Si, and reductive elimination provide an elegant route to pure R_2SiHX derivatives (cf. Section III,C,1 and Table XII, entry 14) (*453*):

$$(Cp')Mn(CO)_3 + Ph_2SiH_2 \rightarrow (Cp')Mn(H)(CO)_2SiHPh_2$$

$$\downarrow \text{halogenation}$$

$$(Cp')Mn(CO)_3 + Ph_2SiHX \xrightarrow{CO} (Cp')Mn(H)(CO)_2SiXPh_2$$

$$(Cp' = \eta^5\text{-}MeC_5H_4)$$

Section IV,A. Some recently determined bond lengths (from X ray, in picometers) are: d(Si–Ti) in $\text{Me}_3\text{SiTi(Cp)}_2\text{Cl}$, 267 (*449*); d(Si–Mo) in $\overline{\text{SiF}_2\text{C}(t\text{-Bu})}{=}\text{CHSiF}_2\text{Mo(CO)}_5$, 261 (*444*); d(Si–Fe) in (*cyclo*-Si_5Me_9)-$\text{SiMe}_2\text{Fe(CO)}_2\text{(Cp)}$, 235 (*438*).

Section IV,F,2. Further X-ray excited photoelectron spectra (ESCA) of RMn(CO)_5 (R = SiCl_3, SiF_3, Me, halogens, etc.) have yielded Mn, C, and O binding energies, reinforcing the earlier conclusion (*265*) that π effects vary little with change in R (*433*).

Section V,B. The first silicon–iridium cluster, $\text{MeSiIr}_3\text{(CO)}_9$, results from the reaction between NaIr(CO)_4 and MeSiCl_3 (*445*). Also, substituted tricobalt clusters of the type $\text{MeSi[Co(CO)}_3]_2[\text{M(CO)}_3\text{(Cp)}]$ (M = Mo, W) have been prepared by heating $\text{H}_2\text{MeSiM(CO)}_3\text{(Cp)}$ with $\text{Co}_2\text{(CO)}_8$ (*455*).

Section V,C. A development of the highest significance is the characterization of stable Ge≡Mn derivatives (*442*); analogous silicon compounds are eagerly awaited.

Finally, recent catalytic studies in which silicon–metal intermediates have been implicated include the photocatalysis of hydrosilation (Si–Fe, *452*; Si–Co, *448*; Si–Rh, *441*) and the double silylation of alkynes with disilanes (Si–Pd, *457*).

REFERENCES

1. Abel, E. W., and Moorhouse, S., *J. Organomet. Chem.* **28**, 211 (1971).

2. Abraham, K. M., and Urry, G., *Inorg. Chem.* **12**, 2850 (1973).

3. Absi-Halabi, M., and Brown, T. L., *J. Am. Chem. Soc.* **99**, 2982 (1977).

4. Adams, R. D., Brice, M. D., and Cotton, F. A., *Inorg. Chem.* **13**, 1080 (1974).

5. Adams, R. D., Cotton, F. A., Cullen, W. R., Hunter, D. L., and Mihichuk, L., *Inorg. Chem.* **14**, 1395 (1975).

6. Ager, D., and Fleming, I., *J. Chem. Soc., Chem. Commun.* P. 177 (1978); Ager, D. J., Fleming, I., and Patel, S. K., *J. Chem. Soc., Perkin Trans. I* p. 2520 (1981).

7. Allinson, J. S., Ph.D. Thesis, University of London (1974).

8. Allinson, J. S., Aylett, B. J., and Colquhoun, H. M., *J. Organomet. Chem.* **112**, C7 (1976).

9. Amberger, E., Mühlhofer, E., and Stern, H., *J. Organomet. Chem.* **17**, P5 (1969).

10. Anderson, D. W. W., Ebsworth, E. A. V., MacDougall, J. K., and Rankin, D. W. H., *J. Inorg. Nucl. Chem.* **35**, 2259 (1973).

11. Anderson, D. W. W., Ebsworth, E. A. V., and Rankin, D. W. H., *J. Chem. Soc., Dalton Trans.* p. 2370 (1973).

12. Andrews, M. A., Kirtley, S. W., and Kaesz, H. D., *Inorg. Chem.* **16**, 1556 (1977).

13. Ang, H. G., and Lau, P. T., *J. Organomet. Chem.* **37**, C4 (1972).

14. Anglin, J. R., Calhoun, H. P., and Graham, W. A. G., *Inorg. Chem.* **16**, 2281 (1977).

15. Archer, N. J., Haszeldine, R. N., and Parish, R. V., *J. Chem. Soc., Chem. Commun.* p. 524 (1971).
16. Archer, N. J., Haszeldine, R. N., and Parish, R. V., *J. Organomet. Chem.* **81**, 335 (1974).
17. Archer, N. J., Haszeldine, R. N., and Parish, R. V., *J. Chem. Soc., Dalton Trans.* p. 695 (1979).
18. Aronsson, B., Lundström, T., and Rundqvist, S., "Borides, Silicides and Phosphides." Methuen, London, 1965.
19. Ash, M. J., Brookes, A., Knox, S. A. R., and Stone, F. G. A., *J. Chem. Soc. A* p. 458 (1971).
20. Aylett, B. J., *J. Inorg. Nucl. Chem.* **15,** 87 (1960).
21. Aylett, B. J., *Prep. Inorg. React.* **2,** 93–138 (1965).
22. Aylett, B. J., *Adv. Inorg. Chem. Radiochem.* **11,** 249 (1968).
23. Aylett, B. J., *Prog. Stereochem.* **4,** 213 (1969).
24. Aylett, B. J., "Organometallic Compounds," 4th. ed., Vol. 1, Part 2. Chapman & Hall, London, 1979.
25. Aylett, B. J., *J. Organomet. Chem. Libr.* **9,** 327 (1980).
26. Aylett, B. J., and Campbell, J. M., *J. Chem. Soc., Chem. Commun.* p. 217 (1965).
27. Aylett, B. J., and Campbell, J. M., *J. Chem. Soc., Chem. Commun.* p. 159 (1967).
28. Aylett, B. J., and Campbell, J. M., *Inorg. Nucl. Chem. Lett.* **3,** 137 (1967).
29. Aylett, B. J., and Campbell, J. M., *J. Chem. Soc. A* p. 1910 (1969).
30. Aylett, B. J., and Campbell, J. M., *J. Chem. Soc. A* p. 1916 (1969).
31. Aylett, B. J., and Campbell, J. M., *J. Chem. Soc. A* p. 1920 (1969).
32. Aylett, B. J., and Colquhoun, H. M., *J. Chem. Res., Synop.* p. 148; *J. Chem. Res., Miniprint* p. 1677 (1977).
33. Aylett, B. J., and Colquhoun, H. M., *J. Chem. Soc., Dalton Trans.* p. 2058 (1977).
34. Aylett, B. J., and Ficheux, M. A., unpublished work.
35. Aylett, B. J., and Hakim, M. J. *J. Chem. Soc. A* p. 1788 (1969).
36. Aylett, B. J., and Sinclair, R. A., *Chem. Ind. (London)* p. 301 (1965).
37. Aylett, B. J., and Young, R. R., unpublished work.
38. Aylett, B. J., Campbell, J. M., and Walton, A., *Inorg. Nucl. Chem. Lett.* **4,** 79 (1968).
39. Aylett, B. J., Campbell, J. M., and Walton, A., *J. Chem. Soc. A* p. 2110 (1969).
40. Aylett, B. J., Eméleus, H. J., and Maddock, A. G., *J. Inorg. Nucl. Chem.* **1,** 187 (1955).
41. Baay, Y. L., and MacDiarmid, A. G., *Inorg. Nucl. Chem. Lett.* **3,** 159 (1967).
42. Baay, Y. L., and MacDiarmid, A. G., *Inorg. Chem.* **8,** 986 (1969).
43. Baird, M. C., *J. Inorg. Nucl. Chem.* **29,** 367 (1967).
44. Bald, J. F., and MacDiarmid, A. G., *J. Organomet. Chem.* **22,** C22 (1970).
45. Ball, R., and Bennett, M. J., *Inorg. Chem.* **11,** 1806 (1972).
46. Ball, R., Bennett, M. J., Brooks, E. H., Graham, W. A. G., Hoyano, J., and Illingworth, S. M., *J. Chem. Soc., Chem. Commun.* p. 592 (1970).
47. Bancroft, G. M., Clark, H. C., Kidd, R. G., Rake, A. T., and Spinney, H. G., *Inorg. Chem.* **12,** 728 (1973).
48. Bancroft, G. M., Mays, M. J., and Prater, B. E., *J. Chem. Soc. A* p. 956 (1970).
49. Basato, M., Fawcett, J. P., and Poë, A., *J. Chem. Soc., Dalton Trans.* p. 1350 (1974).
50. Behrens, H., *Adv. Organomet. Chem.* **18,** 1 (1980).
51. Bennett, M. A., and Patmore, D. J., *Inorg. Chem.* **10,** 2387 (1971).
52. Bennett, M. J., and Simpson, K. A., *J. Am. Chem. Soc.* **93,** 7156 (1971).
53. Bennett, M. J., Graham, W. A. G., Hoyano, J. K., and Hutcheon, W. L., *J. Am. Chem. Soc.* **94,** 6232 (1972).

54. Bennett, M. J., Graham, W. A. G., Smith, R. A., and Stewart, R. P., *J. Am. Chem. Soc.* **95,** 1684 (1973).
55. Bent, H. A., *Chem. Rev.* **61,** 275 (1961).
56. Bentham, J. E., and Ebsworth, E. A. V., *Inorg. Nucl. Chem. Lett.* **6,** 145 (1970).
57. Bentham, J. E., and Ebsworth, E. A. V., *J. Chem. Soc. A* p. 2091 (1971).
58. Bentham, J. E., Cradock, S., and Ebsworth, E. A. V., *J. Chem. Soc. A* p. 587 (1971).
59. Berezhnoi, A. S., "Silicon and Its Binary Systems." Consultants Bureau, New York, 1960.
60. Berg, G. C. van den, and Oskam, A., *J. Organomet. Chem.* **78,** 357 (1974).
61. Berg, G. C. van den, and Oskam, A., *J. Organomet. Chem.* **91,** 1 (1975).
62. Berg, G. C. van den, Oskam, A., and Vrieze, K., *J. Organomet. Chem.* **57,** 329 (1973).
62a. Berry, A. D., and MacDiarmid, A. G., *Inorg. Nucl. Chem. Lett.* **5,** 601 (1969).
63. Berry, A. D., Corey, E. R., Hagen, A. P., MacDiarmid, A. G., Saalfield, F. E., and Wayland, B. B., *J. Am. Chem. Soc.* **92,** 1940 (1970).
64. Bettler, C. R., Sendra, J. C., and Urry, G., *Inorg. Chem.* **9,** 1060 (1970).
65. Bichler, R. E. J., and Clark, H. C., *J. Organomet. Chem.* **23,** 427 (1970).
66. Bichler, R. E. J., Booth, M. R., and Clark, H. C., *J. Organomet. Chem.* **24,** 145 (1970).
67. Bichler, R. E. J., Clark, H. C., Hunter, B. K., and Rake, A. T., *J. Organomet. Chem.* **69,** 367 (1974).
68. Bikovetz, A. L., Kuzmin, O. V., Vdovin, V. M., and Krapivin, A. M., *J. Organomet. Chem.* **194,** C33 (1980).
69. Bird, P., Harrod, J. F., and Khin Aye Than, *J. Am. Chem. Soc.* **96,** 1222 (1974).
70. Blackburn, S. N., Haszeldine, R. N., Parish, R. V., and Setchfield, J. H., *J. Organomet. Chem.* **192,** 329 (1980).
71. Boese, R., and Schmid, G., *J. Chem. Soc., Chem. Commun.* p. 349 (1979).
72. Booth, B. L., and Shaw, B. L., *J. Organomet. Chem.* **43,** 369 (1972).
73. Bradley, G. F., and Stobart, S. R., *J. Chem. Soc., Dalton Trans.* p. 264 (1974).
74. Brookes, A., Howard, J., Knox, S. A. R., Riera, V., Stone, F. G. A., and Woodward, P., *J. Chem. Soc., Chem. Commun.* p. 727 (1973).
75. Brookes, A., Knox, S. A. R., Riera, V., Sosinsky, B. A., and Stone, F. G. A., *J. Chem. Soc., Dalton Trans.* p. 1641 (1975).
76. Brookes, A., Knox, S. A. R., and Stone, F. G. A., *J. Chem. Soc. A* p. 3469 (1971).
77. Brooks, E. H., Elder, M., Graham, W. A. G., and Hall, D., *J. Am. Chem. Soc.* **90,** 3587 (1968).
78. Brown, T. L., Edwards, P. A., Harris, C. B., and Kirsch, J. L., *Inorg. Chem.* **8,** 763 (1969).
79. Burlitch, J. M., and Theyson, T. W., *J. Chem. Soc., Dalton Trans.* p. 828 (1974); Haupt, H.-J., and Neumann, F., *J. Organomet. Chem.* **33,** C56 (1971).
80. Burne, A. R., Ph.D. Thesis, University of London (1971).
81. Burnham, R. A., and Stobart, S. R., *J. Chem. Soc., Dalton Trans.* p. 1269 (1973).
82. Burnham, R. A., and Stobart, S. R., *J. Chem. Soc., Dalton Trans.* p. 1489 (1977).
83. Campbell, J. M., Ph.D. Thesis, University of London (1967).
84. Campbell-Fergusson, H. J., and Ebsworth, E. A. V., *Chem. Ind. (London)* p. 300 (1965); *J. Chem. Soc. A* p. 1508 (1966); p. 705 (1967).
85. Čapka, M., Svoboda, P., Bažant, V., and Chvalovský, V., *Collect. Czech. Chem. Commun.* **36,** 2785 (1971).
86. Cardin, D. J., Keppie, S. A., and Lappert, M. F., *J. Chem. Soc. A* p. 2594 (1970).
87. Cardin, D. J., Keppie, S. A., Lappert, M. F., Litzow, M. R., and Spalding, T. R., *J. Chem. Soc. A* p. 2262 (1971).

88. Cardoso, A. M., Clark, R. J. H., and Moorhouse, S., *J. Organomet. Chem.* **186,** 241 (1980).
89. Carre, F., Cerveau, G., Colomer, E., Corriu, R. J. P., Young, J. C., Ricard, L., and Weiss, R., *J. Organomet. Chem.* **179,** 215 (1979).
90. Cella, J. A., Cargioli, J. D., and Williams, E. A., *J. Organomet. Chem.* **186,** 13 (1980).
91. Cerveau, G., Colomer, E., Corriu, R., and Douglas, W. E., *J. Chem. Soc., Chem. Commun.* p. 410 (1975).
92. Cerveau, G., Colomer, E., Corriu, R., and Douglas, W. E., *J. Organomet. Chem.* **135,** 373 (1977).
93. Cerveau, G., Colomer, E., and Corriu, R., *J. Organomet. Chem.* **136,** 349 (1977); see also Cerveau, G., Chauvière, G., Colomer, E., and Corriu, R. J. P., *ibid.* **210,** 343 (1981).
94. Chalk, A. J., and Harrod, J. F., *J. Am. Chem. Soc.* **87,** 1133 (1965).
95. Chalk, A. J., and Harrod, J. F., *J. Am. Chem. Soc.* **89,** 1640 (1967).
96. Chan, T. M., Connolly, J. W., Hoff, C. D., and Millich, F., *J. Organomet. Chem.* **152,** 287 (1978).
97. Chatt, J., Eaborn, C., and Ibekwe, S. D., *J. Chem. Soc., Chem. Commun.* p. 700 (1966).
98. Chatt, J., Eaborn, C., Ibekwe, S. D., and Kapoor, P. N., *J. Chem. Soc. A* p. 1343 (1970).
99. Chatt, J., Eaborn, C., and Kapoor, P. N., *J. Chem. Soc. A* p. 881 (1970).
100. Cheng, C.-W., and Liu, C.-S., *J. Chem. Soc., Chem. Commun.* p. 1013 (1974).
101. Chipperfield, J. R., Ford, J., and Webster, D. E., *J. Organomet. Chem.* **102,** 417 (1975).
102. Chipperfield, J. R., Hayter, A. C., and Webster, D. E., *J. Chem. Soc., Chem. Commun.* p. 625 (1975); Chipperfield, J. R., Ford, J., Hayter, A. C., and Webster, D. E., *J. Chem. Soc., Dalton Trans.* p. 360 (1976).
103. Chipperfield, J. R., Hayter, A. C., and Webster, D. E., *J. Chem. Soc., Dalton Trans.* p. 921 (1977).
104. Ciriano, M., Green, M., Gregson, D., Howard, J. A. K., Spencer, J. L., Stone, F. G. A., and Woodward, P., *J. Chem. Soc., Dalton Trans.* p. 1294 (1979).
105. Ciriano, M., Howard, J. A. K., Spencer, J. L., Stone, F. G. A., and Wadepohl, H., *J. Chem. Soc., Dalton Trans.* p. 1749 (1979).
106. Clark, H. C., and Hauw, T. L., *J. Organomet. Chem.* **42,** 429 (1972).
107. Clark, H. C., and Rake, A. T., *J. Organomet. Chem.* **74,** 29 (1974).
108. Clark, H. C., and Rake, A. T., *J. Organomet. Chem.* **82,** 159 (1974).
108a. Clark, H. C., Dixon, K. R., and Nicolson, J. G., *Inorg. Chem.* **8,** 450 (1969); see also Wharf, I., and Onyszchuk, M., *Can. J. Chem.* **50,** 3450 (1972).
109. Cleland, A. J., Fieldhouse, S. A., Freeland, B. H., and O'Brien, R. J., *J. Chem. Soc., Chem. Commun.* p. 155 (1971).
110. Cleland, A. J., Fieldhouse, S. A., Freeland, B. H., and O'Brien, R. J., *J. Organomet. Chem.* **32,** C15 (1971).
111. Clemmit, A. F., and Glockling, F., *J. Chem. Soc. A* p. 2163 (1969).
112. Clemmit, A. F., and Glockling, F., *J. Chem. Soc., Chem. Commun.* p. 705 (1970).
113. Clemmit, A. F., and Glockling, F., *J. Chem. Soc. A* p. 1164 (1971).
114. Colomer, E., and Corriu, R. J. P., *J. Chem. Soc., Chem. Commun.* p. 176 (1976).
115. Colomer, E., and Corriu, R. J. P., *J. Organomet. Chem.* **133,** 159 (1977); see also Cerveau, G., Colomer, E., and Corriu, R. J. P., *Angew. Chem., Int. Ed. Engl.* **20,** 478 (1981).
116. Colomer, E., and Corriu, R. J. P., *J. Chem. Soc., Chem. Commun.* p. 435 (1978).

117. Colomer, E., Corriu, R. J. P., and Vioux, A., *J. Chem. Soc., Chem. Commun.* p. 175 (1976).
118. Colomer, E., Corriu, R. J. P., and Vioux, A., *J. Chem. Res., Synop.* p. 168; *J. Chem. Res., Miniprint* p. 1939 (1977).
119. Colomer, E., Corriu, R. J. P., and Vioux, A., *Inorg. Chem.* **18**, 695 (1979).
120. Colomer, E., Corriu, R. J. P., and Young, J. C., *J. Chem. Soc., Chem. Commun.* p. 73 (1977); see also Cerveau, G., Colomer, E., Corriu, R. J. P., and Young, J. C., *J. Organomet. Chem.* **205**, 31 (1981).
121. Colquhoun, H. M., Ph.D. Thesis, University of London (1975).
122. Connolly, J. W., and Hoff, C. D., *J. Organomet. Chem.* **160**, 467 (1978).
123. Corriu, R. J. P., and Douglas, W. E., *J. Organomet. Chem.* **51**, C3 (1973).
124. Corriu, R. J. P., and Moreau, J. J. E., *J. Chem. Soc., Chem. Commun.* p. 278 (1980).
125. Couldwell, M. C., and Simpson, J., *J. Chem. Soc., Dalton Trans.* p. 714 (1976).
126. Couldwell, M. C., Simpson, J., and Robinson, W. T., *J. Organomet. Chem.* **107**, 323 (1976).
127. Cowie, M., and Bennett, M. J., *Inorg. Chem.* **16**, 2321 (1977).
128. Cowie, M., and Bennett, M. J., *Inorg. Chem.* **16**, 2325 (1977).
129. Cradock, S., Ebsworth, E. A. V., and Robertson, A., *J. Chem. Soc., Dalton Trans.* p. 22 (1973).
130. Crozat, M. M., and Watkins, S. F., *J. Chem. Soc., Dalton Trans.* p. 2512 (1972).
131. Cundy, C. S., and Lappert, M. F., *J. Chem. Soc., Chem. Commun.* p. 445 (1972).
132. Cundy, C. S., and Lappert, M. F., *J. Chem. Soc., Dalton Trans.* p. 665 (1978).
133. Cundy, C. S., and Lappert, M. F., *J. Organomet. Chem.* **144**, 317 (1978).
134. Cundy, C. S., Kingston, B. M., and Lappert, M. F., *Adv. Organomet. Chem.* **11**, 253 (1973).
135. Cundy, C. S., Lappert, M. F., Dubac, J., and Mazerolles, P., *J. Chem. Soc., Dalton Trans.* p. 910 (1976).
136. Cundy, C. S., Lappert, M. F., and Yuen, C.-K., *J. Chem. Soc., Dalton Trans.* p. 427 (1978).
137. Curtis, M. D., *Inorg. Nucl. Chem. Lett.* **6**, 859 (1970).
138. Curtis, M. D., *Inorg. Chem.* **11**, 802 (1972).
139. Curtis, M. D., and Greene, J., *J. Am. Chem. Soc.* **100**, 6362 (1978).
140. Curtis, M. D., Greene, J., and Butler, W. M., *J. Organomet. Chem.* **164**, 371 (1979).
141. Dalton, J., *Inorg. Chem.* **10**, 1822 (1971).
142. Dalton, J., *Inorg. Chem.* **11**, 915 (1972).
143. Darensbourg, D. J., *Inorg. Chim. Acta* **4**, 597 (1970).
144. Darensbourg, D. J., and Darensbourg, M. Y., *Inorg. Chem.* **9**, 1691 (1970).
145. Dean, W. K., and Graham, W. A. G., *Inorg. Chem.* **16**, 1061 (1977); Chan, L. Y. Y., Dean, W. K., and Graham, W. A. G., *ibid.* p. 1067.
146. Dean, W. K., Simon, G. L., Treichel, P. M., and Dahl, L. F., *J. Organomet. Chem.* **50**, 193 (1973).
147. Dessy, R. E., Pohl, R. L., and King, R. B., *J. Am. Chem. Soc.* **88**, 5121 (1966); King, R. B., *Acc. Chem. Res.* **3**, 417 (1970).
148. Devarajan, V., and Cyvin, S. J., *Acta Chem. Scand.* **26**, 1 (1972).
149. Ditchfield, R., Jensen, M. A., and Murrell, J. N., *J. Chem. Soc. A* p. 1674 (1967).
150. Dobson, G. R., and Ross, E. P., *Inorg. Chim. Acta* **5**, 199 (1971).
151. Duffy, D. N., and Nicholson, B. K., *J. Organomet. Chem.* **164**, 227 (1978).
152. Durig, J. R., Meischen, S. J., Hannum, S., Hitch, R. R., Gondal, S. K., and Sears, C. T., *Appl. Spectrosc.* **25**, 182 (1971).
153. Durkin, T. R., and Schram, E. P., *Inorg. Chem.* **11**, 1048 (1972).

154. Eaborn, C., "Organosilicon Compounds." Butterworth, London, 1960.
155. Eaborn, C., Harrison, M. R., Kapoor, P. N., and Walton, D. R. M., *J. Organomet. Chem.* **63**, 99 (1973).
156. Eaborn, C., Hitchcock, P. B., Tune, D. J., and Walton, D. R. M., *J. Organomet. Chem.* **54**, C1 (1973).
157. Eaborn, C., Kapoor, P. N., Tune, D. J., Turpin, C. L., and Walton, D. R. M., *J. Organomet. Chem.* **34**, 153 (1972).
158. Eaborn, C., Metham, T. N., and Pidcock, A., *J. Organomet. Chem.* **63**, 107 (1973).
159. Eaborn, C., Metham, T. N., and Pidcock, A., *J. Organomet. Chem.* **131**, 377 (1977).
160. Eaborn, C., Odell, K. J., Pidcock, A., and Scollary, G. R., *J. Chem. Soc., Chem. Commun.* p. 317 (1976).
161. Eaborn, C., Ratcliff, B., and Pidcock, A., *J. Organomet. Chem.* **65**, 181 (1974).
162. Eaborn, C., Ratcliff, B., and Pidcock, A., *J. Organomet. Chem.* **66**, 23 (1974).
163. Eaborn, C., Tune, D. J., and Walton, D. R. M., *J. Chem. Soc., Chem. Commun.* p. 1223 (1972).
164. Eaborn, C., Tune, D., and Walton, D. R. M., *J. Chem. Soc., Dalton Trans.* p. 2255 (1973).
165. Ebsworth, E. A. V., and Fraser, T. E., *J. Chem. Soc., Dalton Trans.* p. 1960 (1979).
166. Ebsworth, E. A. V., and Leitch, D. M., *J. Chem. Soc., Dalton Trans.* p. 1287 (1973).
167. Ebsworth, E. A. V., Edward, J. M., and Rankin, D. W. H., *J. Chem. Soc., Dalton Trans.* p. 1667 (1976).
168. Ebsworth, E. A. V., Edward, J. M., and Rankin, D. W. H., *J. Chem. Soc., Dalton Trans.* p. 1673 (1976).
169. Ebsworth, E. A. V., Marganian, V. M., Reed, F. J. S., and Gould, R. O., *J. Chem. Soc., Dalton Trans.* p. 1167 (1978).
170. Edwards, J. D., Goddard, R., Knox, S. A. R., McKinney, R. J., Stone, F. G. A., and Woodward, P., *J. Chem. Soc., Chem. Commun.* p. 828 (1975).
171. Edwards, J. D., Knox, S. A. R., and Stone, F. G. A., *J. Chem. Soc., Dalton Trans.* p. 545 (1980).
172. Elder, M., *Inorg. Chem.* **8**, 2703 (1969).
173. Elder, M., *Inorg. Chem.* **9**, 762 (1970).
174. Ellis, J. E., *J. Organomet. Chem.* **86**, 1 (1975).
175. Ellis, J. E., and Faltynek, R. A., *J. Chem. Soc., Chem. Commun.* p. 966 (1975).
176. Ellis, J. E., Barger, P. T., and Winzenburg, M. L., *J. Chem. Soc., Chem. Commun.* p. 686 (1977).
177. Emerson, K., Ireland, P. R., and Robinson, W. T., *Inorg. Chem.* **9**, 436 (1970).
178. Etzrodt, G., and Schmid, G., *J. Organomet. Chem.* **169**, 259 (1979).
179. Fawcett, J. P., and Harrod, J. F., *Can. J. Chem.* **54**, 3102 (1976).
180. Fawcett, J. P., and Harrod, J. F., *J. Organomet. Chem.* **113**, 245 (1976).
181. Fieldhouse, S. A., Cleland, A. J., Freeland, B. H., Mann, C. D. M., and O'Brien, R. J., *J. Chem. Soc. A* p. 2536 (1971).
182. Fink, W., *Helv. Chim. Acta* **58**, 1464 (1975); **59**, 606 (1976).
183. Fink, W., and Wenger, A., *Helv. Chim. Acta* **54**, 2186 (1971).
184. Fischer, E. O., Fischer, H., Schubert, U., and Pardy, R. B. A., *Angew. Chem., Int. Ed. Engl.* **18**, 871 (1979).
185. Fleming, I., and Roessler, F., *J. Chem. Soc., Chem. Commun.* p. 276 (1980); Fleming, I., Newton, T. W., and Roessler, F., *J. Chem. Soc., Perkin Trans. I*, p. 2527 (1981).
186. Fritz, G., and Hohenberger, K., *Z. Anorg. Allg. Chem.* **464**, 107 (1980).
187. Gangji, S. H. H., Ph.D. Thesis, University of London (1977).

188. Gansow, O. A., Schexnayder, D. A., and Kimura, B. Y., *J. Am. Chem. Soc.* **94,** 3406 (1972).

189. Gerlach, R. F., Graham, B. W. L., and Mackay, K. M., *J. Organomet. Chem.* **118,** C23 (1976).

190. Gerlach, R. F., Graham, B. W. L., and Mackay, K. M., *J. Organomet. Chem.* **182,** 285 (1979).

191. Gerlach, R. F., Mackay, K. M., and Nicholson, B. K., *J. Organomet. Chem.* **178,** C30 (1979); Gerlach, R. F., Mackay, K. M., Nicholson, B. K., and Robinson, W. T., *J. Chem. Soc., Dalton Trans.* p. 80 (1981).

192. Gingerich, K. A., *J. Chem. Phys.* **50,** 5426 (1969).

193. Gladysz, J. A., Tam, W., Williams, G. M., Johnson, D. L., and Parker, D. W., *Inorg. Chem.* **18,** 1163 (1979).

194. Gladysz, J. A., Williams, G. M., Tam, W., Johnson, D. L., Parker, D. W., and Selover, J. C., *Inorg. Chem.* **18,** 553 (1979).

195. Glockling, F., and Hill, G. C., *J. Chem. Soc. A* p. 2137 (1971).

196. Glockling, F., and Hooton, K. A., *J. Chem. Soc. A* p. 1066 (1967).

197. Glockling, F., and Houston, R. E., *J. Organomet. Chem.* **50,** C31 (1973).

198. Glockling, F., and Irwin, J. G., *Inorg. Chim. Acta.* **6,** 355 (1972).

199. Glockling, F., and McGregor, A., *J. Inorg. Nucl. Chem.* **35,** 1481 (1973).

200. Glockling, F., and Pollock, R. J. I., *J. Chem. Soc., Dalton Trans.* p. 497 (1975).

201. Glockling, F., and Stobart, S. R., *MTP Int. Rev. Sci.: Inorg. Chem.,* Ser. One **6,** 63 (1972).

202. Glockling, F., and Wilbey, M. D., *J. Chem. Soc. A* p. 1675 (1970).

203. Glockling, F., McGregor, A., Schneider, M. L., and Shearer, H. M. M., *J. Inorg. Nucl. Chem.* **32,** 3101 (1970).

204. Goddard, R., and Woodward, P., *J. Chem. Soc., Dalton Trans.* p. 559 (1980).

205. Green, M., Howard, J. A. K., Proud, J., Spencer, J. L., Stone, F. G. A., and Tsipis, C. A., *J. Chem. Soc., Chem. Commun.* p. 671 (1976).

206. Green, M., Spencer, J. L., Stone, F. G. A., and Tsipis, C. A., *J. Chem. Soc., Dalton Trans.* pp. 1519, 1526 (1977).

207. Greene, J., and Curtis, M. D., *J. Am. Chem. Soc.* **99,** 5176 (1977).

208. Greene, J., and Curtis, M. D., *Inorg. Chem.* **17,** 2324 (1978).

209. Grobe, J., and Möller, U., *J. Organomet. Chem.* **31,** 157 (1971).

210. Grobe, J., and Möller, U., *J. Organomet. Chem.* **36,** 335 (1972).

211. Grobe, J., and Walter, A., *J. Organomet. Chem.* **140,** 325 (1977).

212. Grobe, J., Martin, R., and Möller, U., *Angew. Chem., Int. Ed. Engl.* **16,** 248 (1977).

213. Grynkewich, G. W., and Marks, T. J., *Inorg. Chem.* **15,** 1307 (1976).

213a. Hagen, A. P., and MacDiarmid, A. G., *Inorg. Nucl. Chem. Lett.* **6,** 345 (1970).

214. Hagen, A. P., Higgins, C. R., and Russo, P. J., *Inorg. Chem.* **10,** 1657 (1971).

215. Hagen, A. P., McAmis, L., and Stewart, M. A., *J. Organomet. Chem.* **66,** 127 (1974).

216. Hamilton, R. S., and Corey, E. R., *Abstr. 156th Meet., Am. Chem. Soc.* p. INOR 025 (1968).

217. Harris, P. J., Howard, J. A. K., Knox, S. A. R., McKinney, R. J., Phillips, R. P., Stone, F. G. A., and Woodward, P., *J. Chem. Soc., Dalton Trans.* p. 403 (1978).

218. Harris, R. K., and Kimber, B. J., *J. Magn. Reson.* **17,** 174 (1975).

219. Harris, R. K., and Mann, B. E. eds., "NMR and the Periodic Table." Academic Press, New York, 1978.

220. Harris, R. K., Jones, J., and Ng, S., *J. Magn. Reson.* **30,** 521 (1978).

220a. Harrod, J. F., and Chalk, A. J., *in* "Organic Syntheses via Metal Carbonyls" (I. Wender and P. Pino, eds.) Vol. 2, p. 673–704. Wiley (Interscience), New York, 1977.

221. Harrod, J. F., and Smith, C. A., *Can. J. Chem.* **48**, 870 (1970).
222. Harrod, J. F., and Smith, C. A., *J. Am. Chem. Soc.* **92**, 2699 (1970).
223. Harrod, J. F., Gilson, D. F. R., and Charles, R., *Can. J. Chem.* **47**, 2205 (1969).
224. Harrod, J. F., Smith, C. A., and Khin Aye Than, *J. Am. Chem. Soc.* **94**, 8321 (1972).
225. Hart-Davis, A. J., and Graham, W. A. G., *J. Am. Chem. Soc.* **93**, 4388 (1971).
226. Haszeldine, R. N., Malkin, L. S., and Parish, R. V., *J. Organomet. Chem.* **182**, 323 (1979).
227. Haszeldine, R. N., Mather, A. P., and Parish, R. V., *J. Chem. Soc., Dalton Trans.* p. 923 (1980).
228. Haszeldine, R. N., Parish, R. V., and Riley, B. F., *J. Chem. Soc., Dalton Trans.* p. 705 (1980).
229. Haszeldine, R. N., Parish, R. V., and Setchfield, J. H., *J. Organomet. Chem.* **57**, 279 (1973).
230. Haszeldine, R. N., Parish, R. V., and Taylor, R. J., *J. Chem. Soc., Dalton Trans.* p. 2311 (1974).
231. Hayashi, T., Yamamoto, K., and Kumada, M., *J. Organomet. Chem.* **112**, 253 (1976).
232. Hencken, G., and Weiss, E., *Chem. Ber.* **106**, 1747 (1973).
233. Highsmith, R. E., Bergerud, J. R., and MacDiarmid, A. G., *J. Chem. Soc., Chem. Commun.* p. 48 (1971).
234. Hoffmann, R., Howell, J. M., and Rossi, A. R., *J. Am. Chem. Soc.* **98**, 2484 (1976).
235. Höfler, F., *Top. Curr. Chem.* **50**, 129 (1974).
236. Höfler, M., and Scheuren, J., *J. Organomet. Chem.* **55**, 177 (1973).
237. Höfler, M., Scheuren, J., and Spilker, D., *J. Organomet. Chem.* **102**, 205 (1975).
233. Höfler, M., Scheuren, J., and Weber, G., *J. Organomet. Chem.* **78**, 347 (1974).
239. Holtman, M. S., and Schram, E. P., *J. Organomet. Chem.* **187**, 147 (1980).
240. Hönle, W., and von Schnering, H. G., *Z. Anorg. Allg. Chem.* **464**, 139 (1980).
241. Hooton, K. A., *J. Chem. Soc. A* p. 1251 (1971).
242. Howard, J., and Woodward, P., *J. Chem. Soc., Dalton Trans.* p. 59 (1975).
243. Howard, J., Knox, S. A. R., Stone, F. G. A., and Woodward, P., *J. Chem. Soc., Chem. Commun.* p. 1477 (1970).
244. Howard, J. A. K., Knox, S. A. R., McKinney, R. J., Stansfield, R. F. D., Stone, F. G. A., and Woodward, P., *J. Chem. Soc., Chem. Commun.* p. 557 (1976).
245. Howard, J. A. K., Knox, S. A. R., Riera, V., Sosinsky, B. A., Stone, F. G. A., and Woodward, P., *J. Chem. Soc., Chem. Commun.* p. 673 (1974).
246. Howard, J. A. K., Knox, S. A. R., Riera, V., Stone, F. G. A., and Woodward, P., *J. Chem. Soc., Chem. Commun.* p. 452 (1974).
247. Howard, J. A. K., Knox, S. A. R., Stone, F. G. A., Szary, A. C., and Woodward, P., *J. Chem. Soc., Chem. Commun.* p. 788 (1974).
248. Hoyano, J. K., and Graham, W. A. G., *Inorg. Chem.* **11**, 1265 (1972).
249. Hoyano, J. K., and Graham, W. A. G., unpublished observations, quoted in Ref. (225).
250. Hoyano, J. K., Elder, M., and Graham, W. A. G., *J. Am. Chem. Soc.* **91**, 4568 (1969).
251. Hutcheon, W. L., Ph.D. Thesis, University of Alberta, Edmonton (1971), quoted in Ref. (128); see also footnote 2 of Ref. (248).
252. Ingle, W. M., Preti, G., and MacDiarmid, A. G., *J. Chem. Soc., Chem. Commun.* p. 497 (1973).
253. Innorta, G., Foffani, A., and Torroni, S., *Inorg. Chim. Acta* **19**, 263 (1976).
254. Isaacs, E. E., and Graham, W. A. G., *Can. J. Chem.* **53**, 467 (1975).
255. Isaacs, E. E., and Graham, W. A. G., *Can. J. Chem.* **53**, 975 (1975).

256. Jansen, P. R., Oskam, A., and Olie, K., *Cryst. Struct. Commun.* **4**, 667 (1975).
257. Jeitschko, W., *MTP Int. Rev. Sci.: Inorg. Chem., Ser. Two* **5**, 219 (1975).
258. Jetz, W., and Graham, W. A. G., *J. Am. Chem. Soc.* **89**, 2773 (1967).
259. Jetz, W., and Graham, W. A. G., *Inorg. Chem.* **10**, 4 (1971).
260. Jetz, W., and Graham, W. A. G., *Inorg. Chem.* **10**, 1159 (1971).
261. Jetz, W., and Graham, W. A. G., *Inorg. Chem.* **10**, 1647 (1971).
262. Jetz, W., and Graham, W. A. G., *J. Organomet. Chem.* **69**, 383 (1974).
263. Job, R. C., and Curtis, M. D., *Inorg. Chem.* **12**, 2510, 2514 (1973).
263a. Johannesen, R. B., Brinkman, F. E., and Coyle, T. D., *J. Phys. Chem.* **72**, 660 (1968).
264. Johnson, D. L., and Gladysz, J. A., *J. Am. Chem. Soc.* **101**, 6433 (1979); *Inorg. Chem.* **20**, 2508 (1981).
265. Jolly, W. L., Avanzino, S. C., and Rietz, R. R., *Inorg. Chem.* **16**, 964 (1977).
266. Jutzi, P., and Steiner, W., *Angew. Chem., Int. Ed. Engl.* **16**, 639 (1977).
267. Jutzi, P., Kohl, F., and Kruger, C., *Angew. Chem., Int. Ed. Engl.* **18**, 59 (1979).
268. Kerber, R. C., and Pakkanen, T., *Inorg. Chim. Acta* **37**, 61 (1979).
269. Kettle, S. F. A., and Khan, I. A., *J. Organomet. Chem.* **5**, 588 (1966).
270. King, R. B., and Pannell, K. H., *Inorg. Chem.* **7**, 1510 (1968).
271. King, R. B., and Pannell, K. H., *Z. Naturforsch., B: Anorg. Chem., Org. Chem., Biochem., Biophys., Biol.* **24B**, 262 (1969).
272. King, R. B., Pannell, K. H., Bennett, C. R., and Ishaq, M., *J. Organomet. Chem.* **19**, 327 (1969).
273. Kingston, B. M., and Lappert, J. F., *J. Chem. Soc., Dalton Trans.* p. 69 (1972).
274. Kiso, Y., Tamao, K., and Kumada, M., *J. Chem. Soc., Chem. Commun.* p. 105 (1972).
275. Kiso, Y., Tamao, K., and Kumada, M., *J. Chem. Soc., Chem. Commun.* p. 1208 (1972).
276. Kiso, Y., Tamao, K., and Kumada, M., *J. Organomet. Chem.* **76**, 95 (1974).
277. Kiso, Y., Tamao, K., and Kumada, M., *J. Organomet. Chem.* **76**, 105 (1974).
278. Klanberg, F., Askew, W. B., and Guggenberger, L. J., *Inorg. Chem.* **7**, 2265 (1968).
279. Knox, S. A. R., McKinney, R. J., Riera, V., Stone, F. G. A., and Szary, A. C., *J. Chem. Soc., Dalton Trans.* p. 1801 (1979).
280. Knox, S. A. R., McKinney, R. J., and Stone, F. G. A., *J. Chem. Soc., Dalton Trans.* p. 235 (1980).
281. Knox, S. A. R., Phillips, R. P., and Stone, F. G. A., *J. Chem. Soc., Dalton Trans.* p. 658 (1974).
282. Ködel, W., Haupt, H.-J., and Huber, F., *Z. Anorg. Allg. Chem.* **448**, 126 (1979).
283. Köhler, F. H., Hollfelder, H., and Fischer, E. O., *J. Organomet. Chem.* **168**, 53 (1979).
284. Kono, H., and Nagai, Y., *Chem. Lett.* p. 931 (1974); *Chem. Abstr.* **81**, 120736 (1974).
285. Kono, H., Wakao, N., Ito, K., and Nagai, Y., *J. Organomet. Chem.* **132**, 53 (1977).
286. Kristoff, J. S., and Shriver, D. F., *Inorg. Chem.* **13**, 499 (1974).
287. Krogh-Jespersen, K., Chandrasekhar, J., and Schleyer, P. von R., *J. Org. Chem.* **45**, 1608 (1980).
288. Kruck, T., Job, E., and Klose, U., *Angew. Chem., Int. Ed. Engl.* **7**, 374 (1968).
289. Kummer, D., and Furrer, J., *Z. Naturforsch., B: Anorg. Chem., Org. Chem., Biochem., Biophys., Biol.* **26B**, 162 (1971).
290. Kuzmin, O. V., Bikovetz, A. L., Vdovin, V. M., and Krapivin, A. M., *Izv. Akad. Nauk SSSR, Ser. Khim.* p. 2815 (1979).

291. Lappert, M. F., and Speier, G., *J. Organomet. Chem.* **80,** 329 (1974).
292. Lappert, M. F., Miles, S. J., Power, P. P., Carty, A. J., and Taylor, N. J., *J. Chem. Soc., Chem. Commun.* p. 458 (1977).
293. Lee, A. G., *J. Organomet. Chem.* **16,** 321 (1969).
294. Lemanski, M. F., and Schram, E. P., *Inorg. Chem.* **15,** 1489 (1976).
295. Li, S., Johnson, D. L., Gladysz, J. A., and Servis, K. L., *J. Organomet. Chem.* **166,** 317 (1979).
296. Lichtenberger, D. L., and Brown, T. L., *J. Am. Chem. Soc.* **99,** 8187 (1977).
297. Liu, C.-S., and Cheng, C.-W., *J. Am. Chem. Soc.* **97,** 6746 (1975).
298. MacDiarmid, A. G., *Intra-Sci. Chem. Rep.* **7,** 83 (1973).
299. Mackay, K. M., and Stobart, S. R., *J. Chem. Soc., Dalton Trans.* p. 214 (1973).
300. Malisch, W., *J. Organomet. Chem.* **39,** C28 (1972).
301. Malisch, W., *J. Organomet. Chem.* **61,** C15 (1973).
302. Malisch, W., *Angew. Chem., Int. Ed. Engl.* **12,** 235 (1973).
303. Malisch, W., *J. Organomet. Chem.* **77,** C15 (1974).
304. Malisch, W., *Chem. Ber.* **107,** 3835 (1974).
305. Malisch, W., *J. Organomet. Chem.* **82,** 185 (1974).
306. Malisch, W., and Kuhn, M., *Chem. Ber.* **107,** 979 (1974).
307. Malisch, W., and Kuhn, M., *Chem. Ber.* **107,** 2835 (1974).
308. Malisch, W., and Panster, P., *J. Organomet. Chem.* **64,** C5 (1974); *Chem. Ber.* **108,** 2554 (1975).
309. Malisch, W., and Ries, W., *Angew. Chem., Int. Ed. Engl.* **17,** 120 (1978); *Chem. Ber.* **112,** 1304 (1979).
310. Malisch, W., Schmidbaur, H., and Kuhn, M., *Angew. Chem., Int. Ed. Engl.* **11,** 516 (1972).
311. Mance, A. M., and Van Dyke, C. H., *Inorg. Nucl. Chem. Lett.* **15,** 393 (1979).
312. Mann, C. D. M., Cleland, A. J., Fieldhouse, S. A., Freeland, B. H., and O'Brien, R. J., *J. Organomet. Chem.* **24,** C61 (1970).
313. Manojlović-Muir, L., Muir, K. W., and Ibers, J. A., *Inorg. Chem.* **9,** 447 (1970).
314. Marks, T. J., *Abstr., 162nd Meet., Am. Chem. Soc.* p. INOR 23 (1971).
315. Marks, T. J., *J. Am. Chem. Soc.* **93,** 7090 (1971).
316. Marks, T. J., and Seyam, A. M., *J. Organomet. Chem.* **31,** C62 (1971).
317. Marks, T. J., and Seyam, A. M., *Inorg. Chem.* **13,** 1624 (1974).
318. Matyushenko, N. N., *Rev. High-Temp. Mater.* **1,** No. 2 (1978).
319. McLean, R. A. N., *J. Chem. Soc., Dalton Trans.* p. 1568 (1974).
320. McNeese, T. J., Wreford, S. S., Tipton, D. L., and Bau, R., *J. Chem. Soc., Chem. Commun.* p. 390 (1977).
321. McWeeney, R., Mason, R., and Towl, A. D. C., *Discuss. Faraday Soc.* **47,** 20 (1969).
322. Moro, A., Foa, M., and Cassar, L., *J. Organomet. Chem.* **185,** 79 (1980).
323. Morrison, D. L., and Hagen, A. P., *Inorg. Synth.* **13,** 65 (1972).
324. Moss, J. R., and Graham, W. A. G., *J. Organomet. Chem.* **23,** C23 (1970).
325. Muetterties, E. L., and Hirsekorn, F. J., *J. Chem. Soc., Chem. Commun.* p. 683 (1973).
326. Muir, K. W., *J. Chem. Soc. A* p. 2663 (1971).
327. Muir, K. W., and Ibers, J. A., *Inorg. Chem.* **9,** 440 (1970).
328. Nakadaira, Y., Kobayashi, T., and Sakurai, H., *J. Organomet. Chem.* **165,** 399 (1979).
329. Nametkin, N. S., Vdovin, V. M., Poletaev, V. A., Svergun, V. I., and Sergeeva, M. B., *Bull. Acad. Sci. USSR, Div. Chem. Sci. (Engl. Transl.)* p. 2767 (1974).
330. Nasta, M. A., and MacDiarmid, A. G., *J. Organomet. Chem.* **18,** P11 (1969).

331. Nasta, M. A., and MacDiarmid, A. G., *J. Am. Chem. Soc.* **93**, 2813 (1971); **94**, 2449 (1972).

332. Nefedov, O. M., Kolesnikov, S. P., and Ioffe, A. I., *J. Organomet. Chem. Libr.* **5**, 181 (1977).

333. Nesmeyanov, A. N., Anisimov, K. N., Kolobova, N. E., and Khandozhko, V. N., *Izv. Akad. Nauk SSSR, Ser. Khim.* p. 462 (1971); *Chem. Abstr.* **75**, 20543 (1971).

334. Nesmeyanov, A. N., Anisimov, K. N., Kolabova, N. E., and Khandozhko, V. N., *Zh. Obshch. Khim.* **44**, 1287 (1974); *Chem. Abstr.* **81**, 57704 (1974).

335. Nicholson, B. K., and Simpson, J., *J. Organomet. Chem.* **72**, 211 (1974).

336. Nicholson, B. K., and Simpson, J., *J. Organomet. Chem.* **155**, 237 (1978).

337. Nicholson, B. K., Robinson, B. H., and Simpson, J., *J. Organomet. Chem.* **66**, C3 (1974).

338. Nicholson, B. K., Simpson, J., and Robinson, W. T. *J. Organomet. Chem.* **47**, 403 (1973).

339. Nickl, J. J., and Schweitzer, K. K., *Proc. Int. Conf. Chem. Vapor Deposition, 2nd, 1970* p. 297 (1970).

340. Nowotny, H., *MTP Int. Rev. Sci.: Inorg. Chem., Ser. One* **10**, 151 (1972).

341. Ogino, K., and Brown, T. L., *Inorg. Chem.* **10**, 517 (1971).

342. Ojima, I., Inaba, S., Kogure, T., and Nagai, Y., *J. Organomet. Chem.* **55**, C7 (1973).

343. Ojima, I., Kogure, T., Kumagai, M., Horiuchi, S., and Sato, T., *J. Organomet. Chem.* **122**, 83 (1976).

344. Ojima, I., Nihonyanagi, M., and Nagai, Y., *J. Chem. Soc., Chem. Commun.* p. 938 (1972).

345. Okinoshima, H., Yamamoto, K., and Kumada, M., *J. Am. Chem. Soc.* **94**, 9263 (1972).

346. Oliver, A. J., and Graham, W. A. G., *Inorg. Chem.* **10**, 1 (1971).

347. Onaka, S., *Bull. Chem. Soc. Jpn.* **46**, 2444 (1973).

348. Onaka, S., *J. Inorg. Nucl. Chem.* **36**, 1721 (1974).

349. Onaka, S., *Nippon Kagaku Kaishi* p. 255 (1974).

350. Onaka, S., Yoshikawa, Y., and Yamatera, H., *J. Organomet. Chem.* **157**, 187 (1978).

351. Owen, B. B., and Webber, R. J., *AIME, Tech. Publ.* p. 2306 (1948).

352. Pannell, K. H., *J. Organomet. Chem.* **21**, P17 (1970).

353. Pannell, K. H., and Rice, J. R., *J. Organomet. Chem.* **78**, C35 (1974).

354. Pannell, K. H., Wu, C. C., and Long, G. J., *J. Organomet. Chem.* **186**, 85 (1980).

355. Parish, R. V., and Riley, B. F., *J. Chem. Soc., Dalton Trans.* p. 482 (1979).

356. Párkányi, L., Nagy, J., and Simon, K., *J. Organomet. Chem.* **101**, 11 (1975), and references given therein.

357. Patil, H. R. H., and Graham, W. A. G., *Inorg. Chem.* **5**, 1401 (1966).

358. Penfold, B. R., and Robinson, B. H., *Accs. Chem. Res.* **6**, 73 (1973).

359. Piper, T. S., Lemal, D., and Wilkinson, G., *Naturwissenschaften* **43**, 129 (1956).

360. Poate, J. M., Tu, K. N., and Mayer, J. W., "Thin Films." Wiley, New York, 1978.

361. Pomeroy, R. K., *J. Organomet. Chem.* **177**, C27 (1979).

362. Pomeroy, R. K., and Graham, W. A. G., *J. Am. Chem. Soc.* **94**, 274 (1972).

363. Pomeroy, R. K., and Harrison, D. J., *J. Chem. Soc., Chem. Commun.* p. 661 (1980).

364. Pomeroy, R. K., and Wijeskera, K. S., *Can. J. Chem.* **58**, 206 (1980); *Inorg. Chem.* **19**, 3729 (1980).

365. Pomeroy, R. K., Gay, R. S., Evans, G. O., and Graham, W. A. G., *J. Am. Chem. Soc.* **94**, 272 (1972).

366. Powell, C. F., Oxley, J. H., and Blocher, J. M. "Vapor Deposition." Wiley, New York, 1966.

367. Preut, H., and Haupt, H.-J., Z. Anorg. Allg. Chem. **422**, 47 (1976).
368. Purnell, J. H., and Walsh, R., Proc. R. Soc. London, Ser. A **293**, 543 (1966).
369. Rankin, D. W. H., and Robertson, A. J. Organomet. Chem. **85**, 225 (1975).
370. Rankin, D. W. H., and Robertson, A., J. Organomet. Chem. **105**, 331 (1976).
371. Rankin, D. W. H., Robertson, A., and Seip, R., J. Organomet. Chem. **88**, 191 (1975).
372. Razuvaev, G. A., Latyaeva, V. N., Gladyshev, E. N., Krasilnikova, E. V., Lineva, A. N., and Kozina, A. P., Inorg. Chim. Acta **31**, L357 (1978).
373. Redwood, M. E., Reichert, B. E., Schrieke, R. R., and West, B. O., Aust. J. Chem. **26**, 247 (1973).
374. Robiette, A. G., Sheldrick, G. M., Simpson, R. N. F., Aylett, B. J., and Campbell, J. M., J. Organomet. Chem. **14**, 279 (1968).
375. Robinson, W. T., and Ibers, J. A., Inorg. Chem. **6**, 1208 (1967).
376. Ross, E. P., and Dobson, G. R., J. Chem. Soc., Chem. Commun. p. 1229 (1969).
377. Ross, E. P., Jernigan, R. T., and Dobson, G. R., J. Inorg. Nucl. Chem. **33**, 3375 (1971).
378. Saalfield, F. E., McDowell, M. V., DeCorpo, J. J., Berry, A. D., and MacDiarmid, A. G., Inorg. Chem. **12**, 48 (1973).
379. Saalfield, F. E., McDowell, M. V., Gondal, S. K., and MacDiarmid, A. G., Inorg. Chem. **7**, 1465 (1968).
380. Saalfield, F. E., McDowell, M. V., and MacDiarmid, A. G., J. Am. Chem. Soc. **92**, 2324 (1970).
381. Saalfield, F. E., McDowell, M. V., MacDiarmid, A. G., and Highsmith, R. E., Int. J. Mass Spectrom. Ion Phys. **9**, 197 (1972).
382. Sakurai, H., Kamiyama, Y., and Nakadaira, Y., J. Am. Chem. Soc. **98**, 7453 (1976).
383. Sakurai, H., Kamiyama, Y., and Nakadaira, Y., J. Am. Chem. Soc. **99**, 3879 (1977).
384. Sakurai, H., Kamiyama, Y., and Nakadaira, Y., Angew. Chem., Int. Ed. Engl. **17**, 674 (1978).
385. Sakurai, H., Kamiyama, Y., and Nakadaira, Y., J. Organomet. Chem. **184**, 13 (1980).
386. Sakurai, H., Kobayashi, T., and Nakadaira, Y., J. Organomet. Chem. **162**, C43 (1978).
387. Schäfer, H., and MacDiarmid, A. G., Inorg. Chem. **15**, 848 (1976).
388. Schmid, G., Angew. Chem. **84**, 1111 (1972).
389. Schmid, G., Angew. Chem., Int. Ed. Engl. **17**, 392 (1978).
390. Schmid, G., and Balk, H.-J., Chem. Ber. **103**, 2240 (1970).
391. Schmid, G., and Balk, H.-J., J. Organomet. Chem. **80**, 257 (1974).
392. Schmid, G., and Boese, R., Chem. Ber. **105**, 3306 (1972).
393. Schmid, G., and Etzrodt, G., J. Organomet. Chem. **131**, 477 (1977).
394. Schmid, G., and Welz, E., Angew. Chem., Int. Ed. Engl. **16**, 785 (1977).
395. Schmid, G., and Welz, E., Z. Naturforsch., B: Anorg. Chem., Org. Chem. **34B**, 929 (1979).
396. Schmid, G., Bätzel, V., and Etzrodt, G., J. Organomet. Chem. **112**, 345 (1976).
397. Schrieke, R. R., and West, B. O., Inorg. Nucl. Chem. Lett. **5**, 141 (1969).
398. Schrieke, R. R., and West, B. O., Aust. J. Chem. **22**, 49 (1969).
399. Schubert, U., and Rengstl, A., J. Organomet. Chem. **166**, 323 (1979).
400. Schubert, U., and Rengstl, A., J. Organomet. Chem. **170**, C37 (1979).
401. Seyferth, D., Adv. Organomet. Chem. **14**, 97 (1976).
402. Simon, G. L., and Dahl, L. F., J. Am. Chem. Soc. **95**, 783 (1973).
403. Simpson, K. A., Ph.D. Thesis, University of Alberta, Edmonton (1973), quoted in Refs. (128, 256).

404. Smith, R. A., and Bennett, M. J., *Acta Crystallogr., Sect. B* **B33**, 1113 (1977).

405. Smith, R. A., and Bennett, M. J., *Acta Crystallogr., Sect. B* **B33**, 1118 (1977).

406. Sommer, L. H., "Stereochemistry, Mechanism and Silicon." McGraw-Hill, New York, 1965.

407. Sommer, L. H., and Lyons, J. E., *J. Am. Chem. Soc.* **90**, 4197 (1968).

408. Sommer, L. H., Lyons, J. E., and Fujimoto, H., *J. Am. Chem. Soc.* **91**, 7051 (1969).

409. Sosinsky, B. A., Knox, S. A. R., and Stone, F. G. A., *J. Chem. Soc., Dalton Trans.* p. 1633 (1975).

410. Spalding, T. R., *J. Organomet. Chem.* **149**, 371 (1978).

410a. Speier, J. L., *Adv. Organomet. Chem.* **17**, 407 (1979).

411. Svoboda, P., Řeřicha, R., and Hetflejš, J., *Collect. Czech. Chem. Commun.* **39**, 1324 (1974).

412. Szary, A. C., Knox, S. A. R., and Stone, F. G. A., *J. Chem. Soc., Dalton Trans.* p. 662 (1974).

413. Thompson, J. A. J., and Graham, W. A. G., *Inorg. Chem.* **6**, 1365 (1967).

413a. Triplett, K., and Curtis, M. D., *J. Am. Chem. Soc.* **97**, 5747 (1975).

414. Triplett, K., and Curtis, M. D., *Inorg. Chem.* **15**, 431 (1976).

415. Tsipis, C. A., *J. Organomet. Chem.* **187**, 427 (1980).

416. Tsipis, C. A., *J. Organomet. Chem.* **188**, 53 (1980).

417. Uhlig, E., Hipler, B., and Müller, P., *Z. Anorg. Allg. Chem.* **442**, 11 (1978).

418. Vancea, L., and Graham, W. A. G., *Inorg. Chem.* **13**, 511 (1974).

419. Vancea, L., Bennett, M. J., Jones, C. E., Smith, R. A., and Graham, W. A. G., *Inorg. Chem.* **16**, 897 (1977).

420. Vancea, L., Pomeroy, R. K., and Graham, W. A. G., *J. Am. Chem. Soc.* **98**, 1407 (1976).

421. Vyazankin, N. S., Razuvaev, G. A., and Kruglaya, O. A., *Organomet. Chem. Synth.* **1**, 205 (1971).

422. Vyshinskaya, L. I., Vasil'eva, G. A., and Klimova, N. V., *Tr. Khim. Khim. Tekhnol.* p. 79 (1975); *Chem. Abstr.* **85**, 124083 (1976).

423. Ward, R., *MTP Int. Rev. Sci.: Inorg. Chem., Ser. One* **5**, 93 (1972).

424. Watters, K. L., Brittain, J. N., and Risen, W. M., *Inorg. Chem.* **8**, 1347 (1969).

425. Webb, M. J., and Graham, W. A. G., *J. Organomet. Chem.* **93**, 119 (1975).

426. Webb, M. J., Stewart, R. P., and Graham, W. A. G., *J. Organomet. Chem.* **59**, C21 (1973); Webb, M. J., Bennett, M. J., Chan, L. Y. Y., and Graham, W. A. G., *J. Am. Chem. Soc.* **96**, 5931 (1974).

427. Windus, C., Sujishi, S., and Giering, W. P., *J. Am. Chem. Soc.* **96**, 1951 (1974).

428. Windus, C., Sujishi, S., and Giering, W. P., *J. Organomet. Chem.* **101**, 279 (1975).

429. Wong, F. S., and Mackay, K. M., *Inorg. Chim. Acta* **32**, L21 (1979).

430. Yamamoto, K., Hayashi, T., and Kumada, M., *J. Organomet. Chem.* **28**, C37 (1971).

431. Yamamoto, K., Okinoshima, H., and Kumada, M., *J. Organomet. Chem.* **23**, C7 (1970); **27**, C31 (1971).

432. Almeida, J. F., and Pidcock, A., *J. Organomet. Chem.* **208**, 273 (1981).

433. Avanzino, S. C., Chen, H.-W., Donahue, C. J., and Jolly, W. L., *Inorg. Chem.* **19**, 2201 (1980).

434. Blakeney, A. J., and Gladysz, J. A., *J. Organomet. Chem.* **202**, 263 (1980).

435. Blakeney, A. J., and Gladysz, J. A., *J. Organomet. Chem.* **210**, 303 (1981).

436. Blakeney, A. J., and Gladysz, J. A., *Inorg. Chim. Acta Lett.* **53**, L25 (1981).

437. Brinkman, K. C., and Gladysz, J. A., *J. Chem. Soc., Chem. Commun.* p. 1260 (1980).

438. Drahnak, T. J., West, R., and Calabrese, J. C., *J. Organomet. Chem.* **198**, 55 (1980).

439. Ebsworth, E. A. V., Ferrier, H. M., and Fraser, T. E., *J. Chem. Soc., Dalton Trans.* p. 836 (1981).

440. Ebsworth, E. A. V., Fraser, T. E., Henderson, S. G., Leitch, D. M., and Rankin, D. W. H., *J. Chem. Soc., Dalton Trans.* p. 1010 (1981).

441. Faltynek, R. A., *Inorg. Chem.* **20**, 1357 (1981).

442. Gäde, W., and Weiss, E., *J. Organomet. Chem.* **213**, 451 (1981).

443. Holmes-Smith, R. D., Stobart, S. R., Cameron, T. S., and Jochem, K., *J. Chem. Soc., Chem. Commun.* p. 937 (1981).

444. Hseu, T. H., Yun Chi, and Liu, C.-S., *Inorg. Chem.* **20**, 199 (1981).

445. Kruppa, W., and Schmid, G., *J. Organomet. Chem.* **202**, 379 (1980).

446. Radnia, P., and McKennis, J. S., *J. Am. Chem. Soc.* **102**, 6349 (1980).

447. Razuvaev, G. A., Latyaeva, V. N., Vyshinskaya, L. I., Malysheva, A. V., and Vasil'eva, G. A., *Doklady Akad. Nauk S.S.S.R.* **237**, 605 (1977).

448. Reichel, C. L., and Wrighton, M. S., *Inorg. Chem.* **19**, 3858 (1980).

449. Rösch, L., Altnau, G., Erb, W., Pickardt, J., and Bruncks, N., *J. Organomet. Chem.* **197**, 51 (1980).

450. Rybinskaya, M. I., Rybin, L. V., Pogrebnyak, A. A., Nurtdinova, G. V., and Yur'ev, V. P., *J. Organomet. Chem.* **217**, 373 (1981).

451. Sakurai, H., Kamiyama, Y., Mikoda, A., Kobayashi, T., Sasaki, K., and Nakadaira, Y., *J. Organomet. Chem.* **201**, C14 (1980); see also Pannell, K. H., Bassindale, A. R., and Fitch, J. W., *J. Organomet. Chem.* **209**, C65 (1981).

452. Schroeder, M. A., and Wrighton, M. S., *J. Organomet. Chem.* **128**, 345 (1977).

453. Schubert, U., Wörle, B., and Jandik, P., *Angew. Chem., Int. Ed. Engl.* **20**, 695 (1981).

454. Süss-Fink, G., and Reiner, J., *J. Organomet. Chem.* **221**, C36 (1981); Süss-Fink, G., *Angew. Chem., Int. Ed. Engl.* **21**, 73 (1982).

455. Vahrenkamp, H., Steiert, D., and Gusbeth, P., *J. Organomet. Chem.* **209**, C17 (1981).

456. Van Buuren, G. N., Willis, A. C., Einstein, S. W. B., Peterson, L. K., Pomeroy, R. K., and Sutton, D., *Inorg. Chem.* **20**, 4361 (1981); see also Pomeroy, R. K., *J. Organomet. Chem.* **221**, 323 (1981).

457. Watanabe, H., Kobayashi, M., Saito, M., and Nagai, Y., *J. Organomet. Chem.* **216**, 149 (1981).

458. Yun Chi and Liu, C.-S., *Inorg. Chem.* **20**, 3456 (1981).

ADVANCES IN INORGANIC CHEMISTRY AND RADIOCHEMISTRY, VOL. 25

THE ELECTRONIC PROPERTIES
OF METAL SOLUTIONS IN LIQUID AMMONIA
AND RELATED SOLVENTS

PETER P. EDWARDS

University Chemical Laboratory, Cambridge, England

I. Introduction 135
II. The Isolated Solvated Electron in Dilute Solutions 138
III. Electronic Properties of Dilute Solutions 142
   A. Metals in Liquid Ammonia 142
   B. Metals in Amines, Ethers, and Other Solvents 148
IV. Concentrated Solutions and the Nonmetal-to-Metal Transition 168
   A. Models for the Nonmetal-to-Metal Transition 169
   B. Liquid–Liquid Phase Separation and the Nonmetal-to-Metal
      Transition 174
   C. Expanded-Metal Compounds 176
V. Concluding Remarks 178
   Appendix 178
   References 180

I. Introduction

The introduction of sodium metal into anhydrous liquid ammonia produces an intensely colored blue solution in which solvated electrons, sodium cations, and various agglomerates of these species coexist in equilibrium. With increasing metal concentration, the system transforms into a bronze-colored metallic conductor with an equivalent conductance exceeding that of liquid mercury. In the transitional range, cooling of the homogeneous solution can give rise to a remarkable liquid–liquid phase separation in which both dilute (blue) and concentrated (bronze) phases coexist. The underlying experimental simplicity of this system, which exhibits the localization of the fundamental unit of electrical charge in dilute solutions and its itineracy in concentrated solutions, has attracted a considerable amount of study from both chemists and physicists for over a century (37–39, 99, 103, 113, 116, 164, 172).

Although we routinely attribute the discovery of metal–ammonia solutions to the German chemist Weyl in 1863 (172), they were proba-

135

bly first prepared over 50 years earlier by Sir Humphry Davy. Following his discovery and isolation of potassium and sodium in October 1807 (*46*), Davy conducted a long series of experiments on the reaction between potassium and gaseous ammonia to support his contention that potassium was indeed an element and not a hydride of potash, as suggested by Gay-Lussac and Thenard (*136*). Throughout this investigation, Davy had consistently stressed (*136*) the overriding importance of working with dry ammonia, a precaution apparently neglected by the French chemists of the day. An entry (Fig. 1) by Davy for November 14 (or thereabouts) 1808 in his experimental notebook is one of many referring to numerous experiments conducted on the mutual action of potassium and ammonia with each other:

> When 8 grains of potassium are heated in ammoniacal gas—it assumed a beautiful metallic appearance & gradually became of a fine blue colour . . .

Several other entries in this period refer to similar observations. As far as I am aware at the present time, these observations were never published in the scientific literature (*79*), although Davy in the same notebook does make some remarkably perceptive comments regarding these systems (Section V).

The physical properties of these intriguing systems have been studied in serious vein since the 1920s (*111, 164*). At the present time, this interest continues unabated. In its more recent development, the multidisciplinary nature of metal-solution investigations has also spilled over into the wider fields of excess electrons in disordered media (*38, 39, 99, 103*). The recent Colloque Weyl V conference on metal–ammonia solutions and excess electrons in liquids (*39*) brought together almost 200 scientists from all disciplines. The present review

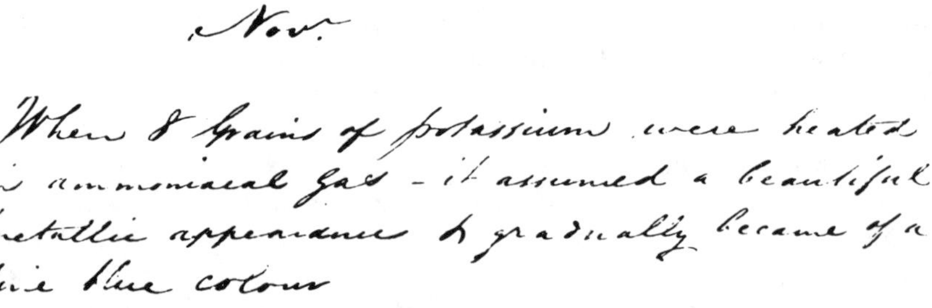
Nov.

When 8 Grains of potassium were heated in ammoniacal gas - it assumed a beautiful metallic appearance & gradually became of a fine blue colour

FIG. 1. An entry from the laboratory notebook of Sir Humphry Davy for the period October 1805–October 1812. The entry is for November 1808; a more precise location is rather difficult, but various entries around the same period suggest November 14 as the most likely date. Reproduced by courtesy of the Royal Institution, London, England.

attempts to cover the electronic properties of metal solutions in the dilute and concentrated ranges, as well as a cursory examination of the nature of the nonmetal-to-metal transition in these systems. In a review some 45 pages long one cannot expect to do justice to the sheer depth and extent of investigation into these systems. In particular, metal–ammonia solutions must be one of only a few systems (if not *the* only one) that can boast reliable and comprehensive optical, conductivity, and magnetic data (among other properties) over something like five orders of magnitude in concentration (*37, 164*).

My aim here is to provide a general background discussion centered around recent advances in studies of metals, not only in ammonia but also in various other solvents. To this author, one of the most important features to emerge from investigations into metal solutions over the last two decades has been the striking advances made possible from studies of metals in nonaqueous solvents other than ammonia. As Dye has suggested (*55*), in any description of metal solutions, we invariably return to the ammonia system for guidance. Certainly in this solvent we have, at the present time, the most data on physical and chemical properties (*164*). In this spirit, the electronic properties of dilute, intermediate, and concentrated metal–ammonia solutions form the underlying framework for this review, and these are discussed at various points in Sections II–IV.

However, in contrast to the ammonia system, metal solutions in amines, ethers, and related solvents are rich in information about *distinguishable* species existing in dilute solutions (*53, 55*). Sections II and III,B will similarly outline certain recent developments in these solvent systems, which have led to a fairly detailed picture of localized-electron states in the dilute range.

Section IV gives an overall view of concentrated solutions and the nonmetal–metal (NM-M) transition. Here again, the majority of data are for metal–ammonia solutions. However, in the past decade a substantial body of experimental data has emerged for the NM-M transition in lithium–methylamine solutions, which allows a direct comparison with the situation existing in metal–ammonia solutions (*60*). This section also considers recent developments in the study of the "expanded-metal" compounds, as typified by $Li(NH_3)_4$, $Ca(NH_3)_6$ (*82*), and $Li(CH_3NH_2)_4$ (*66*), and formed by slow cooling of the concentrated metal solutions.

Throughout this review, metal concentrations are expressed in units of mole percent metal (MPM), where

$$\text{mole percent metal} = \left| \frac{\text{moles metal}}{(\text{moles metal + moles solvent})} \right| \times 100$$

II. The Isolated Solvated Electron in Dilute Solutions

The accumulated evidence for the electrolytic nature of dilute metal solutions is overwhelming—metal atoms introduced into a variety of nonaqueous solvents spontaneously dissociate into localized excess electrons and positive ions (37, 39, 164).

$$Na \rightarrow Na^+ + e^- \tag{1}$$

We recall that the corresponding process (22) in the *gas phase* requires an energy of ~2–4eV! Drawing upon the analogous situation in "conventional" electrolyte systems, e.g., aqueous salt solutions, one infers that both species are "solvated" in the sense that they polarize the surrounding dipolar solvent molecules in their own Coulombic field (101, 104). The idea of solvating a fundamental particle is indeed intriguing. Kraus (109–111) first proposed that the excess electron existed in a cavity in the solvent in which one or more of the ammonia molecules have been excluded, thereby accounting for the extremely low density of the solutions (37). However, it was Gibson and Argo (80) who were the first to apply the description "solvated electron," e_s^-, for the excess electron states in liquid ammonia.†

In the period 1940–1946, Ogg (132) developed the first quantitative theory for the solvated electron states in liquid ammonia. The Ogg description relied primarily on the picture of a particle in a box. A spherical cavity of radius R is assumed around the electron, and the ammonia molecules create an effective spherical potential well with an infinitely high repulsive barrier to the electron. It is this latter feature that does not satisfactorily represent the relatively weakly bound states of the excess electron (9, 103). However, the idea of a potential cavity formed the basis of subsequent theoretical treatments. Indeed, as Brodsky and Tsarevsky (9) have recently pointed out, the simple approach used by Ogg for the excess electron in ammonia forms the basis of the modern theory (157) of localized excess-electron states in the nonpolar, rare-gas systems. [The similarities between the current treatments of trapped H atoms and excess electrons in the rare-gas solids has also recently been reviewed by Edwards (59).]

Jortner, in a milestone contribution (101, 104), elaborated on this

---

† The short-lived hydrated electron counterpart, e_{aq}^-, was discovered in the early 1960s in water and in aqueous electrolyte solutions (7, 90). These hydrated electrons have since been found to play an important role in radiation-induced chemical conversions and in the changes in the physical properties of irradiated aqueous systems (39, 140). Solvated electrons have also been observed in a variety of nonpolar (45) as well as polar liquids. For a recent review, see Kevan and Feng (74).

simple physical picture. He supposed that the electron resided in a cavity of radius R approximately 3.2–3.4 Å in the ammonia. The liquid ammonia in the vicinity of the cavity is polarized by the electron, producing a centrally symmetric potential well of the type

$$V_{(r)} = \begin{cases} \dfrac{-e^2}{\epsilon_{\text{eff}}R} & r < R \\[2mm] \dfrac{-e^2}{\epsilon_{\text{eff}}r} & r \geq R \end{cases} \quad \epsilon_{\text{eff}}^{-1} = \epsilon_\infty^{-1} - \epsilon_0^{-1}$$

(2)

(3)

where ϵ_∞ and ϵ_0 are the high- and low-frequency dielectric constants, respectively. Thus the model combines the idea of Ogg's cavity with a polarization interaction at large distances.

Following a variational solution of the ground (1s) and first excited state (2p) of the electron in this potential well, various other polarization terms are added and a variety of characteristics for the solvated electron (optical transition energy, heat of solution, etc.) can be calculated (101, 105). For illustrative purposes, we shall utilize this simple model because of its obvious transparency in relating certain (macroscopic) features of solvent properties to the energy levels and wave functions for the solvated electron in polar solvents.

The derived potentials, energy levels, and wave functions for the solvated electron in two representative solvents, methylamine and hexamethylphosphoramide (HMPA), are compared in Fig. 2. A characteristic broad optical-absorption band in the near infrared (see Fig. 12, Section III,B,2 for a typical trace) is generally attributed (101, 104) to the Franck–Condon 1s → 2p transition within this potential well. Theory predicts (9, 104, 146, 147) and experiments confirm (146, 152) a high transition probability for this excitation, giving rise to the intense blue color characteristic of these systems. On this model, the solvent is considered only in terms of a continuum for $r > R$, and no account is taken of the peculiarities of the local structure around e⁻—for example, in the first coordination layer of the cavity (105). On a simplified approach, the energy of the optical transition energy might be judged to be coarsely related to the binding energy of the solvated electron in that solvent. However, the exact nature of electron binding and the associated lifetime of the solvated entity in polar solvents is a complex problem (1, 102, 105, 120, 171). Lifetime considerations (146) also have to take into account the pseudobimolecular reaction of electrons in the solvent, e.g., in water,

$$2e_{\text{aq}}^- + 2H_2O \rightarrow H_2 + 2OH^-$$

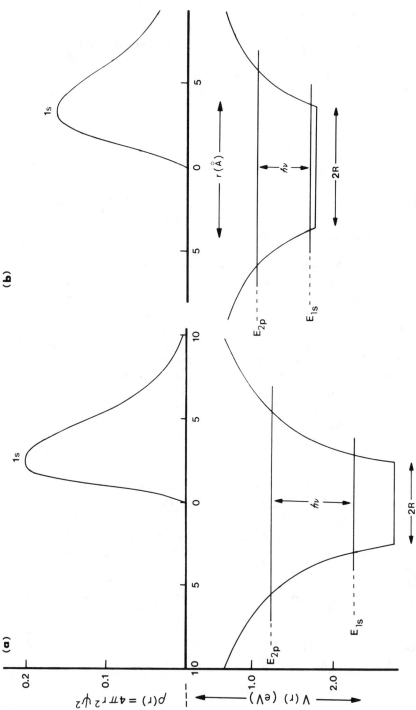

FIG. 2. Wave functions and energy levels for the solvated electron in (a) methylamine (MeA) and (b) hexamethylphosphoramide (HMPA). The potential $V(r)$ and wavefunction are based upon the model of Jortner (101) and computed using values of the optical and static dielectric constants of the two solvents. The optical absorption responsible for the characteristic blue color is marked by $h\nu$ and represents transitions between the 1s and 2p states. The radius of the cavity is ~3 Å in MeA, and ~4.5 Å in HMPA.

(pseudobimolecular rate constant $k = 5 \times 10^9$ liter mol^{-1} sec^{-1}). The relative stability of the spin-paired dimer species in ammonia (Section III,A,2) may well retard the corresponding reaction of electrons in this solvent (54, 146).

For our illustrative purposes, HMPA is taken as representative of a large, bulky, aprotic solvent in which electron and, indeed, any anion solvation is very weak (129, 131). Methylamine is representative of a solvent in which the electron binding is considerably larger (60, 127). Within the cavity, the electronic wave function assumes its maximum value; outside the cavity, the wave function asymptotically approaches zero. Although there is obviously a high degree of charge confinement within the cavity, a considerable amount of charge dispersion exists far beyond the first solvation layer.

All attempts (74, 89) to find a sensible, quantitative relation between the wavelength of maximum absorption (λ_{max}) and typical macroscopic properties of the solvent (i.e., dielectric constant) have so far failed (146). However, the size of the solvent cavity in which the electron is trapped also plays a decisive role (101) in determining the transition energy [Eqs. (2), (3)], and the solvent dependence of λ_{max} might well indicate a variation in cavity size from solvent to solvent. In this spirit, Dorfman and Jou (48) have evaluated cavity radii on the basis of the simple Jortner model for the solvated electron. The values are shown in Fig. 3, which shows a plot of the optical transition energy E_{max} versus

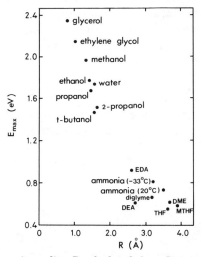

FIG. 3. Values of the cavity radius R calculated, from Jortner (101), to fit the values for E_{max}. [From Dorfman and Jou (48), Fig. 9. Reprinted with permission from Springer-Verlag, Berlin.]

R. The cavity radii in water and alcohol are around 0.9–1.5 Å, and in ammonia and the various amines and ethers between 2.5 and 4 Å.

Considerably more elaborate treatments (1, 40, 105) for calculating the electronic energy levels and wave functions for the excess electron have since been developed and attempt to introduce certain microscopic features of the local molecular environment. Such calculations for electrons in ammonia were first reported by Copeland, Kestner, and Jortner (40). The reader is referred to the paper by Banerjee and Simons (1) and the recent review by Brodsky and Tsarevsky (9) for comprehensive discussions of the current theoretical descriptions for solvated electrons in disordered condensed media (see also Ref. 171).

III. Electronic Properties of Dilute Solutions

Numerous physical and chemical investigations (37–39, 99, 103, 113, 116, 164) of metals in nonaqueous solvents reveal unequivocally that solvated electrons and metal cations interact in the dilute and intermediate concentration range to form various agglomerate species, both paramagnetic and diamagnetic. In this section we attempt to illustrate the rich variety of these localized-electron species in dilute solutions. We begin with a short summary of the situation in metal–ammonia solutions. Our objective here is twofold: (1) to set the scene for our later discussions concerning the transition to the metallic state in more concentrated metal solutions, and (2) to allow a comparison of the situation existing in dilute ammonia solutions to that found in the lower-dielectric-constant amine and ether solvents.

A. Metals in Liquid Ammonia

The overall changes in magnetic and transport properties of metal–ammonia solutions from the dilute to concentrated regimes are shown in Fig. 4.

1. Electron–Cation Interactions

As with any electrolyte, various aggregate species are expected to form as the concentration of solute increases. In particular, both the electrical conductivity and (metal) NMR data (Fig. 4) signal the appearance of neutral species at metal concentrations in excess of 10^{-3} MPM (37), the conductivity via a Morse-like behavior in the equivalent conductance, the magnetic resonance via a finite Knight (contact) shift for the metal nucleus. As expected for a high-dielectric-constant solvent, the ion-pairing association

$$M_s^+ + e_s^- \leftrightarrow (M_s^+, e_s^-) \tag{4}$$

is best viewed as a simple (and very short-lived (26), $\sim 10^{-12}$ sec) ionic aggregation involving a complex in which there is negligible distortion of either the solvated electron or solvated cation wave function. The very small electron spin density at the metal nucleus [judged from the metal Knight shift (13) data to be approximately 1% of the free-atom value] does indeed indicate a weakly interacting association complex (54). In the concentration range where the ion-pairing process is quite clearly indicated by the conductivity and NMR data, the optical absorption peak shows certain minor but significant modifications (Fig. 4). The observed red shift of this band [for a recent compilation of data, see Harris (88)] with solution composition in the range 10^{-3} to 10^{-1}

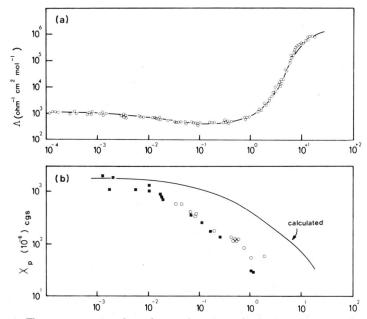

FIG. 4. The concentration dependence of various electronic properties of metal–ammonia solutions. (a) The ratio of electrical conductivity to the concentration of metal-equivalent conductance, as a function of metal concentration (240 K). [Data from Kraus (111).] (b) The molar spin (○) and static (■) susceptibilities of sodium–ammonia solutions at 240 K. Data of Hutchison and Pastor (spin, Ref. 98) and Huster (static, Ref. 97), as given in Cohen and Thompson (37). The spin susceptibility is calculated at 240 K for an assembly of noninteracting electrons, including degeneracy when required (37).

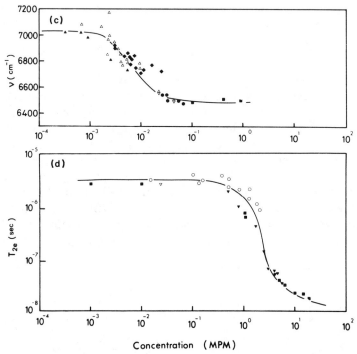

FIG. 4. (continued) (c) The optical properties of sodium–ammonia and potassium–ammonia solutions (208 K), and variation of the near-IR band maxima with concentration. [Data of Douthit and Dye (49), Gold and Jolly (84), Koehler and Lagowski (107), Rubinstein et al.(144), and Quinn and Lagowski (142), as given by Harris (88).] (d) ESR spin–spin relaxation time T_{2e} at 238 K. [Data from Chan et al. (∇, Ref. 31), Essig ($\blacksquare$, Ref. 72), and Hutchison and Pastor ($\bigcirc$, Ref. 98), as given in Schindewolf (148).]

MPM ($-65°C$, 208 K) was originally attributed (49) to ion-pair formation [Eq. (4)]. However, more recent work (144) suggests the shift results from the presence of two absorption centers in the intermediate concentration range: the solvated electron and a second absorber involving a binary combination of solvated electrons, a diamagnetic species. The precise origins of this red shift in the optical band in ammonia solutions are still a matter of some controversy. However, the accumulated evidence for the link between the *red* shift and *electron spin-pairing* is extremely strong (144, 165). In addition, recent precise studies of optical band shapes for the solvated electron in ethylenediamine (see Section III,B,2) suggest a *blue* shift upon ion-pairing (150), as observed for the (M_s^+, e_s^-) species in a variety of solvent systems (65, 151, 153).

Another important property, the apparent molar volume of the solvated electron, is essentially unaffected by the electron–cation and, indeed, the electron–electron interaction (37).

In summary then, the appellation (53, 84, 85) "loose ion-pair" for the solvated electron–metal cation association complex in dilute metal–ammonia solutions is therefore a particularly apt description.

2. Electron–Electron Interactions

The nature of the interaction leading to electron *spin-pairing* is far less clear (54). The rapid decrease in the molar spin susceptibility of metal–ammonia solutions in the concentration range 10^{-3} to 1 MPM below that expected for an assembly of noninteracting electrons (Fig. 4) has been taken as direct evidence for the formation of diamagnetic complexes having an even number of electrons in a singlet ground state (37, 54). For example, at 0.1 MPM, approximately 90% of the electrons in a sodium–ammonia solution are spin-paired, and the diamagnetic state is apparently several times kT (thermal energy) lower in energy than the triplet or dissociated doublet states. In addition, lower temperatures favor the diamagnetic state. Figure 5 shows the experimental data for metal ammonia solutions at 25°C (298 K), 0°C (273 K), and −33°C (240 K), expressed in terms of paramagnetic spin concentrations (88).

In the weak-interaction model (85) developed in the previous section to explain ion-pairing in metal–ammonia solutions, aggregation interactions involving M_s^+ and e_s^- are relatively weak, and leave the isolated solvated electron properties virtually intact. However, a major difficulty (29, 54, 134) arises with the type of model when one considers the precise nature of the corresponding electron spin-pairing interaction in ammonia solutions. It is worth expanding on this issue because it probably remains one of the fundamental dilemmas of metal–ammonia solutions in the dilute range (54).

The mere existence of diamagnetic states in dilute metal–ammonia solutions in itself presents no direct threat to a weak-interaction model; spin–spin interactions could presumably be incorporated into an ion-cluster model (85), where spin pairing occurs via collision complexes or by a long-range interaction between isolated solvated electrons, possibly mediated via an intermediate cation. In a recent study of solvated electron interactions, Schettler and Lepoutre (145) found that the ground singlet state is separated from the lowest triplet state by kT at interunit separations below 10.5 Å. However, we note that the mean distance between metal atoms in the spin-pairing concentration range

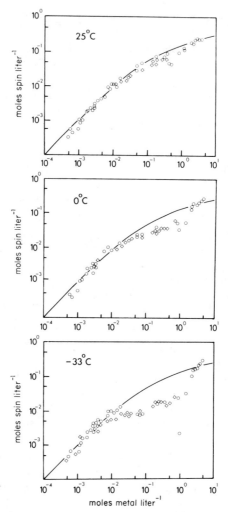

FIG. 5. Electron spin pairing in metal–ammonia solutions at 25°C (298 K), 0°C (273 K), and −33°C (240 K). Paramagnetic spin concentrations for sodium–ammonia and potassium–ammonia solutions. [Experimental data from Refs. *47, 76, 98, 114, 115,* and *159.* Adapted from Harris (*88*). Used with permission.] The solid line indicates the expected spin-pairing behavior for noninteracting electrons (*88*).

(10^{-3} to 10^{-2} MPM) is ~90 to 180 Å! The long-range interaction picture runs into an even more fundamental difficulty when one considers the exceptionally long lifetime (~1 μsec) of the electron–electron spin-pairing interaction required to maintain the electron spin-state lifetimes, as detected by ESR (*134*). Measured electron spin–spin relaxation times (T_{2e}) for solutions in the concentration range 10^{-4} to 10 MPM are shown in Fig. 4d. These ESR linewidths (ΔH) are among the narrowest known (typically in the region 0.02 to 0.03 G) and, particularly for the lighter alkali metals (134), are essentially constant over the composition range in which extensive spin pairing is taking place (37). In the light of the above comments concerning solvated electron interactions, a long-range interaction between solvated electrons in the concentration range 10^{-3} to 10^{-2} MPM might be expected to produce relatively weakly bound diamagnetic units, which would rapidly separate into two individual e_s^- units. This should lead to a considerable uncertainty in the electron spin lifetime and *an electron spin relaxation rate that is markedly composition dependent*. This effect is *not* observed (Fig. 4). The extremely narrow ESR, with its attendant long T_{2e} $(T_{2e} \approx 1/\Delta H)$ in the μsec range (29, 134), then requires (54) that processes of the type

$$e_1 + e_1 \rightleftharpoons e_2 \tag{5}$$

have encounter lifetimes *exceeding* 10^{-6} sec (recall that the corresponding lifetime of electron–cation encounter ion pairs in dilute ammonia solutions is only $\sim 10^{-12}$ sec). Thus the singlet state, once formed, must apparently exist for a microsecond or longer before conversion into a dissociated pair of electrons.

It has been suggested (54) that the exceptionally long lifetime of the singlet e_2 state may still arise from weak, long-range interactions, but with the stringent lifetime requirements being satisfied through additional electron-exchange processes of the type

$$e_1^\alpha + \begin{bmatrix} \text{diamagnetic} \\ \text{cluster} \end{bmatrix} \rightleftharpoons \begin{bmatrix} \text{diamagnetic} \\ \text{cluster} \end{bmatrix} + e_1^\alpha$$

However, this scheme does present major conceptual difficulties and has been criticized recently (160). In particular, the formation of specific diamagnetic (cluster) units at these very low metal concentrations might be considered unrealistic in view of the extremely large (average) electron–electron separation (~ 100 Å). Clustering phenomena are well established for concentrations close to the nonmetal-to-metal transition in these systems (33, 155), but for sodium–ammonia solutions this does not occur until approximately 4 MPM (Section IV).

Even without the complexities introduced via electron-exchange processes of the type outlined above, the fundamental nature of the interaction between two solvated electrons leading to spin pairing has yet to be firmly established. A treatment of the possible bonding in terms of a valence bond approximation has been attempted by Schettler and Lepoutre (145), stressing the diffuse nature (Fig. 2) of the trapping potential for the solvated electron. This treatment takes no account of the (rapidly) varying electron density with metal composition. A super-exchange-type process, involving an intermediary metal cation, is implicit in the recent suggestion of Mott (122) for the spin-paired species in ammonia. While there is some evidence for "ion triples," $e_s^-M_s^+e_s^-$, in a recent study of lanthanoid metals in ammonia (138), the suggestion has also come under some criticism (14). Clearly, charge neutrality *might* demand (133) some cation involvement, but the details of the interaction remain to be firmly established.

Two final points are of interest. First, in the overall context of spin pairing in dilute solutions, and the subsequent "metallization" process at higher concentrations [Fig. 4(a)], it is important to note that as the system moves *through* the nonmetal-to-metal transition (located around 4 MPM), the paramagnetic susceptibility quite rapidly approaches the predicted value for an assembly of degenerate (metallic) electronic states (37) (see also Fig. 5). The possible implications of this observation to the nature of spin pairing in the nonmetallic regime are discussed by Edwards (62). Second, the concept of *null properties* in dilute metal–ammonia solutions—i.e., properties which are essentially insensitive to metal concentration—formed the observational basis (54) for the weak-interaction (85), ion-cluster model (Section III,A,1). In this, complexes are formed from individual units which retain their essential properties in the loosely bound agglomerate species. However, it is amusing to note that the insensitivity of T_{2e} to metal concentration in the dilute regime (Fig. 4) now forms the observational basis for a model that demands a *stable, long-lived* diamagnetic entity in dilute metal–ammonia solutions (54).

The time is indeed ripe for more detailed experimental and theoretical studies of the spin-pairing phenomenon in metal solutions. Indeed, a considerable portion of the experimental data in this controversial area dates back to the late 1950s, and, for some data, even earlier (76).

B. Metals in Amines, Ethers, and Other Solvents

One underlying theme to emerge from the previous section is the lack of specific information regarding *distinguishable* species in dilute metal–ammonia solutions (55). This might well be anticipated to be a natural consequence for agglomerate species in a host solvent with a large dielectric constant ($\epsilon_0 = 16$ at 300 K). In general terms, the prop-

erties of dilute metal–ammonia solutions are then best characterized
(22) in terms of an assembly of isolated fragments (M_s^+, e_s^-) in which the
excess-electron state forfeits any parentage in the electronic states of
the gas-phase alkali atom.

In contrast to ammonia solutions, in which the metal cation effectively assumes a "bystander" role in the formation of aggregate species,
metal solutions in the low-dielectric-constant amines and ethers (ϵ_0
typically <10 at 300 K) show a multitude of metal-dependent species,
both paramagnetic and diamagnetic. The general body of experimental
information requires the presence of at least *three distinct stoichiometries* in solution (51, 53, 55),

$$M_s^+ + e_s^- \rightleftharpoons (M_s^+, e_s^-) \rightleftharpoons (M_s^-), \tag{6}$$

the solvated electron, an aggregate species of e_s^- and M_s^+, and a dielec-
tron species, now well characterized in certain solvents (51, 52) as a
centrosymmetric (solvated) alkali metal anion (M_s^-). The association of
(solvated) cation and electron can no longer be described as a weakly
interacting ion-pair species; in these solvents we have immediately
recognizable changes in the electronic properties of the aggregate
species compared to its constituent partners. A problem of classification now enters (51, 53) when, for example, terms such as "loose" and
"contact" ion pairs and "monomers" are introduced to describe the association complex (143). It is to this problem that we turn to in the
following section. However, at this stage we may safely make the general comment that the majority of aggregate species are more "tightly
bound" than in ammonia solutions—as indeed one would expect from
the decrease in dielectric constant. By this we mean that the
electron–cation interaction can be sufficiently strong to introduce a
noticeable metal dependence into the electronic properties.

This metal dependence also carries over to the spin-paired species. In
the low-dielectric-constant solvents, these can no longer be represented
in terms of weak, long-range interactions possibly mediated by an in-
tervening cation, but are now distinct, new molecular entities (Section
III,B,5).

In most general terms, the species identified in metal solutions are
best described as "matrix-bound," in that the system comprises a solute
(metal) species in a host (solvent) matrix. Before discussing details of
the experimental data, we first attempt a general rationalization (61,
108) of the spectrum of possible matrix-bound states found in metal
solutions.

1. Paramagnetic States: Ion Pairs to Solvated Alkali Metal Atoms

Consider the incorporation of a rubidium atom into a polar solvent and imagine that the valence (5s) electron is removed from it, leaving behind a positive ion Rb^+. This ion will polarize the surrounding solvent so that at *large distances* it produces an electrostatic potential (*108*)

$$U = (e/\epsilon_0)r \tag{7}$$

where ϵ_0 is the static dielectric constant. When (and if) the valence electron is brought back, one of several situations may result (*106, 170*).

(i) It may be energetically favorable for the valence electron to occupy a large centrosymmetric orbit, over most of which Eq. (7) applies. The Coulombic attraction between the positive hole and the bound electron is reduced by the static dielectric constant of the solvent. The solute ion–solvent interaction effectively reduces or screens the Rb^+–valence electron interaction. In terms of a centrosymmetric picture (*3, 4*) for a monomer species (see Fig. 6), this results in a substantial increase in the Bohr radius† (a_H^*) of the 5s-electron wave function. In this situation, the potential is basically Coulombic, and the matrix-bound states are closely related to the corresponding states of the hydrogen atom,‡ i.e., via a modified Schroedinger equation (*5, 108*) in which the potential is given by Eq. (7) and the modified Bohr radius by

$$a_H^* = (\epsilon_0)a_0/m^* \tag{8}$$

where m^* is an effective mass for the excess electron, and a_0 is the Bohr radius of the hydrogen 1s orbital (0.5292 Å). It is important to note that in the limiting situation, matrix-bound states characterized by this screened hydrogenic potential [Eq. (7)], although spatially localized around the Rb^+ core, have *no* recognizable characteristics of the gas-phase Rb atom (*5, 106, 108, 170*). Eigenfunctions and eigenenergies are *solely* dependent upon the characteristics (m^*, ϵ_0) of the host solvent. As such it would be very difficult to differentiate their electronic properties from those of the *isolated* excess electron [see also (iv)]. Such states

---

† Approximately taken as the radius of maximum spin density ($4\pi r^2\psi^2$) of the valence electron.

‡ The related situation in highly excited states of the alkali atoms has recently been reviewed by Metcalf (*121*).

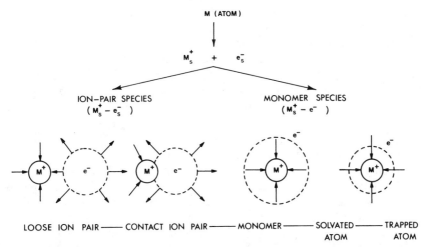

FIG. 6. A schematic representation of several possible (paramagnetic) electron–cation aggregates in metal solutions. [Adapted from Fig. 3 of J. L. Dye (53), with permission from Pergammon Press.]

have indeed been identified (18) in frozen solutions of alkali metals in HMPA (Section III,B,4). An alternative, pictorial representation of this type of weakly interacting state is the "loose" ion-pair species (53), and is also shown in Fig. 6. This latter type of representation emphasizes directional bonding between the constituent partners. We note that in the absence of this directional interaction, there is essentially no distinction between this representation and the large radius monomer state. This aspect is discussed elsewhere (61).

(ii)   The most favorable energy state for the electron may be to occupy an orbit very near to the rubidium cation. In this case, Eq. (7) does not apply and we have the possibility of a complicated state of affairs in the immediate vicinity of the cation (108). If there exists only a weak pertubation by the surrounding matrix (170) (for example, in the rare-gas solids), we have a "trapped atom" situation (Fig. 6). These states are adequately described (100) in terms of a tight-binding Heitler–London (atomic) scheme. Any changes from the gas-phase properties of the Rb atom are generally small and usually accounted for by second-order polarization and dispersion interactions with the host matrix (16).

(iii)   Obviously an intermediate situation (108) between (i) and (ii) may prevail. In this, the 5s wave function is still primarily confined to the neighborhood of the parent cation (Rb$^+$). However, increased overlap with the surrounding solvent requires that the electronic states of the solute are best described in terms of a strongly perturbed atomic scheme (ii), or, alternatively, as a modified hydrogenic state (i). In the centrosymmetric representation of the paramagnetic state (Fig. 6), both these situations are accounted for in terms of the variation of the effective Bohr radius of the valence electron (16). The alternative scheme (53) recognizes a high electron density at the rubidium nucleus in terms of a "contact" ion-pair species (32), in which there is a significant degree of interaction between e_s^- and the outer 5s orbital of Rb^+, resulting in an appreciable atomic character.

(iv)   States (i) to (iii) all represent bound states, in which the excess electron is trapped in the reduced Coulombic field of the parent ion. If the polarization of the host solvent by the metal cation is sufficiently large, the valence electron will escape the Coulombic field and become ionized [Eq. (1)]. This spontaneous ionization† (which is, in fact, the dissolution process) is now comparable to the introduction of an excess electron into the pure solvent via an adiabatic process—for example, via photochemical or radiation processes (146). The properties of this electronic state are now determined by the factors responsible for the solvation of excess electrons in polar solvents, as outlined in Section II.

Having briefly discussed the various possibilities for matrix-bound paramagnetic states in these systems, we now review the main spectroscopic information regarding these distinguishable species.

2. Fluid Solutions

a. Electron spin resonance studies.   Perhaps the most telling information regarding the nature of the paramagnetic species in these systems arises from ESR studies. The ESR spectrum of the isolated solvated electron consists of a single narrow line with a g_e factor close to the freespin value; e.g., in ammonia the resonance has a width of ~0.030 G and $g_e = 2.0016$ (98). Given the possibility of an interaction with an alkali cation (Fig. 6), an appreciable electron (spin) density can exist at the metal nucleus. Provided this electron–cation association complex has a sufficiently long lifetime, then a hyperfine interaction is possible (12, 26, 156), resulting in $(2I + 1)$ lines in the ESR spectrum (I is the nuclear spin of the cation). In fact, the observation (2, 3, 168) of metal hyperfine splitting in metal–amine solutions provided the first unambiguous identification of the atomic stoichiometry [Eq. (4)] of one of the paramagnetic species in metal–amine solutions.

The striking effect of both the solvent and metal on the ESR spectra

_____

† A related ionization process at high metal concentrations, involving the transition from localized to itinerant excess electrons, is discussed in Section IV.

of metal solutions (58) is illustrated in Fig. 7 for solutions of K, Rb, and Cs in three solvents. A simple ion-pairing treatment (21, 26) is quantitatively capable of predicting the form of the ESR spectrum in the wide variety of solvent systems. Briefly, two parameters are critical in determining the spectra obtained (26): the effective lifetime (τ_M) of the (primary) electron–cation encounter process,

$$M_s^+ + e_s^- \rightarrow (M_s^+, e_s^-)$$

calculated from diffusion-controlled encounter theory, and A, the *metal* hyperfine coupling constant, expressed in hertz. The product $P_c = \langle \tau_M A \rangle$ defines the form of the ESR spectrum. For the high-dielectric-constant solvents (NH_3, HMPA, etc.), $P_c \ll 1$ and a "time-averaged" signal of both paramagnetic species (e_s^-; M_s^+, e_s^-) is observed. In the lower-dielectric-constant solvents, $P_c \gg 1$, and the encounter lifetime is sufficiently long to permit the conversion of the "loose" ion pair (which must surely be the species formed in the *initial* encounter process) to a centrosymmetric species representing something like an "expanded" or solvated metal atom (Fig. 6). The latter species has the capability (in a low-dielectric-constant solvent) of a large hyperfine coupling to the metal nucleus. In Fig. 8 we show the established (26) P_c values for dilute solutions of metals in a variety of solvents.

The success of this type of approach is important since it emphasizes

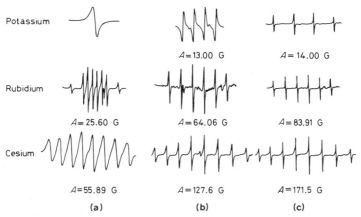

FIG. 7. Electron spin resonance spectra of potassium, rubidium, and cesium in (a) methylamine, (b) ethylamine, and (c) propylamine; adapted from Dye and Dalton (58), with permission. Note the change in scale from pattern to pattern, as indicated by the (metal) hyperfine splittings given. Reprinted with permission from J. L. Dye and L. R. Dalton, *Journal of Physical Chemistry*, **71**, 184 (1967). Copyright 1967 American Chemical Society.

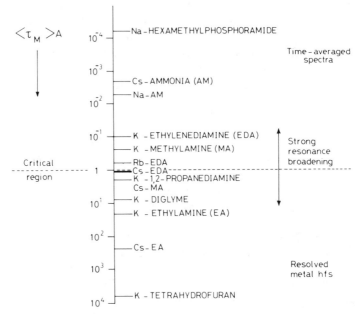

FIG. 8. Classification of ESR spectra (~296 K) of solutions of alkali metals in a variety of nonaqueous solvents: τ_M is the electron–cation encounter lifetime, and A is the metal hyperfine coupling constant, in hertz; THF = tetrahydrofuran, EA = ethylamine, DG = diglyme, MA = methylamine (~220 K), 1,2PDA = 1,2-propanediamine, EDA = ethylenediamine, AM = ammonia (~240 K).

once again that, despite the unique nature of e_s^- and the various types of association complexes formed with solvated metal cations, a standard ion-pairing treatment does indeed give a reasonable semiquantitative description of the magnetic and electrical properties of dilute solutions (26, 37, 164).

The separation (A) of the hyperfine lines in the ESR spectra of metal–amine, and metal–ether solutions represents a direct measure of the average s-electron (spin) density of the unpaired electron at the particular metal nucleus (12, 156). When this splitting is compared to that of the free (gas-phase) atom, we obtain a measure of the "percent atomic character" of the paramagnetic species. The percent atomic character in all these *fluid* systems increases markedly with temperature, and under certain circumstances the paramagnetic species almost takes on "atomic" characteristics (43, 53, 160). Figure 9 shows the experimental data for fluid solutions of K, Rb, and Cs in various amines and ethers, and also for frozen solutions (solid data points) of these metals in HMPA (17). The fluid solution spectra have coupling con-

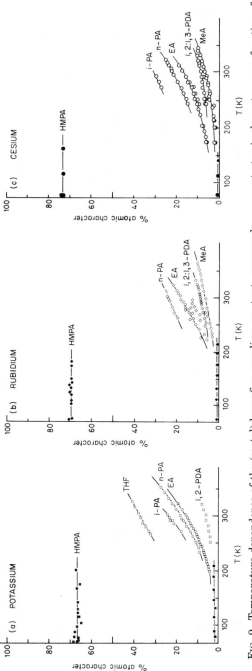

Fig. 9. Temperature dependence of the (metal) hyperfine coupling constant, expressed as percent atomic character, as a function of temperature for solutions of (a) potassium (^{39}K), (b) rubidium (^{85,87}Rb), and (c) cesium (^{133}Cs) in amines, ethers, and HMPA. The experimental data are from a variety of sources, and outlined elsewhere (16). Open symbols denote parameters from fluid solution studies; shaded symbols are from frozen solutions in HMPA. Solvent identification: as in Figure 8, and BuA = butylamine, i-PA and n-PA are isopropylamine and n-propylamine.

stants that are intermediate between values for the high and low atomic character states identified in the frozen HMPA solutions, but tend toward these values both in the high and low temperature extremes, respectively (this correlation between fluid and frozen solution data is discussed in detail in Section III,B,4).

Figure 10b shows the extrapolation of the magnetic data (percent atomic character) for Cs in various solvents (16, 17) to high temperatures, while Figure 10a shows the corresponding g_e factors for Cs–ethylamine solutions (128). Once again, the corresponding data obtained from studies of frozen solutions of the respective metals in HMPA are also included as limiting values at the high and low tem-

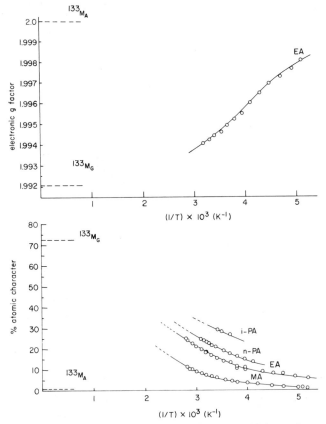

FIG. 10. Magnetic parameters for solutions of cesium-133 in various amines and ethers, and in HMPA. Open symbols are for fluid solution studies (130). Dotted lines on the ordinate show the limiting high- and low-atomic-character states observed in frozen Cs–HMPA solutions (16). Solvent identification as in Fig. 8.

perature extremes. These types of considerations led to the postulate
(30, 161) that the limiting (high-temperature) high atomic character
state in the *fluid* metal–amine and metal–ether solutions is a *solvated*
(Figs. 9, 10) rather than a "free" (i.e., 100% atomic character), alkali
metal atom.

b. *Optical studies.* Optical spectra of fluid metal–amine and
metal–ether solutions contain a band in the near infrared, at first sight
somewhat similar in shape and position to the isolated e_s^- band in these
solvents, and two additional bands at higher energies whose positions
are dependent primarily upon the alkali metal (75, 150). One of these
metal-dependent optical bands has been assigned (119) to solvated alkali metal anions (Section III,B,5), while kinetic studies (75) have
demonstrated that the other, *transient* band (observed via first-detec-
tion techniques, e.g., pulse radiolysis) arises from the paramagnetic
species of atomic stoichiometry, (M_s^+,e_s^-). The position of the (M_s^+,e_s^-)
optical absorption exhibits (28, 148, 151, 153) a distinct *blue* shift from
that of e_s^- observed in the same solvent (compare Section III,A,1 and
Fig. 4 for dilute ammonia solutions). Furthermore, a close correlation
has been established (28, 151) between the magnitude of this shift and
the percent atomic character of the paramagnetic species (M_s^+,e_s^-), as
determined by ESR studies. Figure 11 shows the available optical and
ESR data for K and Cs states in a variety of host matrices (65). This

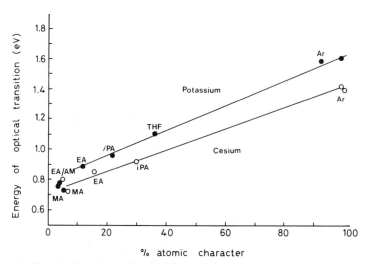

FIG. 11. Correlation of optical and ESR data for monomeric forms of potassium and
cesium in a variety of host matrices. [From Edwards and Catterall (65); used with
permission of Taylor and Francis, Ltd., *Philosophical Magazine.*]

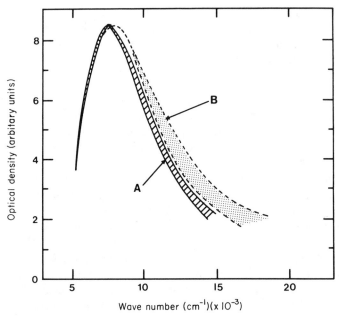

FIG. 12. Infrared optical absorption bands observed in ethylenediamine solutions: (A) spread in band shape of e_s^- observed by pulse radiolysis, (B) spread in band shape observed in alkali metal solutions. [Adapted with permission from W. A. Seddon and J. W. Fletcher, *Journal of Physical Chemistry*, **84**, 1104 (1980). Copyright 1980 American Chemical Society.]

correlation extends over the entire range from ion-pair-like species in the high-dielectric-constant solvents (e.g., NH_3) through the monomeric, solvated atom states to the atomic variants found in the nonpolar rare-gas matrices.

The role of ion pairing in producing a blue shift in the optical spectrum may also extend to systems previously supposed to be metal-independent in dilute solutions. In an excellent and timely review, Seddon and Fletcher (150) have highlighted the recent advances made in the study of optical properties of dilute metal solvent systems, and at the same time reexamined the precise resolution and interpretation of metal-solution spectra. Figure 12 shows the optical spectra (150) attributed to e_s^- in ethylenediamine (EDA). The low-energy profile (A) encompasses the spread in band shape observed in pulse radiolysis studies of pure EDA. The high-energy profile (B) represents the range of published spectra observed in alkali metal solutions in EDA. Two effects are noticeable: (i) a slight shift toward the blue in the band maxima in the *presence* of alkali metal cations, and at the same time, (ii) a significant broadening of the band, especially in the more concentrated alkali-metal solutions. Observation (i) is consistent with the general trend of a blue shift in the band maximum of e_s^- upon ion pairing (Fig. 11), while (ii) may, in part, be the result of ion-pair formation; there also remains the distinct possibility of an underlying contribution from (M_s^-). Clearly, a more quantitative assessment of the optical spectra of dilute metal–solvent systems must now follow from the approach taken by Seddon and Fletcher (*150*). In conclusion, we note also that the absorption spectra of (M_s^+, e_s^-) and (M_s^-) exhibit a certain asymmetry, similar to that observed for the isolated solvated electron.

3. Current Models for Paramagnetic States in Amine and Ether Solutions

Two conceptually different models have been proposed to explain the temperature dependence of the (metal) hyperfine coupling constant and electronic g_e factor in metal–amine and metal–ether solutions.

a. Multistate model. The multistate or dynamic model was originally developed by Hirota *et al.* (*95*) to explain the anomalous linewidth variation of cation hyperfine lines in ion pairs of the type $M^+ R \cdot ^-$, where $R \cdot$ is an organic radical. The model was applied to metal solutions by Dye and co-workers (*43*) and by Catterall, Symons, and Tipping (*30*), and proposes that two or more paramagnetic species, each having different g_e factors and (metal)·hyperfine coupling constants, are in rapid equilibrium with each other. The individual species are assumed to have a structure that, to first order, is insensitive to temperature variations. The observed hyperfine coupling constant A is then represented by

$$A = \sum_i x_i A_i \qquad (9)$$

where x_i and A_i are the mole fraction and hyperfine coupling constant for the ith state, respectively. In fluid solutions, individual states cannot be observed by ESR, and observed properties (A, g_e) represent a weighted average of those of the individual states. The temperature dependence of A then reflects the *changing relative populations of states*. The variations in line widths across the ESR spectra (Fig. 7) are interpreted in terms of the time-dependent modulation of the hyperfine interaction between the unpaired electron and the metal nucleus of the two (or more) magnetically inequivalent states of stoichiometry M.

b. Continuum model. In the continuum model of Bar-Eli and Tuttle (*2, 3*), a *single* centrosymmetric species is invoked to explain the metal hyperfine coupling. The temperature variation in A then arises from the temperature-dependent changes in the solvation structure of the central metal ion. Inherent in the continuum model is a time-dependent modulation of the hyperfine coupling arising from the inter-conversion of nonequivalent states *within* the one state distribution (Fig. 13), which explains the anomalous line-width variation in the ESR. The model was put on a quantitative basis by O'Reilly and Tsang (*135*).

Figure 13 is a schematic representation of the anticipated distribution functions (number of states with a given hyperfine coupling constant) at high and low temperatures for the two models (*17*).

The observation (*153*) of a *single* transient optical-absorption band

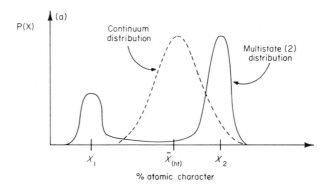

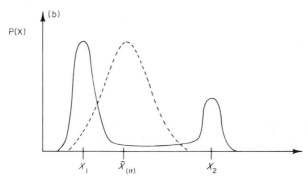

FIG. 13. A schematic representation of the multistate and continuum models for metal–amine and metal–ether solutions at (a) high and (b) low temperatures. The multistate picture shown here assumes, for simplicity, two species of atomic character X_1 and X_2. The continuum model suggests a single species with an atomic character of $\overline{X}_{(ht)}$ and $\overline{X}_{(lt)}$ at the extremes of high and low temperatures, respectively. P(X) is a measure of the number of states with a given percent atomic character.

from the cation–electron aggregate (M_s^+, e_s^-) in several amines and tet-rahydrofuran, and its correlation (151) with ESR results (Fig. 11), would, at first, appear to contradict the multistate picture. However, in the series THF, dimethoxyethane, and the polyglycoldimethyl ethers (glymes), evidence was obtained (150) for the existence of *two* distinct optical bands resulting from the reaction of e_s^- with Na_s^+ to give two or more distinct cation–electron aggregates in equilibrium. This latter result is in accord with ESR multistate models.

The essential difference between the multistate and the continuum model rests upon the temperature dependence of the density of states (Fig. 13). An experimental distinction between the models may also be possible (21) from a study of genuine vitreous (quick-frozen) solutions, i.e., systems that, at low temperature, retain the appropriate fluid solution species existing *at* the freezing point. For these systems, the continuum model predicts a *single* species at low temperature with a low A value, while the multistate picture requires the superposition of spectra from two (or more) species: one (or more) in high abundance with a low A value, and at least one in lower abundance with a high value of A (Fig. 13).

For a variety of reasons (16, 21), we turned our attention to HMPA in the hope of obtaining *genuine* frozen solutions suitable for spectroscopic examination. Indeed, ESR studies of these rapidly frozen solutions revealed a rich variety of signals (16–23, 64, 65).

4. Quenched Metal–HMPA Solutions

A detailed discussion of the analysis and characterization of these signals is given elsewhere (17); here we quote typical results for illustrative purposes. A representative spectrum (21) at high machine amplification of a frozen rubidium–HMPA solution is illustrated in Fig. 14. Identified ^{85}Rb and ^{87}Rb resonances are indicated in the figure and the initial assignment summarized schematically. We have proposed (18, 19) a system of nomenclature that classifies states in terms of their percentage occupation of the outer metal (ns) orbital, labeling alphabetically from the lowest atomic character state (^xM_A, ~1%) to the highest (^xM_I, ~80%). Superscripts denote the mass number of the isotope.

From this type of analysis, it is possible to arrive at a distribution function (21) for the paramagnetic states in these vitreous solids, and a typical plot for ^{85}Rb species at liquid-nitrogen temperatures is shown in Fig. 15. This analysis has also recently been extended (17) to potassium and cesium solutions. The results clearly show the presence of

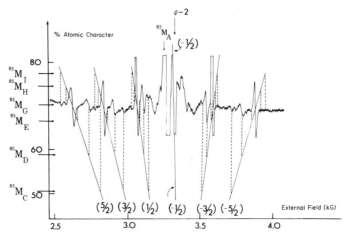

FIG. 14. Electron spin resonance spectrum of a frozen solution of rubidium in HMPA, at high machine amplification. The full lines show the variation of resonant field position with A for $g_e = 1.99800$, and a microwave frequency of 9.1735 GHz. The lines are anchored at the crossovers of the $^{85}M_G$ species ($A = 251.3$ G). Positions of the M_C, M_D, M_E, M_G, M_H, and M_I absorptions are indicated. Reprinted with permission from R. Catterall and P. P. Edwards, *Journal of Physical Chemistry*, **79**, 3010 (1975). Copyright 1975 American Chemical Society.

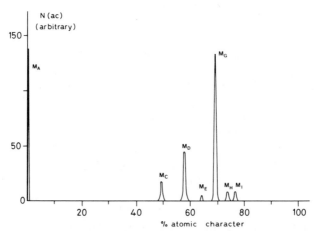

FIG. 15. The experimental distribution function for paramagnetic states identified in frozen rubidium–HMPA solutions at liquid-nitrogen temperatures (*21*). Reprinted with permission from R. Catterall and P. P. Edwards, *Journal of Physical Chemistry*, **79**, 3010 (1975). Copyright 1975 American Chemical Society.

centers differing widely in percent metal (ns) character. In addition, the coupling constants observed in fluid metal–amine and metal–ether solutions are always intermediate between values for the ^xM_G and ^xM_A species in frozen metal–HMPA solutions, but tend toward these values, respectively, in both the high and low temperature extremes. These results obviously confirm the presence of a distribution of paramagnetic states in these quick-frozen solutions. Unfortunately, the underlying problem here is whether these species observed in the vitreous solid are indeed representative of a variety of paramagnetic states in the liquid solution as well (53). This is a difficult question to answer. However, two major features (17) of these studies are important in this regard. The first is the remarkable reproducibility of the results for the frozen solutions (17). The second aspect concerns the strong correlation in the ESR parameters (A and g_e) for both fluid metal–amine and metal-ether solutions, and frozen metal–HMPA solutions. This is highlighted in Fig. 16, where we plot the variation of the electronic g_e factor shift [$\Delta g_e = g_e$ (free atom) $- g_e$] as a function of percent atomic character for the species M_A to M_H identified in frozen HMPA solutions, together with the corresponding data for fluid amine and ether solutions. For all metals studied, the electronic g_e factor moves smoothly away from the free atom value as A moves toward the atomic value. This behavior is indeed strong evidence that the limiting high-temperature paramagnetic state in the fluid solutions is the *solvated* metal atom. In addition, the maximum observed g_e shifts for each metal correlate well (21) with the (metal) spin–orbit coupling constants λ_{so}. Similarly, measured electron spin–lattice relaxation times (24, 25) for the solvated alkali metal atoms in frozen metal–HMPA solutions show a strong correlation with the magnitude of the spin–orbit coupling constant (Table I).

In these vitreous solids, then, the *entire* spectrum of possible

TABLE I

SPIN–LATTICE RELAXATION DATA FOR ALKALI METAL ATOMS[a]

Metal	λ_{so} (eV)[b]	T_{1e} (sec)	T_{2e} (sec)
Na	1.41×10^{-3}	$>5 \times 10^{-2}$	$\sim 1 \times 10^{-8}$
K	4.75×10^{-3}	6.4×10^{-2}	1.3×10^{-8}
Rb	1.95×10^{-2}	1.5×10^{-3}	1.2×10^{-8}
Cs	4.55×10^{-2}	2.0×10^{-4}	1.3×10^{-8}

[a] From Catterall and Edwards (24, 25).
[b] 1 eV $\cong 1.6021 \times 10^{-9}$ J.

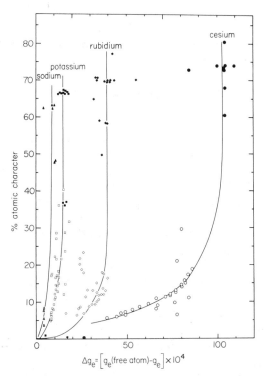

FIG. 16. Correlation of percent atomic character with Δg_e for frozen solutions of sodium, potassium, rubidium, and cesium in HMPA, and for fluid solutions in various amines and ethers. Reprinted with permission from R. Catterall and P. P. Edwards, *Journal of Physical Chemistry,* **79,** 3010 (1975). Copyright 1975 American Chemical Society. Frozen M–HMPA solutions are shaded symbols. Open symbols are fluid amine and ether solutions.

electron–cation aggregates (Section III,B,1) appear to exist. The high-atomic-character states ($^{85}M_C$ to $^{85}M_I$, Fig. 15) are best viewed as solvated metal atoms (Fig. 6).

The low-atomic-character state ($^{85}M_A$) can be interpreted (*19*) in terms of the loose ion-pair picture (Fig. 6), or alternatively, as a large-radius monomeric state [a_H^* estimated (*17*) to be in the region 10–15 Å]. Experimental magnetic parameters and unpaired electron spin densities at the metal nucleus for the ^xM_A species are shown in Table II.

Observed line widths are metal dependent. However, values of $|\psi_{(o)}|^2_M$, the unpaired electron (spin) density at the metal nucleus, *are approximately independent of the metal atom for potassium, rubidium, and cesium species.* As such, the states $^xM_A(X = 39, 85, 133)$ are best described as *true* hydrogenic states in this disordered medium.

TABLE II

EXPERIMENTAL MAGNETIC PARAMETERS FOR THE ^XM_A (X = 39, 85, 133) STATE IN
LOW-TEMPERATURE ALKALI METAL–HMPA GLASSES[a]

Isotope	I	g_e	ΔH (G)	A (G)	$10^{24}\|\psi_{(0)}\|^2_M$ (electrons cm^{-3})[b]
^{39}K	$\frac{3}{2}$	2.0018	4.9 ± 0.3	0.80 ± 0.1	0.073 ± 0.007
^{85}Rb	$\frac{5}{2}$	2.0009	9.4 ± 0.5	1.48 ± 0.1	0.065 ± 0.005
^{133}Cs	$\frac{7}{2}$	1.9994	14.9 ± 0.8	1.85 ± 0.05	0.060 ± 0.006

[a] From Catterall and Edwards (18, 19).
[b] For comparison purposes, $\|\psi_{(0)}\|^2$ atomic for the alkali atoms in question is: ^{39}K,
7.4790 × 10^{24}; Rb, 15.8225 × 10^{24}; ^{133}Cs, 26.4508 × 10^{24} (electrons cm^{-3}).

As outlined in Section III,B,1, these states forfeit any parentage in the
electronic states of the gas-phase (parent) alkali atoms.

5. Diamagnetic States: Ion Triples to Alkali Metal Anions

Perhaps the most striking characteristic of metal–amine and
metal–ether solutions that sets them apart from metal–ammonia solutions is the metal-dependent optical-absorption band that occurs at
higher energies than the infrared band associated with e_s^- (51, 75, 150).
This shifts progressively to higher energies for solutions of Cs, Rb, K,
and Na in a given solvent (Fig. 17). The metal, temperature, and solvent dependence of this band prompted Matalon, Golden, and Ottolenghi (119) to suggest that the species responsible in amine solutions

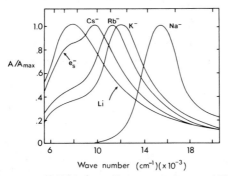

FIG. 17. Optical spectra of lithium, sodium, potassium, rubidium, and cesium in
ethylenediamine with identification of absorption peaks of Na$^-$, K$^-$, Rb$^-$, Cs$^-$ and e_s^-.
Absorption for e_s^- is taken from pulse radiolysis studies. [Taken from Fig. 1 of Dye (51);
used with permission of Verlag-Chemie, *Angew. Chem. Int. Ed. Engl.*]

were alkali metal anions, and that the transitions were similar to the charge-transfer-to-solvent (CTTS) bands of I^- The term alkali metal anion refers to a spherically symmetric species with two electrons in the outer s-like orbital centered on the metal. However, this is only one of several possibilities open to a diamagnetic cluster of two electrons and a cation, and Fig. 18 is a schematic representation of the various species. Indeed in a "good" solvent such as ammonia, we expect (53, 143) this spin-paired species to resemble a loose triple ion, since a wide variety of properties are essentially unaffected by the electron–electron interaction (Section III,A,2). However, in the low-dielectric-constant solvents the evidence for a spherically symmetric alkali metal anion species has accummulated steadily over the last decade (51). In addition, the existence of alkali metal anions in the gas phase has been recognized for some time (50). In 1974, Dye and co-workers (56, 162) isolated and characterized a crystalline salt of the sodium anion. When a saturated solution of sodium metal in ethylamine in the presence of a cation-complexing agent (for example, bicyclic diamino ether, or "crypt") is cooled, gold-colored crystals form, whose structure corresponds to a sodium cation trapped in the crypt and a sodium anion outside (Fig. 19). More recent studies have also revealed the existence of alkali metal anions in thin solid films formed by evaporation of metal–ammonia solutions containing cation-complexing agents (42).

The NMR spectra of Na^-, Rb^-, and Cs^- have now been reported (52, 56). The chemical shift of Na^-, large and diamagnetic, is the same as that calculated for the gaseous anion and is essentially independent of solvent. The apparent absence of a large paramagnetic shift of Na^- upon solvation, and the solvent independence of the peak position, have been cited (56) in support of a model in which the solvent is excluded from the region occupied by the sodium 2p electrons. However, recent studies (20) of the photodetatchment of electrons from K^- and Rb^- in HMPA glasses at low temperature reveal a significant degree of solvation in the alkali anion species. The overall conclusion (20, 52) must be that the alkali metal anion in these systems is a distinct, centrosymmetric species, with the two valence electrons residing in an expanded ns orbital on the metal.

The existence of metal anions in these nonaqueous systems is not confined to elements of group IA of the Periodic Table. Peer and Lagowski (137) have recently presented spectroscopic evidence for the existence of the auride ion Au^- in liquid ammonia, while Dye (51) has recently given an extensive assessment of the stability of crystalline salts of a variety of possible metal anions.

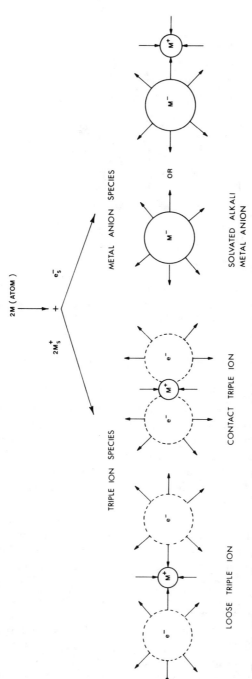

FIG. 18. Diamagnetic electron–cation clusters in metal solutions. [Adapted, in part, from Dye (53); used with permission of Pergamon Press.]

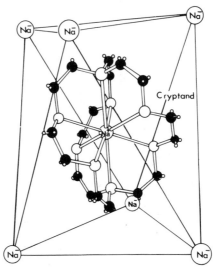

FIG. 19. Structure of the central Na^+,2,2,2-cryptand ion and the placement of the surrounding six Na^- ions in the Na^+,2,2,2-cryptand · Na^- crystal. (From Anions of the alkali metals, J. L. Dye. Copyright © 1977 by Scientific American, Inc. All rights reserved.)

IV. Concentrated Solutions and the Nonmetal-to-Metal Transition

In the preceeding sections, we have seen that the dielectric constant plays an important role in the binding energy of a localized-electron state—either isolated solvated electrons (Section II) or electron–cation aggregates (Section III). It is therefore particularly instructive to study its variation as the metal concentration is increased. Such measurements were carried out on sodium–ammonia solutions at 1.2, 5.4, and 10 GHz by Mahaffey and Jerde (117), and by Breitschwerdt and Radscheit (8). Over the concentration range 0.1 to 1 MPM, the results at 10 GHz (shown in Fig. 20) exhibit a very rapid increase to a value of nearly 100 before falling off to large negative values characteristic of Drude behavior for metals. Such a "dielectric catastrophe" was indeed the basis of the first consideration (34) of the nonmetal-to-metal (NM-M) transition in sodium–ammonia solutions by Herzfeld (93) in 1927. Stated simply, the dielectric constant goes to infinity at the metallic onset; the resultant binding energy on the (isolated) electron vanishes, and the system acquires metallic status! Over the composition range 0.1 to 10 MPM in sodium–ammonia solutions, the electrical conductivity increases by approximately four orders of magnitude (73, 94), and in the vicinity of the saturation point, it reaches a value exceeding that of liquid mercury (equivalent conductivity, −33°C,

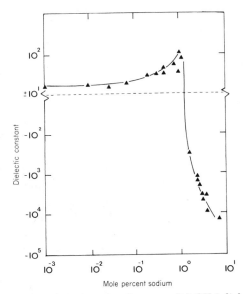

FIG. 20. The real part of the microwave frequency (10 GHz) dielectric constant as a function of sodium metal concentration at 298 K in sodium–ammonia solutions. *Note the break in the ordinate.* [Adapted from Mahaffey and Jerde (*117*); used with permission from the American Physical Society, *Reviews of Modern Physics.*]

Na–NH$_3$, 1.02×10^6 ohm^{-1} cm^2 mol^{-1}; for Hg, 0°C, 0.16×10^6 ohm^{-1} cm^2 mol^{-1}). In the intermediate composition range 1 to 7 MPM, a NM-M transition occurs, and changes in the electronic, thermodynamic, and mechanical properties of the system are equally impressive (*35, 37, 124, 154*). A detailed discussion of the concentration dependence of various properties of metal–ammonia solutions is given in the book by Thompson (*164*). In addition, a recent review (*60*) at Colloque Weyl V also summarizes the available data for lithium–methylamine solutions (*10, 11, 63, 127, 128, 166*).

Our canvas here is to provide a qualitative description of current models for the NM-M transition, developed for both metal–ammonia and metal–methylamine solutions. For this purpose we also draw upon interpretations from other systems in which the transition from localized to itinerant electron regimes is well recognized (*78*).

A. MODELS FOR THE NONMETAL-TO-METAL TRANSITION

1. The Herzfeld Theory of Metallization

In 1927, K. F. Herzfeld published a paper (*93*) in which he proposed a simple criterion for determining when an element or system will exhibit true metallic characteristics. On classical grounds, Herzfeld argued that the characteristic frequency of bound electrons, representing a measure of the force holding the electron in the free atom, is diminished at high electron densities to the value

$$\nu = \nu_0(1 - R/V)^{1/2} \tag{10}$$

where ν_0 is the characteristic frequency in the isolated (low electron concentration) species, R is the molar refractivity, and V is the molar volume.

If $(R/V) = 1$, the resultant force on the localized electron vanishes, the electron is set free, and the system acquires metallic status. The previous statements may be recast in slightly different form using the venerable Clausius–Mossotti relationship (81), where

$$R/V = (n^2 - 1)/(n^2 + 2) \tag{11}$$

and n is the index of refraction, and n^2 is the optical dielectric constant.

When the condition $(R/V) = 1$ is fulfilled, the refractive index/dielectric constant goes to infinity, and we have a NM-M transition. The Herzfeld criterion when applied to metal–ammonia solutions does indeed predict (67, 93) that localized, solvated electrons are set free by mutual action of neighboring electrons at metal concentrations above ~4–5 MPM, and measurements (Fig. 20) similarly indicate a dielectric catastrophe in this concentration range. The simple Herzfeld picture has recently been applied by Edwards and Sienko (70) to explain the occurrence of metallic character in the Periodic Table.

2. The Mott Transition

Mott's major work (124–126) on the transition to the metallic state in simple metals has for a long time been used as a framework for the description of the NM-M transition in metal–ammonia solutions (37, 164, 165).

Consider a monovalent metal such as sodium (125). Each sodium atom carries with it one valence (3s) electron. The lower half of this band is then half filled. As far as a conventional picture is concerned (139), the 3s band is half filled and the valence electrons are therefore itinerant and move freely about the system. If we now increase the internuclear distance, there is a reduction in the width of the 3s band. Ultimately, as common sense would have it (125, 126), the band reduces to the discrete s level of the *isolated* sodium atom. Metallic conduction is no longer possible, although according to the band-model approximation we still have a half-filled band and therefore metallic conduction should still be maintained! The band-model approximation thus breaks down for "narrow" bands in which there is considerable electron–electron correlation (86, 87, 124). In a number of papers beginning in 1949 (124–126), Sir Nevill Mott addressed this fundamental problem. In Mott's view, the transition from metallic to nonmetallic regimes is sharp—indeed, first order: electrons are *either* localized in distinct regions of space, *or* itinerant. Figure 21 contrasts (165) the standard (band picture) view of the conductivity variation with those to Mott's.

Mott first gave an estimate (125) of this critical density at the NM-M transition in terms of the screening properties of an itinerant electron gas. The "Mott criterion" is

$$n_c^{1/3} a_H^* \geqslant 0.25 \tag{12}$$

where n_c is the critical concentration of centers, and a_H^* is an effective radius for the isolated (localized electron) state. From an extensive analysis of experimental data for disordered materials, Edwards and Sienko (68) have recently found that a particular (scaled) form of the Mott criterion exhibits an apparent universality, in that the relation

$$n_c^{1/3} a_H^* = 0.26 \pm 0.05 \tag{13}$$

predicts the critical concentration for the onset of metallic character for systems around the range of $\sim 10^9$ in critical densities (or alternatively

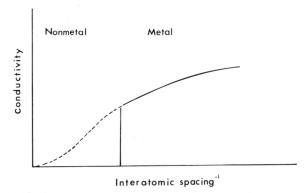

FIG. 21. A comparison of the anticipated changes in conductivity with atomic spacing from a conventional viewpoint (dashed line) and that of Mott. The abscissa is the reciprocal of the interatomic spacing. [Adapted from Thompson (165), with permission.]

600 Å in a_H^*) provided a_H^* is obtained directly from *experimental* parameters that characterize the localized electron state. The data (*64, 68, 69*) for metal solutions in ammonia, methylamine, and HMPA are included in Fig. 22. Inherent in the Mott picture (*124*) is a *major* change in the *thermodynamic properties* of the solutions in the transition region. This important feature is discussed in Section IV,B.

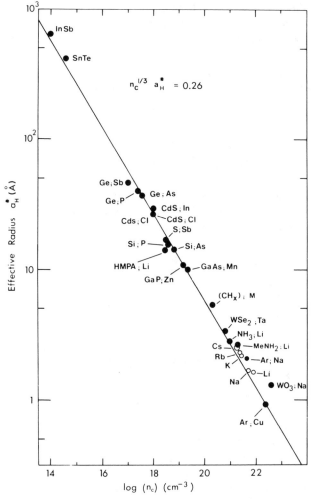

FIG. 22. The nonmetal-to-metal transition: a logarithmic plot of the effective radius a_H^* of the localized-electron state versus the critical (electron) concentration for metallization, n_c, in a variety of systems. [Adapted from Edwards and Sienko (*68*), and used with permission from the American Physical Society, *The Physical Review (Solid State)*.]

3. Inhomogeneous Picture for the Transition

The fundamental basis of Mott's picture for the NM-M transition in metal–ammonia solutions is that the system is essentially homogeneous (and slightly disordered) for temperatures sufficiently far removed from the consolute temperature for phase separation. An alternative model has been advanced by Cohen and Jortner (35, 36). Their description assumes that solutions in the transition range are microscopically inhomogeneous, consisting of highly concentrated regions with metallic conduction and dilute regions with electrolytic conduction. The dilute and metallic domains themselves are of constant and well-defined concentration, 2.3 and 9 MPM, respectively. An increase in the overall metal concentration ultimately leads to a finite fraction of the material being connected as a conduction path leading across the sample (35, 146) (Fig. 23). A percolation problem was then posed, and the approach utilized to account (almost quantitatively) for numerous transport properties (35, 36). In contrast to the Mott picture (123), the NM-M transition here is essentially continuous; the particular interpretation used by Jortner and Cohen (35, 36) puts the percolation threshold near 3.5 MPM. However, the overriding question (122, 123) is whether we have any direct experimental evidence for a microscopic structure in which the local concentration fluctuates between the two well-defined boundaries. Damay and Chieux have recently carried out small-angle neutron scattering (SANS) studies on sodium and lithium solutions in liquid deuteroammonia at concentrations near 4 MPM, from the liquid–liquid critical temperature up to room temperature (44). Their complete SANS data could apparently be attributed to concentration fluctuations on purely thermodynamic grounds for systems close to a critical point. However, the more recent heat capacity studies by Steinberg et al. (158) reopen the important question of the validity

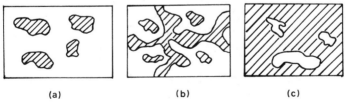

(a)   (b)   (c)

FIG. 23. Illustration of the nonmetal-to-metal transition according to the percolation picture of Cohen and Jortner (see text). Solvent regions of high metal concentration are shaded. Metallic regions grow with increase of metal concentration. (a) Below the percolation threshold there are isolated metallic regions and no conductance. (b) Above the percolation threshold a metallic path crosses the material and conduction occurs. (c) Above a certain critical concentration, the insulating regions are disjoint.

of a percolation hypothesis for the transition region in metal–ammonia solutions.

B. Liquid–Liquid Phase Separation and the Nonmetal-to-Metal Transition

If a solution containing approximately 4 mole percent sodium in ammonia is cooled below −42°C (231 K) a remarkable liquid–liquid phase separation occurs (33, 155). The solution physically separates into two distinct layers—a low-density, bronze metallic phase that floats out on top of a more dense, less concentrated dark-blue phase. The first experimental observation of this striking phenomenon in sodium–ammonia solutions was made by Kraus (109, 110) in 1907; more recent studies have mapped out the phase coexistence curves for a variety of alkali and alkaline earth metals in liquid ammonia, and these are delineated and discussed elsewhere (164).

In 1958, Pitzer (141), in a remarkable contribution that appears to have been the first theoretical consideration of this phenomenon, likened the liquid–liquid phase separation in metal–ammonia solutions to the vapor–liquid condensation that accompanies the cooling of a nonideal alkali metal vapor in the gas phase. Thus, in sodium–ammonia solutions below 231 K we would have a phase separation into an insulating vapor (corresponding to matrix-bound, localized excess electrons) and a metallic (matrix-bound) liquid metal. This suggestion of a "matrix-bound" analog of the critical liquid–vapor separation in pure metals preceeded almost all of the experimental investigations (41, 77, 91, 92) into dense, metallic vapors formed by an expansion of the metallic liquid up to supercritical conditions. It was also in advance of the possible fundamental connection between this type of critical phenomenon and the NM-M transition, as pointed out by Mott (125) and Krumhansl (112) in the early 1960s.

The precise nature of the electronic interactions between centers must obviously change dramatically at the NM-M transition, e.g., from van der Waals type interaction to metallic cohesion (112). These gross changes in electronic properties at the transition are sufficient to noticeably influence the thermodynamic features of the system (86, 87). The conditions therefore appear highly conducive for a *thermodynamic phase transition* to accompany the *electronic transition* at the critical density. In fact, the transition to the metallic state in metal–ammonia solutions is accompanied by a decrease in both enthalpy and entropy (146, 149), and it has been argued convincingly (124, 125) that the phase separation in supercritical alkali metals and metal solutions is indeed a consequence of these changes in thermodynamic functions (Fig. 24).

In a recent communication (70), we have reexamined Pitzer's early hypothesis in the light of the considerable advances made recently both in the theoretical and experimental study of supercritical fluid alkali metals (41, 77), doped semiconductors (78), and metal–ammonia and metal–methylamine (60, 166) solutions. The latter systems are particularly instructive (70). Preliminary data (10) for the lithium–methylamine system suggests that phase separation occurs around 15–16 MPM, compared to approximately 4 MPM in lithium–ammonia solutions (Fig. 25). In both solvent systems, this critical composition also marks the onset of the NM-M transition for $T > T_c$ (critical consolate temperature). Table III is a collation of critical (consolate) densities in lithium–ammonia, sodium–ammonia, and lithium–methylamine solutions, together with the observed critical densities of the pure alkali metals and the corresponding estimates of the Mott criterion ($n_c^{1/3} a_H^*$) for all systems under investigation (70). These condensation phenomena observed in both gaseous and matrix-bound systems appear to be closely related. Furthermore, they are significantly correlated in that, in all cases, critical (metal) densities at the metallic onset ($T > T_c$), or at the phase separation limit ($T < T_c$) are in good agreement with the Mott criterion [Eq. (13)]. Pitzer's suggestion (141),

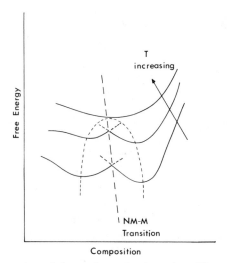

FIG. 24. The electronic and thermodynamic phase transitions at the nonmetal-to-metal transition: a schematic representation of the free energy of a metal–ammonia solution in the temperature range of the miscibility gap, showing the NM-M transition as a function of metal concentration for increasing temperatures.

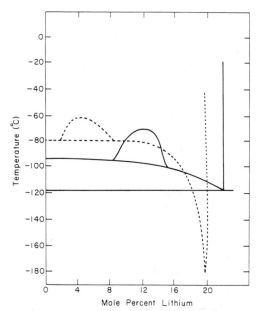

FIG. 25. A comparison of phase relations in the systems lithium–ammonia (– – –) and lithium–methylamine (——). The latter is the tentative phase diagram based on preliminary studies (10).

that phase separation in metal–ammonia solutions is the analog within the liquid ammonia medium of the liquid–vapor separation of the pure metal, does indeed represent a particularly apt description of the phenomenon (70).

C. EXPANDED-METAL COMPOUNDS

The lithium–ammonia phase diagram shows a deep pseudoeutectic (Fig. 25) at approximately 20 MPM and 88.8 K, which signals the appearance of a compound, tetraamminelithium(0) of stoichiometry $Li(NH_3)_4$. A great deal of evidence, both direct and indirect, has now been amassed that shows that a true compound, rather than a simple eutectic solid, does indeed exist at low temperatures (118, 164). Recent studies of metal–ammonia compounds suggest that they may crystallize in unusual structural arrangements (82, 83, 167). For example, the compound $Ca(ND_3)_6$ may possess a novel ND_3 geometry in which the ammonia molecules are nearly planar and have two inequivalent sets of deuterons, with one N–D distance being rather short (~0.9 Å), while the other two are extremely long (~1.4 Å) (82). Whatever the pecu-

TABLE III

THE MOTT CRITERION FOR METAL SOLUTIONS AND EXPANDED FLUID
METALS[a]

System	n_c (cm^{-3})	a_H^* (Å)	$n_c^{1/3}a_H^*$
I. Supercritical alkali metals[b]			
Li	9.43×10^{21}	1.59	0.34
Na	5.48×10^{21}	1.71	0.30
K	2.89×10^{21}	2.16	0.31
Rb	2.40×10^{21}	2.29	0.31
Cs	$1.9 \ \times 10^{21}$	2.52	0.32
II. Metal solutions[c]			
Li–NH$_3$ (209 K)	9.94×10^{20}	2.83	0.28
Na–NH$_3$ (231 K)	9.03×10^{20}	2.88	0.28
Li–MeNH$_2$ (~200 K)	1.85×10^{21}	2.60	0.32

[a] Full details given in Ref. 69.

[b] Values of n_c at the gas–liquid critical point; a_H^* derived from radii corresponding to the principal maxima in the radial distribution functions $[r^2\psi^2(r)]$, using relativistic wave functions.

[c] Values of n_c taken at the concentration of the liquid–liquid phase separation; a_H^* derived from the adiabatic cavity model for the solvated electron (see Section II).

liarities of the local structure in the complexes, the most appealing overall description (154) of the solid compound is that of an "expanded metal." In this, the ammonia simply takes the role of a space-filling diluent, effectively increasing the Li–Li separation relative to that in the pure metal. This dilution effect highlights one of the great attractions (66) of these materials, namely, that via pertinent choice of solvent these low-electron-density materials could be tailored in such a fashion as to move them toward the metal–nonmetal transition.

Edwards, Lusis, and Sienko have recently reported an ESR study (60) of frozen lithium–methylamine solutions which suggests the existence of a compound tetramethylaminelithium(0), Li(CH$_3$NH$_2$)$_4$, bearing all the traits (60) of a highly expanded metal lying extremely close to the metal–nonmetal transition. Specifically, both the nuclear-spin and electron-spin relaxation characteristics of the compound, although nominally metallic, cannot be described in terms of the conventional theories of conduction ESR (6, 15, 71) and NMR in pure metals (60, 96, 169).

The use of cyclic (crown) polyethers and cryptates to enhance metal solubilities also opens up many interesting possibilities for studying expanded-metal compounds in the low-dielectric-constant solvents

(51). In addition, thin solid films obtained by evaporation of metal–ammonia solutions that contain a cation complexing agent (cryptand) have transmission spectra that suggest the existence of expanded metals in these systems (42).

V. Concluding Remarks

The fundamental theme for this review was the attempt to illustrate the "flavor" of current research in this area. If one were to summarize (62, 146) the present status of our knowledge of these remarkable systems, it might be that most of the properties of excess electrons in solution can be interpreted in terms of models that are (qualitatively) easily understood, but quantitatively evaluated only with considerable effort. This is, indeed, an observation pertinent not only to metal solutions, but to inorganic chemistry in general!

We conclude this review as we have started it—with a brief reference to unpublished work by Sir Humphry Davy:

During the course of many laborious experiments (1808–1809) on the action of dry gaseous ammonia with molten potassium (Fig. 1), Davy speculated briefly on the possible reaction at the metal interface, questioning "whether any substances can be formed which will not absorb Ammonia." Unfortunately, Davy's intense activity during this period led to considerable fatigue, and (136) "a most severe fit of illness, which for a time caused an awful pause in his researches, broke the thread of his pursuits, and turned his reflections into different channels."

Appendix

The purpose of this section, added in proof, is to include certain work published since the review was completed.

The first observation of the NMR spectrum of Na^- in a metal solution without added cation-complexing agents has recently been reported (65a). The rationale behind this observation was that the solvent HMPA by itself appeared to fulfill many of the requirements generally sought from macrocyclic complexing agents: high solubility via cation complexation, stability to electron reduction, weak anion solvation, etc. The NMR spectrum of Na^- in fluid Na–HMPA solutions, shown in Fig. 26, exhibits precisely the same chemical shift as that observed for Na^- in solutions of Na in anhydrous methylamine and ethylamine in the presence of 2,2,2-cryptand. The metal anion here is truly "gas-like" in

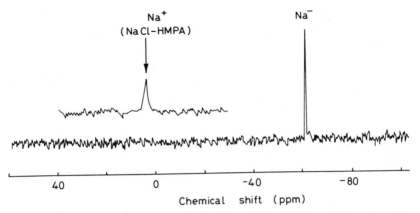

FIG. 26. The ^{23}Na NMR spectrum of Na$^-$ in a solution of sodium in HMPA (*65a*).

that the sodium 2p electrons are effectively shielded from solvent interactions by the presence of two spin-paired electrons in an almost unperturbed 3s orbital on the metal.

The magnetic susceptibility data for metal–ammonia solutions cited earlier (see Fig. 5) have now appeared in the chemical literature (*88a*). For metal–ammonia solutions at $-65°C$, the spin concentration clearly

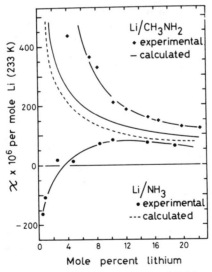

FIG. 27. The magnetic susceptibility of solutions of lithium in ammonia and in methylamine. The broken and solid lines indicate the predicted concentration variations of the susceptibility, based on an electron-gas picture in which electron–electron correlations are neglected (*157a, 157b*).

mimics the analytical concentration of the metal over a wide concentration range. This contrasts markedly with the situation at $-33°C$ and $0°C$ (Fig. 5).

Detailed magnetic susceptibility studies of both fluid and frozen solutions of lithium in methylamine have now been carried out (157a, 157b). The data for lithium–methylamine solutions at $-40°C$ are shown in Fig. 27. The corresponding behavior of lithium–ammonia solutions is also shown. The observed susceptibility of lithium-methylamine solutions is in reasonable agreement with the calculated susceptibility of a free-electron gas (neglecting electron-correlation effects). In contrast, large diamagnetic deviations are observed for lithium–ammonia solutions below ~8 MPM (see Section III,A,2). Clearly, the methylamine solutions do not display the extensive spin-pairing phenomena observed in ammonia solutions at $-40°C$. The electronic properties of the corresponding solid expanded-metal compounds, including lithium 2,1,1-cryptate electride were discussed in the recent ACS symposium (157b) (see also Ref. 113a).

ACKNOWLEDGEMENTS

I would like to thank Dr. R. Catterall, Professor M. J. Sienko, and Professor Sir Nevill Mott for numerous stimulating discussions on the chemistry and physics of metal solutions. I also thank Mrs. I. M. McCabe, librarian at the Royal Institution, London, for her expert assistance in helping me uncover Sir Humphry Davy's early contributions in this area, and Mr. M. Springett for his considerable help in preparing figures. The financial assistance of the SERC, The Royal Society, and NATO is gratefully acknowledged.

REFERENCES

1. Banerjee, A., and Simons, J., J. Chem. Phys. **68,** 415 (1978).
2. Bar-Eli, K., and Tuttle, T. R., Jr., Bull. Am. Phys. Soc. [2] **8,** 352 (1963).
3. Bar-Eli, K., and Tuttle, T. R., Jr., J. Chem. Phys. **40,** 2508 (1964); **44,** 114 (1966).
4. Becker, E., Lindquist, R. H., and Alder, B. J., J. Chem. Phys. **25,** 971 (1956).
5. Bethe, H., in "Solid State Physics" (J. S. Blakemore, ed.), Saunders, Philadelphia, Pennsylvania, 1970.
6. Beuneu, F., and Monod, P., Phys. Rev. B: Condens. Matter **18,** 2422 (1978).
7. Boag, J. W., and Hart, E. J., Nature (London) **197,** 45 (1963).
8. Breitschwerdt, K. G., and Radscheit, H., Ber. Bunsenges. Phys. Chem. **80,** 797 (1976).
9. Brodsky, A. M., and Tsarevsky, A. V., Adv. Chem. Phys. **44,** 483 (1980).
10. Buntaine, J. R., Ph.D. Thesis, Cornell University, Ithaca, New York (1980), on file with University Microfilms, Order No. 8015643.
11. Buntaine, J. R., Sienko, M. J., and Edwards, P. P., J. Phys. Chem. **84,** 1230 (1980).

12. Carrington, A., and McLachlan, A. D., "Introduction to Magnetic Resonance." Harper & Row, New York, 1967.
13. Catterall, R., *in* "Metal-Ammonia Solutions" (J. J. Lagowski and M. J. Sienko, eds.), p. 105. Butterworth, London, 1970.
14. Catterall, R., and Carmichael, I., comments at Colloque Weyl V, *J. Phys. Chem.* **84**, 1128 (1980).
15. Catterall, R., and Edwards, P. P., *Adv. Mol. Relaxation Processes* **7**, 87 (1975).
16. Catterall, R., and Edwards, P. P., *Adv. Mol. Relaxation Interaction Processes* **13**, 123 (1978).
17. Catterall, R., and Edwards, P. P., *Adv. Mol. Relaxation Interaction Processes* (in press).
18. Catterall, R., and Edwards, P. P., *Chem. Phys. Lett.* **42**, 540 (1976).
19. Catterall, R., and Edwards, P. P., *Chem. Phys. Letts.* **43**, 122 (1976).
20. Catterall, R., and Edwards, P. P., *J. Chem. Soc., Chem. Commun.* p. 591 (1980).
21. Catterall, R., and Edwards, P. P., *J. Phys. Chem.* **79**, 3010 (1975).
22. Catterall, R., and Edwards, P. P., *J. Phys. Chem.* **84**, 1196 (1980).
23. Catterall, R., and Edwards, P. P., *Mol. Phys.* **32**, 555 (1976).
24. Catterall, R., and Edwards, P. P., *Solid State Commun.* (in press).
25. Catterall, R., and Edwards, P. P., *Int. Symp. Magn. Reson., 6th, 1977* p. 89 (1977).
26. Catterall, R., Edwards, P. P., Slater, J., and Symons, M. C. R., *Chem. Phys. Lett.* **64**, 275 (1979).
27. Catterall, R., and Mott, N. F., *Adv. Phys.* **18**, 665 (1969).
28. Catterall, R., Slater, J., Seddon, W. A., and Fletcher, J. W., *Can. J. Chem.* **54**, 3110 (1976).
29. Catterall, R., and Symons, M. C. R., *J. Chem. Soc.* p. 13 (1966).
30. Catterall, R., Symons, M. C. R., and Tipping, J. W., *J. Chem. Soc.* p. 1529 (1966); p. 1234 (1967).
31. Chan, S. I., Austin, J. A., and Paez, O. A., *in* "Metal-Ammonia Solutions" (J. J. Lagowski and M. J. Sienko, eds.), p. 425. Butterworth, London, 1970.
32. Chen, K. S., Mao, S. W., Nakamura, K., and Hirota, N., *J. Am. Chem. Soc.* **93**, 6004 (1971).
33. Chieux, P., and Sienko, M. J., *J. Chem. Phys.* **53**, 566 (1970).
34. Cohen, M. H., *Rev. Mod. Phys.* **40**, 839 (1968).
35. Cohen, M. H., and Jortner, J., *J. Phys. Chem.* **79**, 2900 (1975).
36. Cohen, M. H., and Jortner, J., *Phys. Rev. B: Solid State* **13**, 1548 (1976).
37. Cohen, M. H., and Thompson, J. C., *Adv. Phys.* **17**, 857 (1968).
38. Colloque Weyl IV. Electrons in fluids—the nature of metal-ammonia solutions, *J. Phys. Chem.* **79**, No. 26 (1975).
39. Colloque Weyl V. Excess electrons and metal-ammonia solutions, *J. Phys. Chem.* **84**, 1065 (1980).
40. Copeland, D. A., Kestner, N. R., and Jortner, J. J., *J. Chem. Phys.* **53**, 1189 (1970).
41. Cusack, N. E., *Proc. Scott. Univ. Summer Sch. Phys.* **19**, 455 (1978).
42. DaGue, M. G., Landers, J. S., Lewis, H. L., and Dye, J. L., *Chem. Phys. Lett.* **66**, 169 (1979).
43. Dalton, L. R., Rynbrandt, J. D., Hansen, E. M., and Dye, J. L., *J. Chem. Phys.* **44**, 3969 (1966).
44. Damay, P., and Chieux, P., *J. Phys. Chem.* **84**, 1203 (1980).
45. Davis, H. T., and Brown, R. G., *Adv. Chem. Phys.* **31**, 329 (1975).
46. Davy, H., *Philos. Trans. R. Soc. London* **98**, 1 (1808).
47. Demortier, A., Ph.D. Thesis, Lille (1970).

48. Dorfman, L. M., and Jou, F. Y., *in* "Electrons in Fluids" (J. Jortner and N. R. Kestner, eds.), p. 447. Springer-Verlag, Berlin and New York, 1973.
49. Douthit, R. C., and Dye, J. L., *J. Am. Chem. Soc.* **82**, 4472 (1960).
50. Dukel'skii, V. M., Zandberg, E. Ya., and Ionov, N. I., *Dokl. Akad. Nauk. SSSR,* **62**, 232 (1948), cited by J. L. Dye, in Ref. *51.*
51. Dye, J. L., *Angew. Chem., Int. Ed. Engl.* **18**, 587 (1979).
52. Dye, J. L., *Prog. Macrocyclic Chem.* **1**, 63 (1979).
53. Dye, J. L., *Pure Appl. Chem.* **49**, 3 (1977).
54. Dye, J. L., *in* "Metal-Ammonia Solutions" (J. J. Lagowski and M. J. Sienko, eds.), p. 1. Butterworth, London, 1970.
55. Dye, J. L., *in* "Electrons in Fluids" (J. Jortner and N. R. Kestner, eds.), p. 77. Springer-Verlag, Berlin and New York, 1973.
55a. Dye, J. L., *Sci. American* **237**(7), 92 (1977).
56. Dye, J. L., Andrews, C. W., and Ceraso, J. M., *J. Phys. Chem.* **79**, 3076 (1975).
57. Dye, J. L., Ceraso, J. M., Lok, M. T., Barnett, B. L., and Tehan, F. J., *J. Am. Chem. Soc.* **96**, 608 (1974).
58. Dye, J. L., and Dalton, L. R., *J. Phys. Chem.* **71**, 184 (1967).
59. Edwards, P. P., *J. Chem. Phys.* **70**, 2631 (1979).
60. Edwards, P. P., *J. Phys. Chem.* **84**, 1215 (1980).
61. Edwards, P. P., *Chem. Rev.* (to be submitted for publication).
62. Edwards, P. P., *Phys. Chem. Liq.* **10**, 189 (1981).
63. Edwards, P. P., Buntaine, J. R., and Sienko, M. J., *Phys. Rev. B: Condens. Matter* **19**, 5835 (1979).
64. Edwards, P. P., and Catterall, R., *Can. J. Chem.* **55**, 2258 (1977).
65. Edwards, P. P., and Catterall, R., *Philos. Mag. [Part] B* **39**, 81, 371 (1979).
65a. Edwards, P. P., Guy, S. C., Holton, D. M., and McFarlane, W., *J. Chem. Soc., Chem. Commun.* p. 1185 (1981).
66. Edwards, P. P., Lusis, A. R., and Sienko, M. J., *J. Chem. Phys.* **72**, 3103 (1980).
67. Edwards, P. P., and Schroer, W., unpublished work.
68. Edwards, P. P., and Sienko, M. J., *Phys. Rev. B: Solid State* **17**, 2575 (1978).
69. Edwards, P. P., and Sienko, M. J., *J. Am. Chem. Soc.* **103**, 2967 (1981).
70. Edwards, P. P., and Sienko, M. J., *Acc. Chem. Research* **15**, 87 (1982).
71. Elliott, R. J., *Phys. Rev.* **96**, 266 (1954).
72. Essig, W., Ph.D. Dissertation, University of Karlsruhe (1978), cited by Schindewolf (*148*).
73. Even, U., Swennumson, R. D., and Thompson, J. C., *Can. J. Chem.* **55**, 2240 (1977).
74. Feng, Da-Fei, and Kevan, L., *Chem. Rev.* **80**, 1 (1980).
75. Fletcher, J. W., and Seddon, W. A., *J. Phys. Chem.* **79**, 3055 (1975).
76. Freed, S., and Sugarman, N., *J. Chem. Phys.* **11**, 354 (1943).
77. Freyland, W., *J. Non-Cryst. Solids* **35/36**, 1313 (1980).
78. Friedman, L. R., and Tunstall, D. P., eds., "The Metal Non-Metal Transition in Disordered Systems." Scottish Universities Summer School in Physics, Edinburgh University, Edinburgh, 1978.
79. Fullmer, J. Z., "Sir Humphry Davy's Published Works." Harvard Univ. Press, Cambridge, Massachusetts.
80. Gibson, G. E., and Argo, W. L., *J. Am. Chem. Soc.* **40**, 1327 (1918).
81. Glasstone, S., "Textbook of Physical Chemistry." Van Nostrand-Reinhold, Princeton, New Jersey, 1940.
82. Glaunsinger, W. S., *J. Phys. Chem.* **84**, 1163 (1980).
83. Glaunsinger, W. S., White, T. R., von Dreele, R. B., Gordon, D. A., Marzke, R. F., Bowman, A. L., and Yarnell, J. L., *Nature (London)* **271**, 414 (1978).

84. Gold, M., and Jolly, W. L., *Inorg. Chem.* **1**, 818 (1962).
85. Gold, M., Jolly, W. L., and Pitzer, K. S., *J. Am. Chem. Soc.* **84**, 2264 (1962).
86. Goodenough, J. B., *in* "New Developments in Semiconductors" (R. R. Wallace, R. Hams, and M. J. Zuckermann, eds.), p. 107. Nordhoff Int., Gröningen, Leyden, 1971.
87. Goodenough, J. B., *in* "The Robert A. Welch Foundation Conferences on Chemical Research, XIV. Solid State Chemistry," Chapter 3. Houston, Texas, 1970. This article outlines similar considerations for narrow d-band materials.
88. Harris, R. L., Ph.D. Thesis, University of Texas at Austin (1979).
88a. Harris, R. L., and Lagowski, J. J., *J. Phys. Chem.* **85**, 856 (1981).
89. Hart, E. J., ed., "Solvated Electron," American Chemical Society, Washington, D.C., 1965.
90. Hart, E. J., and Boag, J. W., *J. Am. Chem. Soc.* **84**, 4090 (1963).
91. Hensel, F., *Ber. Bunsenges. Phys. Chem.* **80**, 786 (1976).
92. Hensel, F., *Can. J. Chem.* **55**, 2225 (1977).
93. Herzfeld, K. F., *Phys. Rev.* **29**, 701 (1927).
94. Hirasawa, M., Nakamura, Y., and Shimoji, M., *Ber. Bunsenges. Phys. Chem.* **82**, 815 (1978).
95. Hirota, N., and Kreilick, R., *J. Am. Chem. Soc.* **88**, 614 (1966).
96. Holcomb, D. F., *Proc. Scott. Univ. Summer Sch. Phys.* **19**, 251 (1978).
97. Huster, E., *Ann. Phys. (Leipzig)* [5] **33**, 477 (1938).
98. Hutchison, C. A., and Pastor, R. C., *J. Chem. Phys.* **21**, 1959 (1953).
99. International Conference on Electrons in Fluids, Banff, Alberta, Canada, 1976, *Can. J. Chem.* **55**, 1796 (1977).
100. Jen, C. K., Bowers, V. A., Cochran, E. L., and Foner, S. N., *Phys. Rev.* **126**, 1749 (1962).
101. Jortner, J., *J. Chem. Phys.* **30**, 839 (1959).
102. Jortner, J., *Ber. Bunsenges. Phys. Chem.* **75**, 696 (1971).
103. Jortner, J., and Kestner, N. R., eds., "Electrons in Fluids," Colloque Weyl III. Springer-Verlag, Berlin and New York, 1973.
104. Jortner, J., Rice, S. A., and Wilson, E. G., *in* "Metal-Ammonia Solutions" (G. Lepoutre and M. J. Sienko, eds.), p. 222. Benjamin, New York, 1964.
105. Kestner, N. R., *in* "Electrons in Fluids" (J. Jortner and N. R. Kestner, eds.), p. 1. Springer-Verlag, Berlin and New York, 1973.
106. Knox, R. S., "Theory of Excitons." Academic Press, New York, 1963.
107. Koehler, W. H., and Lagowski, J. J., *J. Phys. Chem.* **73**, 2329 (1969).
108. Kohn, W., *Solid State Phys.* **5**, 258 (1957).
109. Kraus, C. A., *J. Am. Chem. Soc.* **29**, 1557 (1907).
110. Kraus, C. A., *J. Am. Chem. Soc.* **30**, 1323 (1908).
111. Kraus, C. A., *J. Am. Chem. Soc.* **43**, 749 (1921); see also *J. Chem. Educ.* **30**, 83 (1953).
112. Krumhansl, J. A., *in* "Physics of Solids at High Pressures" (C. T. Tomizuka and R. M. Emrick, eds.), p. 425. Academic Press, New York, 1965.
113. Lagowski, J. J., and Sienko, M. J., eds., "Metal-Ammonia Solutions," Colloq. Weyl II, IUPAC. Butterworth, London, 1970.
113a. Landers, J. S., Dye, J. L., Stacy, A. M., and Sienko, M. J., *J. Phys. Chem.* **85**, 1096 (1981).
114. Lelieur, J. P., Ph.D. Thesis, Orsay (1972).
115. Lelieur, J. P., and Rigny, P., *J. Chem. Phys.* **59**, 1142 (1973).
116. Lepoutre, G., and Sienko, M. J., eds., "Metal-Ammonia Solutions," Colloq. Weyl I. Benjamin, New York, 1964.

117. Mahaffey, D. W., and Jerde, D. A., *Rev. Mod. Phys.* **40**, 710 (1968).
118. Mammano, N., *in* "Metal-Ammonia Solutions" (J. J. Lagowski and M. J. Sienko, eds.), p. 367. Butterworth, London, 1970.
119. Matalon, S., Golden, S., and Ottolenghi, M., *J. Phys. Chem.* **73**, 3098 (1969).
120. McHale, J., and Simons, J., *J. Chem. Phys.* **67**, 389 (1977).
121. Metcalf, H. J., *Nature (London)* **284**, 127 (1980).
122. Mott, N. F., *J. Phys. Chem.* **84**, 1199 (1980).
123. Mott, N. F., *J. Phys. Chem.* **79**, 2915 (1975).
124. Mott, N. F., "Metal-Insulator Transitions." Taylor & Francis, London, 1974.
125. Mott, N. F., *Philos. Mag.* [8] **6**, 287 (1961).
126. Mott, N. F., *Proc. Phys. Soc. London, Ser. A* **62**, 416 (1949).
127. Nakamura, Y., Horie, Y., and Shimoji, M., *J. Chem. Soc. Faraday Trans.* **70**, 1376 (1974).
128. Nakamura, Y., Toma, T., and Shimoji, M., *Phys. Lett. A* **60**, 373 (1977).
129. Nauta, H., and van Huis, C., *J. Chem. Soc., Faraday Trans.* **68**, 647 (1972).
130. Nicely, V. A., Breit-Rabi analysis of experimental values of L. R. Dalton. M.S. Thesis, Michigan State University (1966) (unpublished).
131. Normant, H., *Angew Chem., Int. Ed. Engl.* **6**, 1046 (1967).
132. Ogg, R. A., *J. Am. Chem. Soc.* **68**, 155 (1946); *J. Chem. Phys.* **14**, 114, 295 (1946); *Phys. Rev.* **69**, 243, 668 (1946).
133. Onsager, L., *Rev. Mod. Phys.* **40**, 709 (1968) (comments following the paper by J. C. Thompson, *ibid.* p. 704).
134. O'Reilly, D. E., *J. Chem. Phys.* **41**, 3729 (1964).
135. O'Reilly, D. E., and Tsang, T., *J. Chem. Phys.* **42**, 3333 (1965).
136. Paris, J. A., "The Life of Sir Humphry Davy." Henry Colburn and Richard Bentley, New Burlington Street, London, 1832.
137. Peer, W. J., and Lagowski, J. J., *J. Am. Chem. Soc.* **100**, 6260 (1978).
138. Peer, W. J., and Lagowski, J. J., *J. Phys. Chem.* **84**, 1110 (1980).
139. Phillips, C. S. G., and Williams, R. J. P., "Inorganic Chemistry." Oxford Univ. Press, London and New York, 1966.
140. Pikaev, A. K., "Solvated Electron in Radiation Chemistry." Nauka, Moscow, 1966.
141. Pitzer, K. S., *J. Am. Chem. Soc.* **80**, 5046 (1958).
142. Quinn, R. K., and Lagowski, J. J., *J. Phys. Chem.* **73**, 2326 (1969).
143. Reichardt, C., *in* "Solvent Effects in Organic Chemistry" (H. F. Ebel, ed.), Monogr. Mod. Chem. Ser., Verlag-Chemie, Weinheim, 1979 (for a wider discussion of the problem).
144. Rubinstein, G., Tuttle, T. R., and Golden, S., *J. Phys. Chem.* **77**, 2872 (1973).
145. Schettler, P. D., Jr., and Lepoutre, G., *J. Phys. Chem.* **79**, 2823 (1975).
146. Schindewolf, U., *Angew. Chem., Int. Ed. Engl.* **17**, 887 (1978).
147. Schindewolf, U., and Schulte-Frohlinde, D., *Ber. Bunsenges. Phys. Chem.*, **75**, 607 (1971).
148. Schindewolf, U., *Z. Phys. Chem. (Wiesbaden)* [N.F.] **112**, 153 (1978).
149. Schindewolf, U., and Werner, M., *J. Phys. Chem.* **84**, 1123 (1980).
150. Seddon, W. A., and Fletcher, J. W., *J. Phys. Chem.* **84**, 1104 (1980).
151. Seddon, W. A., Fletcher, J. W., and Catterall, R., *Can. J. Chem.* **55**, 2017 (1977).
152. Seddon, W. A., Fletcher, J. W., and Sopchyshyn, F. C., *Can. J. Chem.* **56**, 839 (1978).
153. Seddon, W. A., Fletcher, J. W., and Sopchyshyn, F. C., *Chem. Phys.* **15**, 377 (1976).
154. Sienko, M. J. *in* "Metal-Ammonia Solutions" (G. Lepoutre and M. J. Sienko, eds.), p. 23. Benjamin, New York, 1964.

155. Sienko, M. J., and Chieux, P., *in* "Metal-Ammonia Solutions" (J. J. Lagowski and M. J. Sienko, eds.) p. 339. Butterworth, London, 1970.
156. Slichter, C. P., "Principles of Magnetic Resonance." Harper & Row, New York, 1963.
157. Springett, B. E., Jortner, J., and Cohen, M. H., *J. Chem. Phys.* **48**, 2720 (1968).
157a. Stacy, A. M., Ph.D. Thesis, Cornell University, Ithaca, New York (1981).
157b. Stacy, A. M., Edwards, P. P., and Sienko, M. J., "ACS Symposium on Electronic Properties of Inorganic Solids, Las Vegas, March 1982," to be published *J. Solid State Chem.*
158. Steinberg, V., Voronel, A., Linsky, D., and Schindewolf, U., *Phys. Rev. Lett.* **45**, 1338 (1980).
159. Suchannek, R. G., Naiditch, S., and Kleijnot, O. J., *J. Appl. Phys.* **38**, 690 (1967).
160. Symons, M. C. R., *Chem. Soc. Rev.* p. 337 (1976).
161. Symons, M. C. R., *J. Phys. Chem.* **71**, 172 (1967).
162. Tehan, F. J., Barnett, B. L., and Dye, J. L., *J. Am. Chem. Soc.* **96**, 7203 (1974).
163. Teherani, T. H., Peer, W. J., Lagowski, J. J., and Bard, A. J., *J. Am. Chem. Soc.* **100**, 7768 (1978).
164. Thompson, J. C., "Electrons in Liquid Ammonia." Oxford Univ. Press (Clarendon), London and New York, 1976.
165. Thompson, J. C., *in* "The Chemistry of Non-Aqueous Solvents" (J. J. Lagowski, ed.), Vol. 2, p. 265. Academic Press, New York, 1967.
166. Toma, T., Nakamura, Y., and Shimoji, M., *Philos. Mag.* [8] **33**, 181 (1976).
167. von Dreele, R. B., Glaunsinger, W. S., Bowman, A. L., and Yarnell, J. L., *J. Phys. Chem.* **79**, 2992 (1975).
168. Vos, K. D., and Dye, J. L., *J. Chem. Phys.* **38**, 2033 (1963).
169. Warren, W. W., Jr., *Phys. Rev.* **3**, 3708 (1971).
170. Weber, S., Rice, S. A., and Jortner, J., *J. Chem. Phys.* **42**, 1907 (1965).
171. Webster, B., *J. Phys. Chem.* **84**, 1070 (1980).
172. Weyl, W., *Ann. Phys. (Leipzig)* **121**, 601 (1864).

METAL BORATES

J. B. FARMER

Borax Research Limited, Chessington, Surrey, England

I. Introduction 187
II. Nomenclature 188
III. Structural Characteristics of Metal Borates 188
   A. Hydrated Metal Borates 189
   B. Anhydrous Metal Borates 195
   C. Boron–Oxygen Bond Lengths 198
   D. Structure Elucidation by Spectroscopic Techniques 199
IV. Aqueous Solutions of Metal Borates 200
   A. Boron Species in Aqueous Solution 201
   B. Soluble Metal Borate Complexes 207
   C. Heterogeneous Systems 208
V. Preparation and Properties of Metal Borates 211
   A. Alkali Metal, Ammonium, and Silver Borates 211
   B. Group II Metal Borates 216
   C. Group III Metal Borate Hydrates 222
   D. Group IV Metal Borate Hydrates 223
   E. Transition-Metal Borates 223
VI. Concluding Remarks 225
   References 225

I. Introduction

The chemistry of metal borates is surprisingly intricate. Early work was essentially of a synthetic nature, and it was the development of spectroscopic and X-ray crystallographic techniques that demonstrated the complexity of boron systems. During the past decade, crystal structures of metal borates have been determined at a rapid rate, and several novel types of borate anions have come to light. Concurrently, great advances in the understanding of the nature of borate ions in solution have been made through use of Raman, infrared (IR), and nuclear magnetic resonance (NMR) spectroscopic techniques.

This review is concerned with the hydrated metal borates, their interactions in solution, and their synthesis. Section III outlines the known structures of the hydrates and includes details of a few anhydrous borates to show the relationship of their basic structural units to those of the partially and completely hydrated forms. Section IV reviews the work on polyborate ions in solution and the effect that cations have on their equilibria. Section V is a brief account of the properties of some metal borates. Literature up to July 1979 is included.

Because of the great volume of literature on boron–oxygen compounds, it has been necessary to be highly selective in the choice of material for inclusion. Thus the properties of boric acid, and formation of sulfatoborates, perborates, arsenatoborates, metal borate–organic complexes, glasses, and melts are omitted. The chemistry of the lanthanide and actinide borates has not been considered; although these borates represent a current active field of research, an account of their properties would necessitate a large extension of the text. Data on all these aspects of borate chemistry are reviewed in Mellor's "Comprehensive Treatise of Inorganic and Theoretical Chemistry" (*60, 307, 392*).

II. Nomenclature

For many years, confusion has arisen over the ambiguous naming of metal borates. Typically, borates of the type $M_2B_4O_7$ or $M_2O \cdot 2B_2O_3$ have been referred to as both tetraborates and diborates. With more structural data available, the use of such names is obviously misleading. The silver borate $Ag_2O \cdot 4B_2O_3$ has been labeled a tetraborate but its structure shows that it is built up of triborate (a B_3O_3 skeleton) and pentaborate (a B_5O_6 skeleton) units. In this article, formulas are written either as the empirical ratio of oxides or as the structural formulas where appropriate. Thus borax, $Na_2B_4O_7 \cdot 10H_2O$, appears either as $Na_2O \cdot 2B_2O_3 \cdot 10H_2O$ (or in its abbreviated form $1:2:10$) or $Na_2[B_4O_5(OH)_4] \cdot 8H_2O$.

III. Structural Characteristics of Metal Borates

The ability of boron to coordinate to three and four oxygen atoms enables a wide range of theoretical structural entities to be formulated. An interesting history of early postulates in which only three-coordinate boron was considered has been outlined by Bokii and Kravchenko (*56*). Several authoritative reviews of the structural chemistry of borates have been published (*56, 77, 173, 182, 381, 408, 429*), but the most recent and reliable is that by Christ and Clark (*80*). In their paper they have considered structures determined from X-ray or neutron diffraction studies of single crystals, rather than include interpretations of spectroscopic or dehydration characteristics of borates. They have also amended the rules affecting formation of hydrated polyborate ions originally proposed by Christ in 1960 (77) and classified the borate structures using a shorthand notation describing the size of a boron–oxygen ring and the number of trigonal to tetrahedrally coordinated boron atoms; the space available does not permit an adequate description of this system. The arrangement of neighboring ions in anhydrous borates and glasses has been discussed in terms of the "structon theory" (181, 182).

The most common units found in borate structures are shown in Fig. 1. Many of the known borate structures can be rationalized by assigning appropriate charges, protons, or hydroxyl groups (which change the coordination of boron from trigonal to tetrahedral) to these basic units. Thus the BO_3 unit can become BO_3^{3-}, $B(OH)_3$, or $B(OH)_4^-$, respectively. Polymerization by elimination of one water molecule between two hydrated units results in chain formation, and further water elimination gives sheets or networks. These operations are illustrated in Fig. 2 for the triborate ring unit.

A. HYDRATED METAL BORATES

Table I lists many of the known hydrated borate structures, except for those containing silicon. The compounds are arranged according to structural similarity under the classification of Christ and Clark (80)

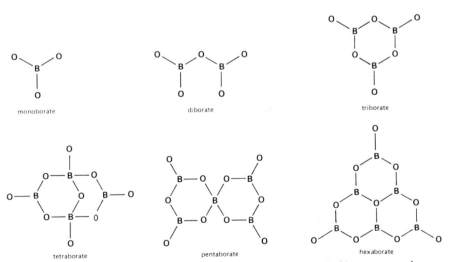

FIG. 1. Basic units of borate structures. All trigonally coordinated boron atoms shown can theoretically become tetrahedrally coordinated.

TABLE I

Structures of Metal Borates

Structural formula	References	Structural formula	References
Isolated B(OH)₃ and BO₃³⁻			
$B(OH)_3$	Sassolite (446)	$Al_3O_6(BO_3)$	(388)
$\alpha\text{-}Li_3BO_3$	(393)	$CaAlO(BO_3)$	(364)
$Mg_3(BO_3)_2$	(39), Kotoite (354)	$LiCdBO_3$	(205)
$Ca_3(BO_3)_2$	(418)	$LiMnBO_3$	(59)
$AlBO_3$	(420)	$FeCoO(BO_3)$	(421)
Tl_3BO_3	(332)	$Mg(Mg_{0.5}Ti_{0.5})O(BO_3)$	Warwickite (300)
$Zn_3(BO_3)_2$	(32, 138)	$Mg_3MnO_2(BO_3)$	Pinakiolite (300)
$Ni_3(BO_3)_2$	(326)	$Mg_3O(OH)_5(BO_3)nH_2O$	Wightmanite (300)
$FeBO_3$	(112)	$Al_6(OH)_3(BO_3)_5$	Jeremejevite (159)
$Mn_3(BO_3)_2$	(58), Jimboite (355)	$Mg_3(OH)F(BO_3)$	Fluoborite (404)
$Co_3(BO_3)_2$	(39)	$Be_2(OH)(BO_3)$	Hambergite (452)
$Mg_3F(BO_3)_3$	(66)	$Ca_{19}Mg_4(CO_3)_4Cl(OH)_7(BO_3)_7 \cdot H_2O$	Sakhaite (316)
BO₂⁻ chain			
$LiBO_2$	(449)	$Sr(BO_2)_2$	(46)
$Ca(BO_2)_2$	(275)		
Isolated B(OH)₄⁻ and BO₄⁵⁻			
$LiB(OH)_4$	(178)	$Sr[B(OH)_4]_2$	(Monoclinic) (228)
$NaB(OH)_4 \cdot 2H_2O$	(44)	$Sr[B(OH)_4]_2$	(Triclinic) (321)
$\alpha\text{-}CsB(OH)_4 \cdot 2H_2O$	(459)	$Ba[B(OH)_4]_2$	(227)
$\alpha\text{-}Ca[B(OH)_4]_2$	(Monoclinic) (320)	$Ba[B(OH)_4]_2 \cdot H_2O$	(253)
$\alpha'\text{-}Ca[B(OH)_4]_2$	(Orthorhombic) (457)	$Ba[B(OH)_4]_2 \cdot 2H_2O$	(322)
$\beta\text{-}Ca[B(OH)_4]_2$	Frovolite (122)	$Na_2Cl[B(OH)_4]$	Teepleite (128, 130)
$Ca[B(OH)_4]_2 \cdot 2H_2O$	(425, 426), Hexahydroborite (384)	$CuCl[B(OH)_4]$	Bandylite (89, 129, 130)
$Mn_3(OH)_4(PO_4)[B(OH)_4]$	Seamanite (301)	$MgCa_4(CO_3)_4[B(OH)_4]_2 \cdot 4H_2O$	Carboborite (437)
$Mg_3(OH)_2(SO_4)_2[B(OH)_4]_2$	Sulfoborite (190)	$MgAlBO_4$	Sinhalite (125)
$Ca_2(AsO_4)[B(OH)_4]$	Cahnite (334)		
Chain BO(OH)₂⁻			
$Ca[BO(OH)_2]_2$	Vimsite (383)		
Network BO₂⁻			
$\gamma\text{-}HBO_2$	Cubic metaboric acid-I (447)		

190

	Isolated B$_2$O$_4$(OH)$^{3-}$ and B$_2$O$_5^{4-}$		
Mg$_2$(OH)[B$_2$O$_4$(OH)]	Ascharite, szaibelyite (406)	Co$_2$B$_2$O$_5$	(40)
Mn$_2$(OH)[B$_2$O$_4$(OH)]	Sussexite (406)	Sr$_2$B$_2$O$_5$	(31)
Na$_2$B$_2$O$_5$	(219)	CaMgB$_2$O$_5$	Mg kurchatovite (438, 440)
Mg$_3$B$_2$O$_6$	(43), Saunite (405)	CaMnB$_2$O$_5$	Mn kurchatovite (439)
	Isolated B$_2$O(OH)$_6^{2-}$		
Ca[B$_2$O(OH)$_6$] · 2H$_2$O	Pentahydroborite (120, 122)	Mg[B$_2$O(OH)$_6$]	Pinnoite (235, 328)
	Isolated B$_3$O$_3$(OH)$_3$ and B$_3$O$_6^{3-}$		
B$_3$O$_3$(OH)$_3$	α-Metaboric acid III (330)	Na$_3$B$_3$O$_6$	(276)
Ba$_3$(B$_3$O$_6$)$_2$	(297)	K$_3$B$_3$O$_6$	(361)
	Isolated B$_3$O$_3$(OH)$_4^-$ and B$_3$O$_5$(OH)$_{12}^{2-}$		
Na[B$_3$O$_3$(OH)$_4$]	Ameghinite (104)	Na$_4$[B$_3$O$_5$(OH)$_{12}$]	(94)
	Chain B$_3$O$_4$(OH) and sheet B$_3$O$_5^-$		
B$_3$O$_4$(OH)(OH$_2$)	β-Metaboric acid-II (448)	CsB$_3$O$_5$	(242)
LiB$_3$O$_5$	(218)		
	Isolated B$_3$O$_3$(OH)$_5^{2-}$ and B$_3$O$_4$(OH)$_4^{3-}$		
Ca[B$_3$O$_3$(OH)$_5$] · H$_2$O	Meyerhofferite (79, 85)	CaMg[B$_3$O$_3$(OH)$_5$]$_2$ · 6H$_2$O	Inderborite (250)
Ca[B$_3$O$_4$(OH)$_3$] · 2H$_2$O	(85, 86)	Zn[B$_3$O$_4$(OH)$_3$] · H$_2$O	(324)
Ca[B$_3$O$_3$(OH)$_5$] · 4H$_2$O	Inyoite (83, 85)	K$_3$[B$_3$O$_4$(OH)$_4$] · 2H$_2$O	(319)
Mg[B$_3$O$_3$(OH)$_5$] · 5H$_2$O	Inderite (92, 343)	Rb$_5$[B$_3$O$_4$(OH)$_4$] · 2H$_2$O	(458)
	Kurnakovite (91, 339)	Ca$_2$Cl[B$_3$O$_4$(OH)$_4$]	Solongoite (441)
	Chain B$_3$O$_4$(OH)$_3^{3-}$		
Ca[B$_3$O$_4$(OH)$_3$] · H$_2$O	Colemanite (81, 85)	CaMg[B$_3$O$_4$(OH)$_3$]$_2$ · 3H$_2$O	Hydroboracite (344)
	Sheet B$_3$O$_5$(OH)$^{2-}$		
Ca[B$_3$O$_4$(OH)]	(85, 88), Fabianite (224)		
	Isolated B$_3$O$_3$(OH)$_6^{3-}$		
Ca$_3$[B$_3$O$_3$(OH)$_6$]$_2$ · 2H$_2$O	Nifontovite (121)		
	Network B$_3$O$_6^{3-}$		
Cu$_2$[B$_3$O$_6$]$_2$	(284)	Ca$_3$(B$_3$O$_6$)$_2$	(280)

(table continues)

TABLE I (Continued)

Structural formula	References	Structural formula	References
	Isolated B₃O₃(OH)₄ · OB(OH)₃³⁻		
Ca₂[B₃O₃(OH)₄ · OB(OH)₃]Cl · 7H₂O	Hydrochloroborite (80)		
	Isolated B₃O₃(OH)₅ · OB(OH)₃³⁻		
Ca₂[B₃O₃(OH)₅ · OB(OH)₃]	Uralborite (367)		
	Isolated B₂O₅(OH)₄²⁻ and B₂O₄(OH)₂²⁻		
Na₂[B₂O₅(OH)₄] · 8H₂O	Borax (267, 302)	Mn[B₂O₅(OH)₄] · 7H₂O	(42)
Na₂[B₂O₅(OH)₄] · 3H₂O	Tincalonite (143)	Rb₂Sr[B₂O₅(OH)₄]₂ · 8H₂O	(191)
K₂[B₂O₅(OH)₄] · 2H₂O	(277)	(NH₄)₂Ca[B₂O₅(OH)₄]₂ · 8H₂O	(41)
Mg[B₂O₅(OH)₄] · 7H₂O	Hungchaoite (424)	Ca₅Mn₂(OH)₄[B₂O₇(OH)₂]	Roweite (299)
	Chain B₂O₆(OH)²⁻		
Na₂[B₂O₆(OH)₂] · 3H₂O	Kernite (82, 90, 144)	Na₂[B₂O₆(OH)₂]	(289)
	Network B₂O₇²⁻		
Li₂B₄O₇	(236, 306)	CdB₄O₇	(183)
ZnB₄O₇	(283)		
	Isolated B₄O₆(OH)₆⁶⁻		
Ca·Mg(CO₃)₂[B₄O₆(OH)₆]	Borcarite (443)		
	Isolated B₅O₆(OH)₄⁻		
NaB₅O₆(OH)₄	(288)	K[B₅O₆(OH)₄] · 2H₂O	(29, 451), Santite (252)
Na[B₅O₆(OH)₄] · 3H₂O	Sborgite (294)	Cs[B₅O₆(OH)₄] · 2DMSO	(131)
β-NH₄[B₅O₆(OH)₄] · 2H₂O	(290)	(NH₄)₃[B₅O₆(OH)₃ · OB₅O₆(OH)₂ · OB₅O₆(OH)₃] · 4H₂O	Ammonioborite (293)
	Isolated B₅O₆(OH)₆³⁻		
NaCa[B₅O₆(OH)₆] · 5H₂O	Ulexite (142)		
	Chain B₅O₇(OH)₂⁻		
NH₄[B₅O₇(OH)₂] · H₂O	Larderellite (291)		

Formula	Mineral (Ref.)	Structure type	Synthetic/other (Ref.)	Mineral (Ref.)
$Na_2[B_5O_7(OH)_3] \cdot 2H_2O$	Ezcurrite (70)	Chain $B_5O_7(OH)_3^{3-}$		
$NaCa[B_5O_7(OH)_4] \cdot 3H_2O$	Probertite (347)	Chain $B_5O_7(OH)_4^{-}$		
$Na_2[B_5O_8(OH)] \cdot 2H_2O$	Nasinite (95)	Sheet $B_5O_8(OH)^{2-}$	$K_4[B_5O_8(OH)] \cdot 2H_2O$ (274)	
$Na_2[B_5O_8(OH) \cdot OB_3O_4(OH)] \cdot 2H_2O$	Biringuccite (93)			
$Ca[B_5O_8(OH) \cdot B(OH)_3] \cdot 3H_2O$	Gowerite (225)	Sheet $B_5O_8(OH)^{2-}$ + isolated $B(OH)_3$	$Sr_2[B_5O_8(OH)]_2B(OH)_3 \cdot H_2O$	Veatchite (87, 346)
$Na_3[B_5O_8(OH)_2] \cdot H_2O$	(287)	Sheet $B_5O_8(OH)_2^{3-}$	$Na_2Ca_3Cl(SO_4)_2[B_5O_8(OH)_2]$	Heidornite (80)
$HKMg_2[B_5O_5(OH)_3 \cdot OB(OH)_2]_2 \cdot 4H_2O$	Kaliborite (97)	Chain $B_5O_7(OH)_3 \cdot OB(OH)_2^{3-}$		
$Mg[B_6O_7(OH)_6] \cdot 2H_2O$	Aksaite (108, 167)	Isolated $B_6O_7(OH)_6^{2-}$	$Co[B_6O_7(OH)_6] \cdot 7H_2O$ (380)	
$Mg[B_6O_7(OH)_6] \cdot 3H_2O$	(107, 139)		$Ni[B_6O_7(OH)_6] \cdot 5H_2O$ (378)	
$Mg[B_6O_7(OH)_6] \cdot 4H_2O$	(106)		$Ni[B_6O_7(OH)_6] \cdot 7H_2O$ (380)	
$Mg[B_6O_7(OH)_6]_4 \cdot 4.5H_2O$	Macallisterite (105)		$K_2Co[B_6O_7(OH)_6]_{12} \cdot 4H_2O$ (379)	
$Na_6Mg[B_6O_7(OH)_6]_4 \cdot 10H_2O$	Rivadavite (109)			
$Na_2Mg[B_6O_6(OH)_4]_{12} \cdot 4H_2O$	Aristarainite (140)	Chain $B_6O_9(OH)_4^{-}$		
$Sr[B_6O_9(OH)_2] \cdot 3H_2O$	Tunellite (84)	Sheet $B_6O_9(OH)_2^{2-}$		
$Sr[B_6O_9(OH) \cdot OB_2O(OH)_3]$	Strontioborite (67)	Sheet $B_6O_9(OH) \cdot OB_2O(OH)_3^{-}$		
$Ca[B_6O_9(OH) \cdot OB_2O(OH)_3]$	(455)		$SrCa[B_6O_9(OH)_2 \cdot OB_6O_9(OH) \cdot OB_2O(OH)_3] \cdot 5H_2O$	Strontioginorite (223)
$Ca_4Mg[B_6O_7(OH)_6 \cdot OAsO_3]_2 \cdot 14H_2O$	Teruggite (103)	Isolated $B_6O_7(OH)_6 \cdot OAsO_3^{5-}$		

193

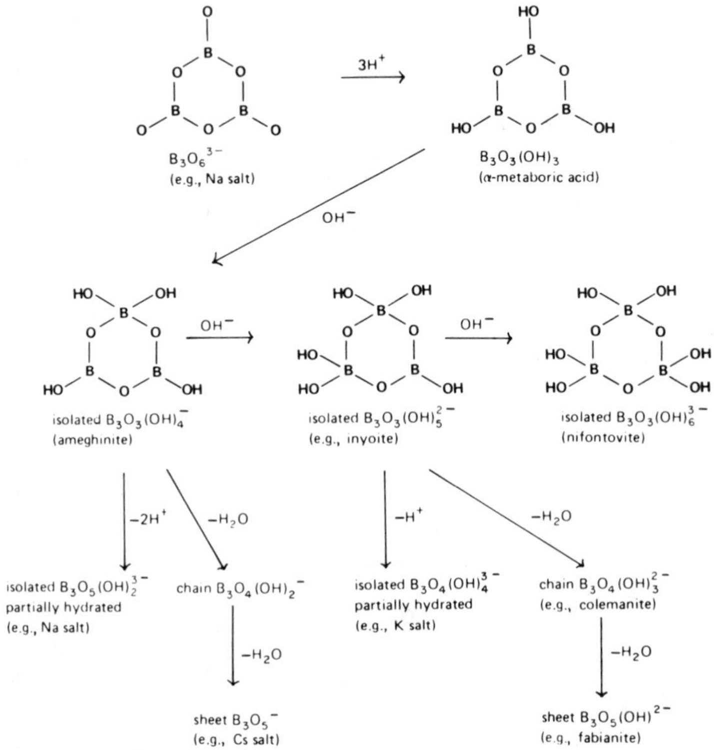

FIG. 2. Structural transformations of the triborate unit. (The use of protons and hydroxyl groups is schematic and does not represent chemical reaction.)

and include some anhydrous borates that are obvious extensions of the hydrated forms.

Of recent interest has been the determination of structures containing "partially hydrated" polyborate ions such as found in $Na_3B_5O_5(OH)_2$ (94), K_3 or $Rb_3[B_3O_4(OH)_4] \cdot 2H_2O$ (319, 458), and roweite ($Ca_2Mn_2(OH)_4[B_4O_7(OH)_2]$) (299), in which oxygen atoms arranged tetrahedrally around boron atoms are protonated and the triangular oxygens are not.

The most common ion of the triborate unit is $B_3O_3(OH)_5^{2-}$, and it is usually found associated with calcium and magnesium. Triborates of calcium $Ca_2B_6O_{11} \cdot xH_2O$ have been reviewed (85). Modification of triborate groups with attachment of a side tetrahedron leads to B_4 species. Uralborite, $Ca_2[B_3O_3(OH)_5 \cdot OB(OH)_3]$, is an example (367) and should be written as shown rather than with the ambiguous shorter formula of $Ca_2[B_4O_4(OH)_8]$. Christ and Clark (80) have doubted the proposed structure of $Mg[B_3O_3(OH)_4 \cdot OB(OH)_2] \cdot 6H_2O$ for $Mg[B_4O_4(OH)_6] \cdot 6H_2O$ (6) on the grounds that particular bond distances are too short to permit such a formulation.

The tetraborate unit $B_4O_5(OH)_4^{2-}$ was first found in borax (302) and is also present in the "octaborates" $M_2^IM^{II}B_8O_{14} \cdot 12H_2O$ (149). The polymerization of these ions to give, e.g., kernite, is not a simple elimination of water but involves the breaking of a B–O bond. The resulting polymeric ion consists of six-membered rings each possessing one trigonal and two tetrahedral borons, and the rings are joined by sharing tetrahedral boron atoms (82, 90, 144). The structure of the calcium borate $Ca_2[B_8O_{13}(OH)_2]$ has been reported as containing tetraborate units (442). Most tetraborates have their structure based on a double ring system, but borcarite $Ca_4Mg(CO_3)_2[B_4O_6(OH)_6]$ provides the one example so far of an eight-membered ring B_4O_4 (443) in which all the boron atoms are tetrahedrally coordinated in the partially hydrated unit.

The hexaborate anions have a unique structure, in that the central oxygen atom is coordinated to three borons. This arrangement was first found in the structure of tunellite (84).

A basic unit that has not been mentioned here is the nonaborate. This has mutually linked six-membered rings (Fig. 3) and has been reported in the form of a modified sheet in preobrazheskite $HMg_3[B_9O_{12}(OH)_4 \cdot (O_2B(OH)_2)_2]$ (348).

B. Anhydrous Metal Borates

The structural chemistry of anhydrous metal borates is also complex; but, unlike the hydrated analogs, polyborate ions can be composed of dissimilar basic units.

The isolated, planar BO_3^{3-} ion (orthoborate) occurs in many of these

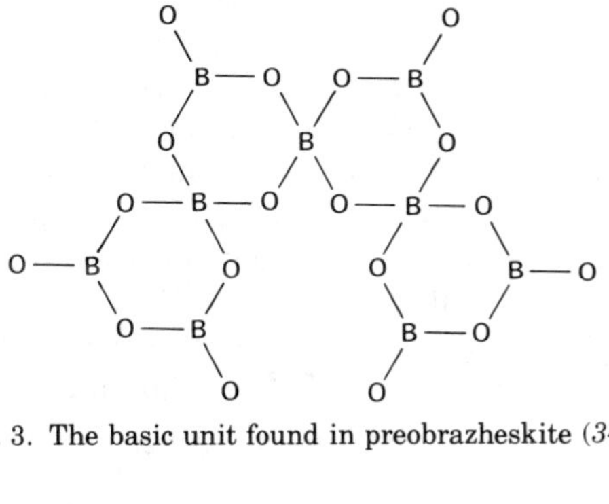

Fig. 3. The basic unit found in preobrazheskite (348).

compounds. In addition to the examples given in Table I, many compounds containing rare-earth elements have been examined. X-Ray crystallographic data on many of these borates have related $M^{III}BO_3$ to the calcite, aragonite, and vaterite polymorphs of $CaCO_3$. The low- and high-temperature forms of lanthanum borate $LaBO_3$ (5, 47) and lutetium borate $LuBO_3$ (22) possess these isolated BO_3^{3-} anions, as do the more recently determined structures of

$Li_3M_2(BO_3)_3$ where   M = Eu (18), Nd (10), and Pr (13),

$Li_6M(BO_3)_3$ where   M = Yb (11) and Ho (19),

$M_3Nd(BO_3)_4$ where   M = Ga (34) and Al (179), and

$Sr_3M_2(BO_3)_4$ where   M = Pr (325), La (14), Nd (7), and Er (9)

The so-called metaborates of empirical formula BO_2^- can take several forms. The infinite linear $(BO_2)_n^{n-}$ ion is found in the low-temperature form of $LiBO_2$ (449), $Ca(BO_2)_2$ (275), and $Sr(BO_2)_2$ (46). At high temperatures and pressures, the lithium compound possesses a sphalerite-like structure with boron in four-coordination (278). The triborate isolated $B_3O_6^{3-}$ group is found in the barium (297), sodium (276), and potassium (361) compounds, and the framework $B_3O_6^{3-}$ group, in which all the boron atoms are tetrahedral, exists for a high-pressure calcium metaborate phase (280) and copper metaborate (284). Two other anhydrous calcium metaborates are known to give the series

CaB_2O_4-I   chain BO_2^- (275)

CaB_2O_4-II   chain $B_2O_4^{2-}$ (368, 450)

CaB_2O_4-III   network $B_6O_{12}^{6-}$ (279)

CaB_2O_4-IV   network $B_3O_6^{3-}$ (280)

The relation between these polymorphs has been summarized (280). Thallium metaborate (413) and the rare-earth metaborates $Sm_2(B_2O_4)_3$, $Gd_2(B_2O_4)_3$ (17), and EuB_2O_4 (269) have the $B_2O_4^{2-}$ chains found in CaB_2O_4-II. In addition to the isolated and framework $B_3O_6^{3-}$ groups, the chain polymer of $B_3O_6^{3-}$ (two tetrahedral and one trigonal boron atoms) is present in the lanthanum metaborate LaB_3O_6 (444). A new type of structure has been found for the metaborates $CoM(BO_2)_5$, where M = Sm (15), La (16), Nd (2), and Ho (12). In these compounds, the boron unit consists of infinite chains of $B_5O_{10}^{5-}$ with each composed of three BO_4 tetrahedra and two BO_3 triangles. Neodymium aluminum metaborate $NdAl_{2.07}[B_4O_{10}]O_{0.6}$ (335) has endless chains of $B_2O_5^{4-}$. Zinc metaborate $Zn_4O(BO_2)_6$ has an unusual structure with all boron atoms tetrahedrally coordinated and situated at the vertices of a Federow's truncated octahedron to form a framework resembling the anion of sodalite (57).

The network $B_4O_7^{2-}$ tetraborate ion consisting of two tetrahedral and two trigonal boron atoms is found in the Li ($236, 306$), Zn (283), and Cd (183) compounds. Other "tetraborates" do not necessarily contain the tetraborate group but are composed of pentaborate and triborate units giving the empirical unit $B_8O_{14}^{n-}$. The sodium borate $Na_2O \cdot 2B_2O_3$ has a three-dimensional network constructed from this combination of units and has one nonbridging oxygen atom (240). Its potassium analog is formed from tetraborate and triborate groups linked by planar BO_3 triangles (239). The borates $Ag_2O \cdot 4B_2O_3$ (233), $BaO \cdot 2B_2O_3$ (45), and $BaO \cdot 4B_2O_3$ (245) possess linked pentaborate and triborate groups, while the strontium and lead ($46, 232, 329$) $MO \cdot 2B_2O_3$ compounds have all the boron atoms tetrahedrally coordinated and an oxygen common to three tetrahedra. The structure of calcium tetraborate has recently been reported as containing a network of $B_8O_{14}^{4-}$ ions (454).

Compounds with empirical formulas that suggest the presence of triborate or hexaborate groups are found to be composites. Thus, the calcium borate $Ca_2B_6O_{11}$ has a symmetric pentaborate unit connected to a framework by additional boron tetrahedra (453). The low-temperature form of sodium borate, α-$Na_2O \cdot 3B_2O_3$, consists of two separate interpenetrating frameworks of alternating pentaborate and tetraborate groups as shown in Fig. 4 (241), while the high-temperature modification is formed from linked pentaborate, triborate, and BO_4 groups (238). Cesium enneaborate $Cs_2O \cdot 9B_2O_3$ (244) has two independent networks of linked but different triborate units.

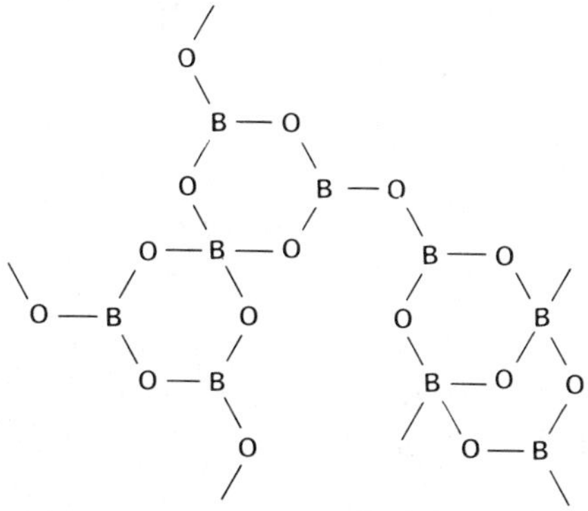

FIG. 4. Schematic representation of the borate ion in the framework structure of α-$Na_2O \cdot 3B_2O_3$ (241).

Potassium pentaborate exists in three forms, of which two have been studied crystallographically. The high-temperature α (237) and low-temperature β (234) forms consist of two separate three-dimensional interlocking networks of pentaborate groups $B_5O_8^-$ differing only in close packing of the matrices. The mineral hilgardite $Ca_2B_5O_9Cl \cdot H_2O$, despite being a hydrate, similarly has a structure based on a network of pentaborate groups with six extra-annular oxygen atoms of $B_5O_9^{3-}$ shared with other groups (141). Another type of pentaborate ion is found in the bismuth borate $Bi_3O(B_5O_{11})$, but in this case, the $B_5O_{11}^{7-}$ ion is isolated (419).

The largest isolated borate anion found so far in an anhydrous borate belongs to the lead borate $6PbO \cdot 5B_2O_3$ (246). The ion $B_{10}O_{21}^{12-}$ is built up from two tetraborate groups linked by two BO_3 triangles. Pentapotassium enneakaidecaborate $5K_2O \cdot 19B_2O_3$ (243) possesses four types of units with interconnected pentaborate, triborate, BO_4 tetrahedra, and BO_3 triangles.

Anhydrous aluminum borates have rings incorporating aluminum atoms. The aluminate $2SrO \cdot Al_2O_3 \cdot B_2O_3$ consists of a six-membered ring of two AlO_4 tetrahedra and one BO_3 triangle (305). $3Li_2O \cdot Al_2O_3 \cdot 2B_2O_3$ has infinite chains of $Al_2(BO_3)_4^{6-}$ in which the rings are formed by two AlO_4 tetrahedra and two BO_3 triangles, interlinked in one direction by two other BO_3 triangles (8).

C. Boron–Oxygen Bond Lengths

Wells (433) has quoted ranges for the boron–oxygen bond lengths of trigonally and tetrahedrally coordinated boron as

$$BO_3: \quad 1.28 \rightarrow 1.43 \text{ Å, average } 1.365 \text{ Å};$$

$$BO_4: \quad 1.43 \rightarrow 1.55 \text{ Å, average } 1.475 \text{ Å}.$$

The isolated ion $B_3O_6^{3-}$ found as sodium (276) and potassium (361) compounds has an exocyclic B—O length of 1.28 and 1.33 Å respectively and an endocyclic B—O length of 1.433 and 1.398 Å respectively. The short BO_2—O^- bond is characteristic for nonbridging atoms attached to rings in anhydrous borates; its value in $Na_2O \cdot 2B_2O_3$ is 1.295 Å (240). In isolated BO_3^{3-} ions, the B–O distance is typically 1.37 Å, but a notable exception is found in $Pr_2Sr_3(BO_3)_4$ (325) where three of the triangles have an average B—O length of 1.46 Å and the fourth is as low as 1.23 Å; this has not been explained.

Boron–oxygen bond lengths in tetrahedrally coordinated boron vary from about 1.4 to 1.55 Å; synthetic $Ca[B_3O_3(OH)_5] \cdot 2H_2O$ possesses these extremes (85). Long B–O bonds are also found if an oxygen is coordinated to three boron atoms, as for example in aristarainite $Na_2Mg[B_6O_8(OH)_4]_2 \cdot 4H_2O$, where the distance is 1.517 Å (140).

Related boron structures exhibit similar bond distances. The structure of sodium perborate has been recently refined (74). It consists of the cyclic anion $[(HO)_2B \cdot (O_2)_2 \cdot B(OH)_2]^{2-}$ with two peroxy bridges. The terminal groups have a B—OH length of 1.442 Å and the ring B—O distance is 1.495 Å. The boron atom in boron triethanolamine $B(OC_2H_4)_3N$ is tetrahedrally surrounded by oxygen and nitrogen atoms, with a B—O length of 1.431 Å (127). Potassium boromalate (282) has a tetrahedral arrangement about the boron atom with the oxygens directly bonded to malic acid; the B–O bond lengths average 1.438 and 1.508 Å.

Hückel theory has been used to correlate bond lengths in planar BO_3 triangles with π-bond order (98), but with the marked deviation from predicted values for $B_3O_6^{3-}$, the treatment is of limited applicability. Other studies have similarly related π-electron charge density with the geometry of borates ($360, 395$). The results indicate that the bond order in boron tetrahedra is slightly lower than unity, and slightly greater than unity in BO_3 triangles; the π-electron charge on boron does not vary significantly among borate structures.

D. STRUCTURE ELUCIDATION BY SPECTROSCOPIC TECHNIQUES

Numerous NMR ($61, 194–196, 331, 445$), Raman ($63–65, 222, 434$), and IR (252) studies of borate glasses and melts have been made and related to the structural units found in crystalline borates, and to the species such as $B_3O_{4.5}$, $B_8O_{13}^{2-}$, and $B_4O_7^{2-}$ postulated by Krogh-Moe (231) in his investigations into the melting-point depression of sodium borate melts. Borate glass structures have been recently reviewed ($164, 338$) and will not be discussed here.

In the examination of crystalline borates, caution has been exercised over the structural interpretations of spectroscopic studies. Certain features pertaining to the boron coordination and water/hydroxyl entities can be deduced, but absolute configurations of molecules require more exacting techniques.

The sodium tetraborates, borax (the decahydrate) and tincalconite (the pentahydrate), have been studied by proton ($33, 281$), boron-11 ($101, 111$), and sodium-23 ($102, 111$) magnetic resonance, and the data is compatible with X-ray studies. Similar agreement with crystal structures has been found with the ^{11}B NMR of the hexaborate ion in tunellite (100) and pentaborates ($26, 254$). Wegener et al. ($192, 430$)

measured the proton magnetic resonance of several potassium borates and have differentiated between water of crystallization and hydroxide groups. The BO_2^- chain in crystalline calcium metaborate $Ca(BO_2)_2$ has been observed by ^{11}B NMR (229), as has the BO_3^{3-} triangle in $Sr_3(BO_3)_2$ (327). The relative merits of NMR techniques for borate structure determination have been reviewed (62, 230).

Infrared spectroscopy is the favored technique for characterizing borates, and several compilations of data have been made (171, 333, 417, 423, 431, 432). The intense absorption at about 970 cm^{-1} is characteristic of tetrahedrally coordinated boron, as measured for the magnesium borates $MgO \cdot B_2O_3 \cdot nH_2O$ (153), and bands at 700 to 900 and 1200 to 1500 cm^{-1} can correspond to triangular boron coordination. Absorptions in the region 400 to 700 cm^{-1} have been attributed to chain structures (423).

With the many crystal structures now known, more reliance can be put on speculative interpretations of IR spectra. The "octoborates" $M_2^IO \cdot M^{II}O \cdot 4B_2O_3 \cdot 12H_2O$ have spectra identical to borax and accordingly are assigned the anion structure $B_4O_5(OH)_4^{2-}$ (155, 157, 158). In the lead borate system, species such as $B_5O_9(OH)^{4-}$, $B_4O_5(OH)_4^{2-}$, $B_5O_7(OH)_3^{2-}$, and $B_3O_3(OH)_4^-$ are claimed (385, 386).

Raman spectroscopy has seldom been used for elucidating structures of crystalline solids. However, with IR, it has proved valuable for observing isostructural rare-earth borates (110) and for measurement of phonon coupling in the zinc borate $Zn_4O(BO_2)_6$ (304).

IV. Aqueous Solutions of Metal Borates

The mode of dissolution of metal borates in aqueous solution is complex. Hydrolysis of the borate anion can result in completely different boron species that are stable only under particular conditions of pH, temperature, and concentration.

The parent acid, H_3BO_3, functions as a weak acid in aqueous solution, possessing an ionization constant of about 6×10^{-10} at 25°C in dilute solution. As its concentration increases, the ionization constant increases markedly. Kolthoff (220) investigated this effect by electrical conductivity and emf measurements, obtaining values of 4.6×10^{-10} and 408×10^{-10} for boric acid concentrations of 0.1 M and 0.75 M respectively at 18°C. These results were attributed to the formation of tetraboric acid.

Other studies into the apparent ionization constant of boric acid and interpretation of data are given by Sprague (392). Of greater relevance to metal borate synthesis and stability is the variation in polyborate ions in solution with pH and concentration.

A. BORON SPECIES IN AQUEOUS SOLUTION

1. Monomeric Boric Acid and Borate Ion

The presence of monomeric H_3BO_3 and $B(OH)_4^-$ in aqueous solutions has been confirmed by spectroscopic techniques. Infrared (126, 415–417) and Raman (176, 247) spectra of boric acid solutions show similar absorptions to crystalline H_3BO_3 (176), for which a planar BO_3 arrangement has been found (446). The monoborate ion $B(OH)_4^-$ has similarly been identified by vibrational spectroscopy (119, 161, 176); its expected tetrahedral structure has been confirmed by comparison of its spectra with that of teepleite $NaB(OH)_4 \cdot NaCl$ (213, 342) and bandylite $Cu[B(OH)_4]_2 \cdot CuCl$ (342), which are known to contain monomeric tetrahedral BO_4 units.

2. Polyborate Ions

The formation of polymeric boron ions in solution is well established. The significant increase in solubility in terms of B_2O_3 in aqueous mixtures of borax and boric acid relative to the individual components is explained by such polyborate formation. The comprehensive reviews by Nies (307) and Sprague (392) reveal that, despite the many investigations over the last 50 years, the identity of the borate ions involved has still not been completely resolved.

a. *Potentiometric investigations.* Of the techniques available for studying polyborate equilibria, potentiometry has received greatest attention. However, interpretation of data from the sodium hydroxide–boric acid system has led to conflicting postulated species, e.g.,

$B_5O_8^-$ and $B_4O_7^{2-}$ (389, 390)

BO_2^-, $HB_2O_4^-$, $B_4O_7^{2-}$, and $HB_4O_7^-$ (255)

$B_5O_8^-$, BO_2^-, $HB_2O_4^-$, $B_4O_7^{2-}$, and $HB_4O_7^-$ (256)

$B_5O_8^-$, HBO_2, and HB_2O_4 (73)

HBO_2, BO_2^-, $B_5O_8^-$, and $B_4O_7^{2-}$ (268)

A phenomenon resulting from polyborate formation is that at a certain ratio M_2O/B_2O_3 the pH of the metal borate solution is unaffected by dilution. For M = Na, this mole ratio is 0.412 and the solution pH is 8.91 (286). This composition was called the isohydric point and the related data used to fit a model based on the presence of $HB_5O_9^{2-}$ species (71); in this early paper (1949) it was claimed that tetracoordinated boron atoms cannot exist in aqueous solution. More recent studies (54, 55, 72) have shown that the isohydric reaction involves hydrolysis of $HB_5O_9^{2-}$:

$$HB_5O_9^{2-} + 6H_2O \rightleftharpoons 3H_3BO_3 + 2BO(OH)_2^- \qquad pK = 4.53$$

The extensive investigations by Ingri et al. (185–189) provided the first reliable quantitative description of borate equilibrium, and several of the postulated species have subsequently been verified by vibrational spectroscopy. Ingri proposed that the polyborates in solution would possess structural units similar to those in crystalline metal borates, i.e., $B(OH)_4^-$, $B_3O_3(OH)_4^-$, $B_3O_3(OH)_5^{2-}$, $B_4O_5(OH)_3^-$, $B_4O_5(OH)_4^{2-}$, and $B_5O_6(OH)_4^-$ (see Section III for details of structures). At this point it should be noted that potentiometry cannot afford information on "degrees of hydration" and that previously described species such as $B_5O_8^-$ can be considered as identical to $B_5O_6(OH)_4^-$:

$$B_5O_8^- + 2H_2O \rightleftharpoons B_5O_6(OH)_4^-$$

$$HBO_2(aq) + H_2O \rightleftharpoons H_3BO_3$$

$$B_4O_7^{2-} + 2H_2O \rightleftharpoons B_4O_5(OH)_4^{2-}$$

Ingris' potentiometric titrations at 25°C in 3 M $NaClO_4$ indicated that at boron concentrations less than 0.025 M only H_3BO_3 and $B(OH)_4^-$ are present; at higher concentrations, $B_3O_3(OH)_4^-$ and $B_3O_3(OH)_5^{2-}$ (189), and at the upper concentration range of 0.4 to 0.6 M boron, either $B_4O_5(OH)_3^-$ or $B_5O_6(OH)_4^-$ are the predominant polymeric species (185). Figure 5 illustrates the distribution of boron species over a range of pH at 25°C.

The presence of certain neutral salts increases the acidity of boric acid as determined potentiometrically (187, 358, 375, 391, 422), and the effect has been attributed to the hydration energies of ions (375, 391, 422). In 3 M KBr solutions containing up to 0.4 M boron, H_3BO_3, $B(OH)_4^-$, $B_3O_3(OH)_4^-$, and $B_4O_5(OH)_4^{2-}$ were consistent with experimental results, but in the presence of $NaClO_4$, NaBr, or LiBr, the inclusion of $B_3O_3(OH)_5^{2-}$ improved the solution analysis (187).

Later investigations in alkali metal chloride and sulfate media suggested that $B_3O_3(OH)_4^-$ and $B_4O_5(OH)_4^{2-}$ were the major polymeric species, with $B_5O_6(OH)_4^-$ and $B_3O_3(OH)_5^{2-}$ being present in small concentrations (391). This same study included systems at temperatures up to

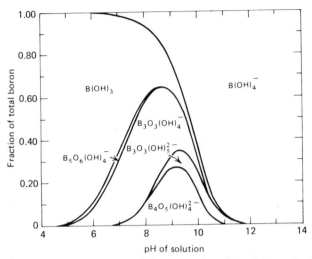

FIG. 5. Variation in the distribution of boron species with solution pH at 25°C. Total boron concentration is 0.4 M (188).

90°C (Table II). Other work has gone beyond this temperature range (50–290°C, Ref. 295), and new species are postulated: in 1 M KCl solutions, boron at high concentrations can exist as $B_2(OH)_7^-$ [equivalent to $B_2O(OH)_5^-$] and $B_3(OH)_{10}^-$ [$B_3O_3(OH)_4^-$] with lesser amounts of either $B_4(OH)_{14}^{2-}$ [$B_4O_5(OH)_4^{2-}$] or $B_5(OH)_{18}^{3-}$ [$B_5O_6(OH)_6^{3-}$].

b. Other studies. Recent investigations by a variety of methods have not yet clarified the situation. Solution pH techniques in concentrated solution have indicated that boron is present mainly as $B_5O_6(OH)_4^-$ (303), as $B_3O_3(OH)_4^-$ (23), or as $B_3O_3(OH)_5^{2-}$ and $B_3O_3(OH)_4^-$ (357); no evidence for the formation of monoborate ions possessing a greater charge than $B(OH)_4^-$ was found by using a lead amalgam electrode (362). Ion exchange analysis suggests $B_3O_3(OH)_4^-$, $B_4O_5(OH)_4^{2-}$, $B_4O_5(OH)_5^-$, and $B_5O_6(OH)_4^{2-}$ (30). Infrared and ion exchange studies indicate the presence of $B_5O_6(OH)_4^-$ at pH 7.5 to 8.5, $B_4O_5(OH)_4^{2-}$ at pH 9.2, and $B(OH)_4^-$ at pH > 10 (126). Cryoscopic measurements based on the depression of freezing point of sodium sulfate decahydrate yield two sets of equilibria involving either $B_4O_5(OH)_4^{2-}$ or a combination of trimeric and pentameric anions (193).

c. Spectroscopic techniques. Nuclear magnetic resonance studies of ^{11}B in solutions of boric acid and metaborates $MB(OH)_4$ are consistent with the presence of H_3BO_3 and tetrahedral $B(OH)_4^-$ in dilute solution (99, 160, 175, 180, 315, 387, 428). In aqueous pentaborate solutions two

TABLE II
FORMATION CONSTANTS OF BORATE IONS AND pH FOR THEIR MAXIMUM CONCENTRATIONS[a]

Reaction	Salt media	At 25°C			At 90°C		
		Formation constant	Optimum pH	Maximum % concentration	Formation constant	Optimum pH	Maximum % concentration
$H_3BO_3 + H_2O \rightleftharpoons B(OH)_4^- + H^+$	3 M KCl	1.76×10^{-9}			2.93×10^{-9}		
	3 M NaCl	2.73×10^{-9}			5.53×10^{-9}		
	5 M NaCl	3.83×10^{-9}			16.5×10^{-9}		
$3H_3BO_3 \rightleftharpoons B_3O_3(OH)_4^- + 2H_2O + H^+$	3 M NaCl	1.81×10^{-7}	8.1	46	9.94×10^{-8}	7.9	28
	3 M NaCl	2.78×10^{-7}	7.8	43	2.38×10^{-7}	7.8	42
	5 M NaCl	6.98×10^{-7}	7.6	44	5.15×10^{-7}	7.5	42
$3H_3BO_3 \rightleftharpoons B_3O_3(OH)_5^{2-} + H_2O + 2H^+$	3 M KCl	4.60×10^{-18}	9.8	2	8.13×10^{-18}	9.7	2
	3 M NaCl	8.63×10^{-17}	9.7	11	1.78×10^{-17}	9.5	1
	5 M NaCl	9.19×10^{-16}	9.6	27	4.35×10^{-17}	8.4	2
$4H_3BO_3 \rightleftharpoons B_4O_5(OH)_4^{2-} + 3H_2O + 2H^+$	3 M KCl	1.73×10^{-15}	9.3	26	2.87×10^{-15}	8.8	30
	3 M NaCl	5.85×10^{-15}	8.9	27	1.93×10^{-15}	8.7	9
	5 M NaCl	2.24×10^{-14}	8.7	24	4.78×10^{-15}	8.2	4
$5H_3BO_3 \rightleftharpoons B_5O_6(OH)_4^- + 5H_2O + H^+$	3 M KCl	1.75×10^{-7}	6.9	3	6.25×10^{-7}	7.1	12
	3 M NaCl	1.83×10^{-7}	6.7	2	9.58×10^{-8}	6.6	1
	5 M NaCl	3.65×10^{-7}	6.3	2	3.94×10^{-7}	6.4	3

[a] Constants determined in 0.4 M boron solutions; concentrations expressed as percent of total boron. Data from Spessard (391).

resonances were observed (298) and the variation in their relative intensities with concentration were interpreted as indicating:

$$OH^- + H_3BO_3 \overset{fast}{\rightleftharpoons} B(OH)_4^-$$

$$2H_3BO_3 + B(OH)_4^- \overset{fast}{\rightleftharpoons} B_3O_3(OH)_4^- + 3H_2O$$

$$4H_3BO_3 + B(OH)_4^- \overset{slow}{\rightleftharpoons} B_5O_6(OH)_4^- + 6H_2O$$

The two resonances were believed to arise from $B_5O_6(OH)_4^-$ and the rapid exchange among H_3BO_3, $B(OH)_4^-$, and $B_3O_3(OH)_4^-$. At 80 MHz a third signal appeared at high boron concentrations, attributed to another polyborate species (387). Aqueous solutions of sodium tetraborate display a single resonance at 14 MHz (298) and 80 MHz (387). Other qualitative studies have also indicated the presence of polyborate ions in concentrated solution (99, 175).

Vibrational spectroscopy has proved to be the most useful technique yet for identifying which boron species are present in solution. Close agreement between solution spectra and those of crystalline borates of known structure have confirmed the presence of hitherto postulated polyborate ions. Details of the IR spectra of the $Na_2O-B_2O_3-H_2O$ system at 26°C with absorptions assigned to polyborate species are shown in Table III (416, 417). In a more recent study (126), the major ions in 0.5 M boron solution were identified as $B_5O_6(OH)_4^-$, $B_4O_5(OH)_4^{2-}$, and $B(OH)_4^-$.

Recent laser Raman investigations (271, 285, 318) have led to simi-

TABLE III

INFRARED SPECTRA OF SODIUM BORATE SOLUTIONS[a]

Mole ratio $B_2O_3:Na_2O$	pH	Absorption bands (cm^{-1})				Frequency assignment
1	11	940	1125	1230		1430	$B(OH)_4^-$
1.5	11	950	1120	1200	1330	1430	$B_3O_3(OH)_5^{2-}$
1.7	10.5	960		1220	1330	1430	$B_3O_3(OH)_5^{2-}$
2	9		1010	1200	1330	1430	$B_4O_5(OH)_4^{2-}$
2.3	8.5		1020	1150	1335	1420 1450	$B_4O_5(OH)_4^{2-}$
2.9	8		1020	1200	1330	1450	$B_3O_3(OH)_4^-$
3.5	7.5		1020	1185	1330	1450	$B_5O_6(OH)_4^-$
4.6	6.5		1020	1100	1340 1180	1440	$B_5O_6(OH)_4^-$
5	6.5		1115	1210	1370	1450	$B_5O_6(OH)_4^-$

[a] From Valyashko and Vlasova (417).

lar conclusions on the state of boron in solution at various pH. No evidence for $B_3O_3(OH)_5^{2-}$ was found. The frequencies assigned to the borates are given in Table IV. The relative intensities of absorptions measured by Maya (285) were used to calculate equilibrium constants; the derived values compared favorably with those determined by Ingri by potentiometric titration (187):

	Maya (285)	Ingri (187)
$3H_3BO_3 \rightleftharpoons B_3O_3(OH)_4^- + H^+ + 2H_2O$	pK = 6.6	pK = 6.7
$5H_3BO_3 \rightleftharpoons B_5O_6(OH)_4^- + H^+ + 5H_2O$	pK = 6.9	pK = 6.6
$4H_3BO_3 \rightleftharpoons B_4O_5(OH)_4^{2-} + 2H^+ + 3H_2O$	pK = 15.6	pK = 14.7

3. Rate of Formation of Polyborate Ions

"Temperature-jump" techniques have been employed to investigate the mode of formation of polyborates (27). The reaction

$$H_3BO_3 + OH^- \rightleftharpoons B(OH)_4^-$$

was too fast to measure (predicted rate 10^{10} mol^{-1} sec^{-1}), but overall rate constants for the formation of $B_3O_3(OH)_4^-$ and $B_5O_6(OH)_4^-$ were obtainable:

$$2H_3BO_3 + B(OH)_4^- \rightleftharpoons B_3O_3(OH)_4^- + 3H_2O \quad k = 2.8 \times 10^3 \text{ mol}^{-2} \text{ sec}^{-1} \text{ at pH 7.4}$$

$$2H_3BO_3 + B_3O_3(OH)_4^- \rightleftharpoons B_5O_6(OH)_4^- + 3H_2O \quad k = 2 \times 10^2 \text{ mol}^{-2} \text{ sec}^{-1} \text{ at pH 7.4}$$

The former value agrees well with the value of 3.5×10^3 found in pressure-jump measurements (317).

Slow changes in borate species have been observed by IR spectroscopy (415, 417). Absorptions by borate solutions prepared by the addi-

TABLE IV

ASSIGNED FREQUENCIES (cm^{-1}) OF RAMAN SPECTRA OF POLYBORATE IONS IN AQUEOUS SOLUTION[a]

$B_3O_3(OH)_4^-$		$B_4O_5(OH)_4^{2-}$		$B_5O_6(OH)_4^-$	
From (271)	From (285)	From (271)	From (285)	From (271)	From (285)
430 w (dp)		385 w (dp)		525 s (p)	527 (p)
458 w (dp)		447 w (dp)			763 (p)
490 w (dp)		565 s (p)	567 (p)	917 w	914 (dp)
609 s (p)	613 (p)				
995 w	995 (p)				

[a] s, sharp; w, weak; p, polarized; dp, depolarized. From Maeda et al. (271).

tion of sodium hydroxide to boric acid, or hydrochloric acid to borax, indicated that over a period of 1 month, the structure of the solution altered, eventually reaching the equilibria appropriate for stable boron species at that pH. Borax solutions at pH 9 to 9.5 provided a spectrum that changed little with time. It is proposed that the relative concentration of four-coordinate boron increases with pH and that the rate of transformation is affected more by cleavage of bridging bonds, $B-O-B$, than by the conversion of trigonal to tetrahedral boron coordination.

B. Soluble Metal Borate Complexes

Until recently, very little had been reported on the important area of metal borate complexation in aqueous solution. The effect of salts on the ionization of boric acid (358, 375) has been mentioned above, and subsequent research suggests that complexation of borate with, for example, calcium ions can account for the enhanced acidity of H_3BO_3. Literature on cationic complexes of boron was reviewed in 1970 (376).

Potentiometric, cryoscopic, and conductometric measurements of the reaction between Fe^{3+} and $B_4O_7^{2-}$ have revealed the presence of $FeB_4O_7^-$ (373), and in oxalate media, the cationic complexes $FeBO_2^{2+}$ and $Fe(BO_2)_2^+$ were assumed to be present (370). The instability constant for $FeBO_2^{2+}$ was estimated as 3.12×10^{-9} by spectrophotometric techniques (68). The appearance of IR absorption bands at 1260, 1105, 860, and 815 cm^{-1} was attributed to the formation of an iron borate complex during oxidation of elemental iron in borate solutions (118).

Instability constants for the formation of $Al(BO_2)_2^+$, $AlBO_2^{2+}$, and $Al(BO_2)_6^{3-}$ have been evaluated (374), although the reliability of the results is in question. Other metals that have been examined in borate solutions include lead (372), cobalt (270, 371), zinc, and cadmium (369), providing evidence for $Pb(BO_2)_3^-$, $PbBO_2^+$, $Co(BO_2)_4^{2-}$, $Ni(BO_2)_3^-$, $Zn(BO_2)_4^{2-}$, and $Cd(BO_2)_4^{2-}$. Silver is believed to form a weak neutral complex $AgBO_2$ (177).

The interaction between sodium and alkaline-earth metal ions and borate has attracted recent attention, particularly from the point of view of association of ions in seawater. Several studies (69, 114, 168, 169, 340) have shown that the boron content of seawater ($4-5 \times 10^{-4}$ M) is too low to support appreciable concentrations of polyborate species. The increase in acidity of boric acid in the presence of metal ions results from ion-pair formation:

$$Na^+ + B(OH)_4^- \rightleftharpoons NaB(OH)_4^0$$

$$M^{2+} + B(OH)_4^- \rightleftharpoons MB(OH)_4^+$$

Measurement of association by density (427) and potentiometric (69, 133, 340) methods are consistent with this scheme. Dissociation constants over the temperature range 10–50°C (where T is the absolute temperature) are (340)

$$pK \ NaB(OH)_4^0 = 0.22 \text{ at } 25°C$$

$$pK \ MgB(OH)_4^+ = 1.266 + 0.001204T$$

$$pK \ CaB(OH)_4^+ = 1.154 + 0.002170T$$

$$pK \ SrB(OH)_4^+ = 1.033 + 0.001738T$$

$$pK \ BaB(OH)_4^+ = 0.942 + 0.001850T$$

and the calculated value for the apparent ionization constant of boric acid in seawater was found to be 1.86×10^{-9} (69, 340). The same data were used to indicate that 44% of the borate ions present in seawater are complexed with sodium, magnesium, and calcium (69).

C. HETEROGENEOUS SYSTEMS

The first comprehensive list of solubility data on borates was compiled by Teeple in 1929 (407). Since then, several hundred aqueous systems have been examined, covering most combinations of reactants over a range of temperatures. Of particular interest is the mode of dissolution of borates and the conditions under which they can be synthesized.

1. Dissolution of Borates

Most borates dissolve in strong acid (e.g., HCl, HNO_3) solution with formation of the weaker boric acid. For colemanite, hydroboracite, ulexite, and inyoite (calcium-, magnesium-, and sodium-containing minerals) the rate of solution increases with acid concentration but passes through a maximum as a protective layer of allegedly mainly boric acid encompasses the borate sample (184). In alkalis (Na_2CO_3 or NaOH), the solution rates are affected by the solubility of protective films of carbonate (200) and solution rates of metal oxides (273). The influence of solubility of by-products of dissolution on solution rates has also been observed in studies on the effect of anions (199) and cations (198) of magnesium and calcium minerals; ferric ions substantially accelerate borate decomposition (198). More recent work on the solubility rates of boric acid in aqueous metal chloride solutions (365) and anhydrous borax in aqueous/organic media (136) or metal nitrate/chloride solutions (137) has shown that dissolution rates increased with increasing water activity and are affected by surface sorption of cations.

The mode of dissolution of metal borates was followed by Svarcs *et al.* in the late 1950s using cryoscopic and electrical conductivity techniques (207–209). The mobilities of anions at 25°C in aqueous solutions of alkali metal borates were found to be 39–40 mho · cm^2, decreasing to 15 in 3% H_3BO_3. This latter value was interpreted as indicating formation of pentaborate ions. Calcium, strontium, and barium metaborates decompose in aqueous solution to give boric acid. However, the mobility of anions in borate $MB_6O_{10} \cdot xH_2O$ (M = Mg, Ca, Sr) was measured as 34 mho · cm^2 and 10.25 in 3% H_3BO_3; the proposed mechanism for hydrolysis was as follows

$$MB_6O_{10} \rightleftharpoons M^{2+} + B_6O_{10}^{2-}$$

$$M^{2+} + B_6O_{10}^{2-} + 3H_2O \rightleftharpoons M^{2+} + B_4O_7^{2-} + 2H_3BO_3$$

$$M^{2+} + B_4O_7^{2-} + 2H_3BO_3 + 3H_2O \rightleftharpoons M^{2+} + 2BO_2^- + 4H_3BO_3$$

$$M^{2+} + 2BO_2^- + 4H_3BO_3 + 3H_2O \rightleftharpoons M^{2+} + 6H_3BO_3$$

Other papers by the same authors (208, 399, 400, 403) discuss the partial and complete hydrolysis of many borates and give details of various resulting boron species.

The relationship between the composition and structure of borates and their decomposition in aqueous solution has been reviewed (78, 226, 414, 417). Borates of the alkali and alkaline-earth metals give an alkaline reaction in solution, as the borates formed by hydrolysis possess a lower boron-to-metal ratio than in the initial material (414).

The pH of an aqueous solution brought into contact with borates can determine the decomposition products both in solution and in the solid phase (414). Conversely, pH plays an important role in the synthesis of metal borates precipitated from aqueous media.

2. Synthesis of Borates

Valyashko and Gode (414) have summarized the conditions necessary for formation of borates of sodium, potassium, ammonium, calcium, and magnesium. The effect of varying the cation is to alter the type of borate that is stable in contact with the aqueous solution. The stability ranges for solid phases in the sodium and calcium systems are shown in Table V.

TABLE V

Stability Ranges for Sodium and Calcium Borates[a]

Solid phase	pH Range for stable solid phase at 25°C	
	$Na_2O-B_2O_3-H_2O$ System	$CaO-B_2O_3-H_2O$ System
H_3BO_3	Up to 6.2	Up to 5.5
$CaO \cdot 3B_2O_3 \cdot 4H_2O$		5.5–7.4
$Na_2O \cdot 5B_2O_3 \cdot 10H_2O$	6.2 → 7.2	
$Na_2O \cdot 2B_2O_3 \cdot 10H_2O$	7.2 → 13.7	
$2CaO \cdot 3B_2O_3 \cdot 13H_2O$		7.4 → 9.6
Metaborates	Over 13.7	Over 9.6

[a] From (414).

The pH ranges over which the solid phases are stable agree reasonably well with the stability ranges of the polyborate ions $B_5O_6(OH)_4^-$, $B_4O_5(OH)_4^{2-}$, $B_3O_3(OH)_5^{2-}$ outlined in Section IV,A; thus it can be expected that to precipitate a borate of a particular structure, it should be prepared from a solution containing those same borate ions.

The effect of calcium ions on polyborate equilibria is significant, but it is only recently that the detailed analysis of the ionic species in the $CaO-B_2O_3-H_2O$ system has been made (133, 377).

A general observation is that the tendency for precipitation of borates possessing a small polyborate ring increases with pH. Further, higher temperatures lead to products of increased B_2O_3/M_xO ratio and lower hydration.

The effect of neutral salts (e.g., NaCl) on the composition of borates precipitated from, or in equilibrium with, aqueous solutions doubtless arises from a reduction in water activity, metal borate complexation, and a shift in polyborate equilibria (Sections IV,A, B). The "indifferent or inert component" method has frequently been used for the synthesis of borates. Potassium and sodium chlorides can be used to enhance the precipitation of specific nickel (48), aluminum (51), iron (49), and magnesium (151) borates. In the $K_2O-B_2O_3-H_2O$ system at 25°C (248), the presence of potassium chloride results in a reduced boric acid crystallization curve, lower borate solubilities, lower pH, and an extended $B_2O_3:K_2O$ range over which the pentaborate crystallizes.

Hence the synthesis of borates is governed by the concentration of cations (whether "neutral" or active), pH, and temperature. These parameters affect the nature of boron ions in solution that are directly related to the composition of precipitated species.

V. Preparation and Properties of Metal Borates

This section briefly outlines the preparation and properties of the major metal borates, with particular emphasis on their hydrates. Recent and comprehensive reviews of the various group metal borates are found in Mellor (60, 307), from which additional data and references covering the entire field can be obtained. For brevity, the borates are referred to in terms of the mole ratio $M_2O : B_2O_3 : H_2O$ or $M_2O : B_2O_3$ of their composition.

A. ALKALI METAL, AMMONIUM, AND SILVER BORATES

These have been reviewed by Nies (307), and the commercially important borates have been reviewed by Doonan and Lower (113).

1. Lithium Borates

Solubilities of lithium borates in the system $Li_2O-B_2O_3-H_2O$ at 10–80°C (341), 100°C (35), and up to 400°C (36) have shown the presence of compounds with $Li_2O : B_2O_3 : H_2O$ mole ratios of 1 : 1 : 1, 1 : 1 : 4, 1 : 1 : 16, 1 : 2 : 2, 1 : 2 : 3, 2 : 1 : 1, 2 : 5 : 7, and 1 : 5 : 10.

a. Metaborates. The two principal borates are the 1 : 1 : 4 and 1 : 1 : 16 compounds, having the structural formulas $LiB(OH)_4$ and $LiB(OH)_4 \cdot 6H_2O$ respectively. Their IR and Raman spectra are recorded (210, 272), and recent studies (115, 116) on their thermal decomposition show that the higher hydrate loses water in four stages: the first H_2O molecule of $LiB(OH)_4 \cdot 6H_2O$ is lost at 30–50°C, the next five at 50–120°C, and the hydroxyl water at 120–160°C and at temperatures higher than 165°C. The transition point between the two hydrates in water is 36.9°C.

b. Tetraborates. The lithium borate $Li_2O \cdot 2B_2O_3 \cdot 3H_2O$ is the most stable borate of the 1 : 2 : x series. In contact with aqueous solutions, it is stable from −3 to 150°C, and can be dehydrated to afford the 1 : 2 : 2, 1 : 2 : 1, and anhydrous 1 : 2 : 0 compounds.

c. Pentaborates. Lithium pentaborate $Li_2O \cdot 5B_2O_3 \cdot 10H_2O$ has a congruent solubility above 40.5° or 37.5°C, and is stable in its aqueous solution up to 140°C, whereupon the 1 : 5 : 2 borate is formed. The "decaborate" $2Li_2O \cdot 5B_2O_3 \cdot 7H_2O$ loses two water molecules at 160°C and becomes anhydrous at 220°C (35).

2. Sodium Borates

Numerous studies have been made into the phase equilibria in the
$Na_2O-B_2O_3-H_2O$ system at $0-100°C$ (e.g., *135, 216, 217, 308*), and
more recently at $150°$ (*96*). A partial phase diagram for this system
covering the range of mole ratio Na_2O/B_2O_3 to 0.5 in solution is shown
in Fig. 6. Over 25 borates of sodium in the absence of other cations have
been identified (*307*).

a. Metaborates. The borates of composition $1:1:x$, where $x = 12, 8,$
4, and 1, have been isolated and found to be stable in contact with
highly alkaline solutions at -3 to $11.5°C$, $11.5-53.6°C$, $53.6-105°C$,
and above $105°C$, respectively. The thermal decomposition (*397*) and IR
(*211*) of the $1:1:8$ compound are consistent with its structural formula
of $NaB(OH)_4 \cdot 2H_2O$ (*44, 210*). The $1:1:4$ borate is prepared by heating
a slurry of the tetrahydrate above $54°C$, crystallization from hot
metaborate solutions, or dehydration of $1:1:8$ *in vacuo*; it loses its
water slowly at room temperature. Aqueous solutions of sodium
metaborates absorb carbon dioxide from the atmosphere to form borax
and sodium carbonate.

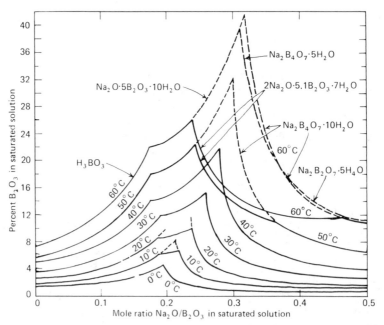

FIG. 6. Solubility isotherm for the system $Na_2O-B_2O_3-H_2O$. From Doonan and Lower
(*113*).

b. Tetraborates. The decahydrate $Na_2O \cdot 2B_2O_3 \cdot 10H_2O$ exists in nature as borax and is commercially the most important of the borates. Its crystal structure shows the borate ion to be $B_4O_5(OH)_4^{2-}$ *(267, 302)*. Borax crystallizes from aqueous solution at temperatures up to 60.7°C as monoclinic prisms, but its crystal habit can be changed by the addition of various substances. The dehydration of borax has been extensively studied. In aqueous solution it dehydrates to the pentahydrate above 60°C. One of the latest investigations *(314)* reports the appearance of a single DTA endothermic peak commencing at 115°C and ending at 238°C, with a minimum at 145°C. Dehydration was complete at 350–450°C, and at 245°C, the rate of dehydration followed the equation

$$\log(b - m) = -8.5 \times 10^{-4}t + C$$

where m is the amount of water lost in t minutes, b is the quantity of water initially present, and C is a constant. Aqueous solutions of borax have a pH of about 9.0, which is almost invariable with small additions of acid or alkali.

The pentahydrate $Na_2O \cdot 2B_2O_3 \cdot 5H_2O$ has the same polyborate ion as borax. It crystallizes from aqueous solution above 60°C but is also unstable with respect to kernite in solution above 40°C. In accord with its structure, $Na_2[B_4O_5(OH)_4]3H_2O$, the first three water molecules are lost at a temperature of 140 or 150°C, with the anhydrous product formed at over 400°C.

In kernite $Na_2O \cdot 2B_2O_3 \cdot 4H_2O$, the borate anion contains parallel infinite chains of $[B_4O_6(OH)_2]^{2-}$. The formation of this ion by B–O bond cleavage (Section III,A) accounts for its slow rate of crystallization and dissolution in aqueous media. Kernite is the stable phase in contact with its solutions from 58 to 95°C.

c. Pentaborates. Ezcurrite $2Na_2O \cdot 5.1B_2O_3 \cdot 7H_2O$, nasinite $2Na_2O \cdot 5B_2O_3 \cdot 5H_2O$, and biringuccite $2Na_2O \cdot 5B_2O_3 \cdot 3H_2O$ are "dibasic" pentaborates and are stable phases in the $Na_2O–B_2O_3–H_2O$ system at 40–94°C, 90–195°C, and 195–240°C, respectively. The most common sodium pentaborate is the 1:5:10 compound, also known as the mineral sborgite. It is stable in contact with its own solution between 2 and 59.5°C, but at temperatures near boiling the 2:9:11 salt is deposited. The 1:5:10 borate has the structural formula $Na[B_5O_6(OH)_4] \cdot 3H_2O$ *(294)*.

3. Potassium Borates

a. Metaborates. The solubilities in the potassium metaborate–water system are shown in Table VI. Omitted from this table are the

TABLE VI

The Potassium Metaborate – Water System[a]

Temperature (°C)	Wt % $K_2O \cdot B_2O_3$ in solution	Solid phase[b]
−20	27.1	Eutectic α-1:1:8 and ice
−18	26.6	α-1:1:8
11	39.0	α-1:1:8
24	43.8	Transition α-1:1:8 to 1:1:2$\frac{2}{3}$
0	35.5	1:1:2$\frac{2}{3}$ (metastable)
30	44.4	1:1:2$\frac{2}{3}$
180	68.8	1:1:2$\frac{2}{3}$
195	73.0	Transition 1:1:2$\frac{2}{3}$ to 1:1:$\frac{2}{3}$
200	73.1	1:1:$\frac{2}{3}$
237	73.4	1:1:$\frac{2}{3}$
250	74.0	Transition 1:1:$\frac{2}{3}$ to 1:1:0

[a] From Nies (307).
[b] The solid phase is expressed as the mole ratio $K_2O : B_2O_3 : H_2O$.

1:1:9 and β-1:1:8 metaborates that are stable at 0°C. The 1:1:2$\frac{2}{3}$ borate loses two water molecules at 177–253°C and the remainder at 253–309°C (460), consistent with its structure $K_3[B_3O_4(OH)_4] \cdot 2H_2O$ (319). Further evidence for this structure was obtained by the appearance of IR absorptions appropriate for trigonal and tetrahedral boron atoms (212).

b. *Tetraborates.* Potassium tetraborate tetrahydrate $K_2O \cdot 2B_2O_3 \cdot 4H_2O$ has a structure similar to that of borax, i.e., $K_2[B_4O_5(OH)_4] \cdot 2H_2O$ (277). Two of its water molecules are lost at 112–195°C, the third at 195–220°C, and the last at 240–420°C (398). The trihydrate 1:2:3 is stable in solution from 50 to over 100°C, and the di- and monohydrates can be prepared from the tetrahydrate at 90 and 140–200°C respectively.

c. *Pentaborates.* The 2:5:5 compound is known as Auger's potassium borate. It has the structural formula $K_2[B_5O_8(OH)] \cdot 2H_2O$ (274) and is stable in solution at temperatures down to 28°C. Its dehydration at 140 and 300–400°C affords the 2:5:3 and 2:5:1 compounds. The major pentaborate of potassium is the α-1:5:8, with the formula $K[B_5O_6(OH)_4] \cdot 2H_2O$ (29, 451). It is stable over a wide range of temperatures, and on dehydration loses water of crystallization at 120–170°C, eventually yielding the anhydrous form at 440°C (398). Recently, the β-1:5:8 compound has been isolated from the α form at 149°C (38).

The hydrolysis of trimethyl borate with water in the presence of potassium alkoxides has afforded polyborates of composition $1:12:7$, $1:5:4$, $1:3:4$, $2:5:5$, and $1:2:2$ (172). Infrared (174) and NMR (172) studies have indicated that these products contain water only in the form of hydroxyl groups.

4. Rubidium Borates

a. Metaborates. The rubidium metaborates of composition $Rb_2O \cdot B_2O_3 \cdot xH_2O$ have been isolated from the system $Rb_2O-B_2O_3-H_2O$. The $1:1:8$ compound is stable in contact with its solution from -19 to $20°C$. At higher temperatures, the $1:1:2\frac{2}{3}$ (isomorphous with the potassium salt) compound crystallizes out (461) and there is evidence for the $1:1:0.5$ and $1:1:\frac{2}{3}$ compounds. The latter is stable in solution from 190 to $220°C$, above which temperature the anhydrous salt is formed.

b. Tetraborates. The regions of stability in aqueous solutions for rubidium tetraborates are:

$1:2:5$	$20-104°C$
$1:2:3$	$104-172°C$
$1:2:2$	$172-244°C$
$1:2:1$	$244-320°C$

The same derivatives can be obtained by dehydrating the $1:2:5$ compound in air.

c. Pentaborates. The pentaborates of rubidium, $Rb_2O \cdot 5B_2O_3 \cdot xH_2O$, have been the subject of recent investigations (37). In the aqueous system at $100°C$, the α-$1:5:8$, β-$1:5:8$, $1:5:4$, $2:5:6$, and $2:5:5$ pentaborates were detected. Thermogravimetric analyses showed that the α-$1:5:8$ compound dehydrates continuously, dehydration of the $1:5:4$ compound gives the $1:5:2$ and $1:5:1$ compounds, and the $2:5:3$ and $2:5:1$ compounds are obtained from heating the $2:5:5$ and $2:5:6$ forms.

5. Cesium Borates

a. Metaborates. Four hydrates of cesium metaborate have been identified. The α-$1:1:8$ form, the crystal structure of which shows the formula α-$CsB(OH)_4 \cdot 2H_2O$ (459), is stable in aqueous solution at 50 to $78°C$. Below $50°C$, the β form is stable, and above $78°C$, the $1:1:2$ compound is the stable phase ($214, 215$). The latter compound can be prepared from the β-1 : 1 : 8 form by dehydration over phosphorous pentoxide.

b. *Tetraborates.* The transition point in aqueous solution between the cesium tetraborates 1 : 2 : 5 and 1 : 2 : 2 is 111°C (*214*). Three water molecules are lost on heating the 1 : 2 : 5 compound in air between 105 and 165°C, and the remainder between 230 and 400°C.

c. *Pentaborates.* The known cesium pentaborates are represented by the 1 : 5 : 8 and 2 : 5 : 7 compositions. Thermal dehydration of the former borate yields the 1 : 5 : x (x = 4, 3, 2, 1) compounds.

The cesium borates obtained by hydrolysis of trimethyl borate in organic media in the presence of $CsOC(CH_3)_3$ have been studied by IR (*132*) and tentatively assigned the formulas $Cs[B_{12}O_{13}(OH)_{11}]$, $Cs[B_4O_4(OH)_5]$, $Cs[B_5O_6(OH)_4]$, $Cs[B_3O_3(OH)_4]$, and $Cs_3[B_5O_6(OH)_6]$.

6. Ammonium Borates

The major ammonium borates are the tetraborate $(NH_4)_2O \cdot 2B_2O_3 \cdot 4H_2O$, and three pentaborates $(NH_4)_2O \cdot 5B_2O_3 \cdot 8H_2O$, larderellite $(NH_4)_2O \cdot 5B_2O_3 \cdot 4H_2O$, and ammonioborite $(NH_4)_2O \cdot 5B_2O_3 \cdot 5\frac{1}{3}H_2O$. The 1 : 5 : 8 compound exists as both rhombic (α) and monoclinic (β) forms, and, on heating to about 100°C, yields ammonioborite (*409, 410*). Isothermal studies (*411*) on the $(NH_4)_2O-B_2O_3-H_2O$ system have indicated that over the temperature range 0–50°C, the 1 : 5 : 8 and 1 : 2 : 4 compounds are metastable phases with respect to the 1 : 5 : 5$\frac{1}{3}$ and a 1 : 2 : 2.5 borate. Crystal structure determinations of the 1 : 5 : x compounds show that they contain true pentaborate groups (Table I).

7. Silver Borates

The $Ag_2O-B_2O_3-H_2O$ system has been investigated over the temperature range 0–100°C, and the borates $Ag_2O \cdot 2B_2O_3 \cdot 2H_2O$ and $2Ag_2O \cdot 5B_2O_3 \cdot 5H_2O$ obtained as stable phases (*356*). The 1 : 2 : 2.5 compound can be prepared by addition of silver nitrate to sodium borate solutions (*206*).

B. GROUP II METAL BORATES

1. Beryllium Borates

The borate chemistry of beryllium is somewhat limited. The mineral hambergite $Be_2BO_3(OH)$ can be synthesized, and studies of the $BeO-B_2O_3-H_2O$ system under hydrothermal conditions have produced

$2BeO \cdot B_2O_3 \cdot 3H_2O$, $BeO \cdot B_2O_3 \cdot H_2O$,   and $BeO \cdot 2B_2O_3 \cdot 1.2H_2O$
(*262, 359*).

2. Magnesium Borates

Many magnesium borate systems have been investigated; a few of the major studies are listed in Table VII. The main magnesium borates stable in contact with aqueous solutions are the $1:1:3$ (pinnoite), $1:3:x$, $2:1:1$, and $2:3:15$ (inderite) compounds.

Pinnoite $MgO \cdot B_2O_3 \cdot 3H_2O$ has a structure represented by $Mg[B_2O(OH)_6]$ (*235, 328*). It is formed in the $MgO-B_2O_3-H_2O$ system over the temperature range 45–200°C, and can be synthesized by warming solutions containing borax with high concentrations of magnesium chloride; the formation temperature can be lowered to 25°C in saturated KCl solution (*151*). Pinnoite is unstable in water at room temperature, and on dehydration, it loses two water molecules at 140–300°C and the third at 300–600° (*1*).

The $1:3:x$ compounds ($x = 4.5, 5, 6, 7$ and 7.5) separate from the $MgO-B_2O_3-H_2O$ systems at 25–200°C or from $MgO-B_2O_3-MgCl_2-H_2O$

TABLE VII

MAGNESIUM BORATE AQUEOUS SYSTEMS

Aqueous system	Temperature (°C)	Reference	Species in contact with solution, expressed as the mole ratio $MgO:B_2O_3:H_2O$
$MgO-B_2O_3-H_2O$	25	(*28, 251*)	$1:2:9, 1:3:7.5, 1:2:8.5, 2:3:15$
	35	(*28*)	$1:3:7.5, 2:3:15$
	45	(*349*)	$1:3:7.5, 1:1:3, 2:3:15$
	70	(*349*)	$1:3:7.5, 1:1:3$
	83	(*28*)	$1:3:7.5, 1:1:3$
	100–200	(*313*)	$1:3:6, 1:3:5, 1:3:4, 1:1:3, 1:2:2.5$
	150	(*366*)	$2:1:1, 1:2:2.5$
	100–700	(*162*)	$2:1:1$
$MgO-MgCl_2-B_2O_3-H_2O$	25	(*28, 350*)	$2:3:3, 3:1:11$
	70	(*21*)	$1:3:7.5, 1:3:5, 1:1:3, 2:1:3$
	83	(*28*)	$1:1:3, 2:1:1$
$MgO-NaCl-B_2O_3-H_2O$	25	(*350*)	$1:3:7.5, 2:3:13, 2:3:15$
$MgCl_2-Na/KCl-KOH-B_2O_3-H_2O$	25	(*150*)	$2:3:15$ (Except at high NaCl concentration)

systems. Macallisterite $MgO \cdot 3B_2O_3 \cdot 7.5H_2O$ can also be prepared by the reaction between magnesium chloride and potassium borates at pH 5.6 (152, 249). The compounds with x = 5, 6, 7, and 7.5 possess isolated $B_6O_7(OH)_6^{2-}$ hexaborate ions (Table I). Their dehydration (4, 124, 261, 312, 396) and IR (259, 312, 313, 394) characteristics are well established.

Inderite $2MgO \cdot 3B_2O_3 \cdot 15H_2O$ crystallizes at pH 7–9 in most magnesium systems below 70°C, although high concentrations of NaCl can limit its formation. A direct synthesis is the reaction of magnesium chloride with potassium borate at pH 7.9 (146, 152). Inderite has the structural formula $Mg[B_3O_3(OH)_5] \cdot 5H_2O$ (92, 343), and its thermal dehydration occurs in four stages with formation of the 2:3:1, 1:2:9, and 1:1:3 compounds (3, 123, 124).

The $2MgO \cdot B_2O_3 \cdot H_2O$ borate is known as ascharite. It can be prepared from the $MgO–B_2O_3–H_2O$ system at temperatures of 150–450°C. The presence of magnesium chloride at a concentration of 25% extends its aqueous stability range down to 83°C. Ascharite is almost insoluble in water, and a solubility of 0.0085% B_2O_3 at 25°C has been reported (456). Its dehydration has been studied (123, 165).

3. Calcium Borates

Calcium borates are widespread in nature, and some are used for the commercial production of boric acid and its salts. As a result, extensive investigations have been made into their occurrence, decomposition, and dissolution in aqueous media. Some of the aqueous phase studies are given in Table VIII.

TABLE VIII

CALCIUM BORATE AQUEOUS SYSTEMS

Aqueous system	Temperature (°C)	Reference	Species in contact with solution, expressed as the mole ratio $CaO : B_2O_3 : H_2O$
$CaO–B_2O_3–H_2O$	25	(251)	1:1:6, 1:3:4, 2:3:13
	30	(204)	1:1:6, 1:3:4, 2:3:13, 6:5:3
	45	(352)	1:1:4, 1:3:4, 2:3:9
	95	(311)	2:1:1
	200–400	(170)	2:1:1, 2:2:1, 2:3:1, 3:9:1
$CaO–MgCl_2–B_2O_3–H_2O$	25	(350, 351)	1:1:4, 1:1:6, 1:3:x, 2:3:9, 2:3:13
$CaO–NaCl–B_2O_3–H_2O$	25	(350, 353)	1:1:6, 1:3:4, 1:3:5, 2:3:13, $Na_2O \cdot 2CaO \cdot 5B_2O_3 \cdot 16H_2O$

a. Metaborates. The two common metaborates of calcium are the $1:1:4$ and $1:1:6$ compounds. They are formed in the $CaO-B_2O_3-H_2O$ system at a pH greater than 9.6, the $1:1:6$ at temperatures of 30°C or less and the $1:1:4$ at over 30°C. The presence of calcium *(154)* or magnesium *(350, 351)* chlorides extends the temperature range for $1:1:4$ metaborate down to less than 20°C. Alternatively, these hydrates can be prepared from calcium chloride–sodium metaborate mixtures *(257)*. The IR *(210)* and dehydration *(20, 462)* properties of the $1:1:6$ and $1:1:4$ borates are consistent with their structural formulas $Ca[B(OH)_4]_2 \cdot 2H_2O$ *(425, 426)* and $Ca[B(OH)_4]_2$ *(320, 457)* respectively. The $1:1:2$ hydrate is formed by dehydrating the hexahydrate at 105°C *(266)*, but unlike the higher hydrates, its structure is represented by the formula $Ca_2[B_3O_3(OH)_5 \cdot OB(OH)_3]$ *(367)*. The structures of calcium metaborate minerals have been reviewed *(382)*.

b. Borates $CaO \cdot 3B_2O_3 \cdot xH_2O$. The $1:3:4$ compound (nobleite) is a stable phase in the $CaO-B_2O_3-H_2O$ system at a pH of 5.5 to 6.5 between 25 and 60°C. It dehydrates via formation of a $1:3:3$ borate *(462)*. The pentahydrate can be prepared by the reaction between calcium chloride, borax, and boric acid at 60°C *(147)*; in nature it occurs as gowerite. The IR spectra of the tetrahydrate and pentahydrate have been recorded *(259)*.

c. Borates $2CaO \cdot 3B_2O_3 \cdot xH_2O$. This series of borates has structures showing the stepwise polymerisation of isolated $B_3O_3(OH)_5^{2-}$ ions. The structures have been discussed by Christ and Clark *(85)* and are shown in Table I. The lowest member, $2:3:1$, occurs in the $CaO-B_2O_3-H_2O$ system above 200°C *(170)* and is identical with the mineral fabianite. At elevated temperatures it is also formed by the hydrothermal reactions between the $2:3:7$, $1:3:5$, or $2:7:8$ borates with boric acid. Colemanite, $2:3:5$, is a commercially important borate; it exhibits interesting ferroelectric *(166)* and optical *(363)* properties. Meyerhofferite, $2:3:7$, the synthetic $2:3:9$, and inyoite, $2:3:13$, are all structurally similar, possessing isolated $B_3O_3(OH)_5^{2-}$ ions. Meyerhofferite is formed by heating aqueous solutions of borax and sodium hydroxide with calcium chloride at 85°C *(156)*. The $2:3:9$ and $2:3:13$ borates are stable species in the $CaO-B_2O_3-H_2O$ system and can be respectively prepared, for example, by reacting calcium chloride with ammonium pentaborate solutions or calcium iodate with borax solutions at 30°C *(435)*. The effect of sodium chloride on the $CaO-B_2O_3-H_2O$ system is to widen the pH range for $2:3:13$ compound formation, from 7.2–9.3 to 8.0–9.6. The thermal dehydration properties of colemanite, meyerhofferite, and inyoite have recently been reinvestigated *(221)*.

4. Strontium Borates

The chemistry of strontium borates is restricted. Investigations (*203, 258*) into the $SrO-B_2O_3-H_2O$ systems have revealed the presence of the stable species $1:1:4$, $1:3:5$, and $1:3:9$ at 30°C; $1:1:4$, $1:3:5$, and $1:3:6$ at 100°C; and $1:1:1$ and $1:4:2$ at 190°C.

Strontium metaborate $SrO \cdot B_2O_3 \cdot 4H_2O$ has been well characterized, the crystal structures of both monoclinic (*228*) and triclinic (*321*) forms indicating the formula $Sr[B(OH)_4]_2$. It is readily formed by reacting soluble strontium salts with sodium metaborate solutions at a pH in excess of 11.

The $SrO \cdot 3B_2O_3 \cdot 4H_2O$ compound, occurring as tunellite, can be synthesized from the reaction between ammonium pentaborate and strontium salts. It was the first borate to be identified as possessing the hexaborate anion (Section III,A). The pentahydrate, $1:3:5$, is stable in the $SrO-B_2O_3-H_2O$ system within the temperature range 0–100°C. Several routes to its preparation have been reported (*206*), and its dehydration has recently been studied (*197*).

The strontium analog of colemanite, $2SrO \cdot 3B_2O_3 \cdot 5H_2O$, can be prepared from the reaction between strontium iodate and borax in water at 65°C (*436*). Like colemanite, it possesses ferroelectric properties.

5. Barium Borates

Very few studies of hydrated barium borates have been published.

The metaborates, $1:1:x$, are the most common borates of barium. $BaO \cdot B_2O_3 \cdot 1\frac{2}{3}H_2O$ precipitates from a boiling solution of barium chloride and boric acid treated with sodium hydroxide to a pH of over 11.4 (*260*); it dehydrates completely at 300°C. The $1:1:4$, $1:1:5$, and $1:1:6$ metaborates have the formula $Ba[B(OH)_4]_2 \cdot xH_2O$ (Table I). The tetrahydrate is stable at 30°C in the $BaO-B_2O_3-H_2O$ system (*202*) and can be separated from sodium borate/barium chloride solutions at pH 11; at a higher pH and in dilute solution, the pentahydrate crystallizes out.

$BaO \cdot 3B_2O_3 \cdot 4H_2O$ is a stable phase in the $BaO-B_2O_3-H_2O$ (*202*) and $BaCl_2-Na_2B_4O_7-H_2O$ (*201*) systems at 30°C, and is also formed by dehydration of the pentahydrate (*323*) or hexahydrate.

6. Zinc Borates

The addition of alkali to solutions containing zinc salts with boron usually leads to formation of amorphous precipitates unless conditions are carefully controlled. Apart from the borate $ZnO \cdot B_2O_3 \cdot H_2O$ formed in aqueous solutions of borax and zinc nitrate, the best known zinc borates belong to the $2:3:x$ series.

The $2ZnO \cdot 3B_2O_3 \cdot 3.5H_2O$ compound is the lowest hydrate under normal conditions and is prepared by the reaction between aqueous zinc sulphate and boric acid at temperatures above 70°C (309), or between zinc oxide and boric acid above 100°C (310). It retains its water up to 260°C. At lower temperatures, zinc salts react with borate solutions to give the $2:3:7$ and $2:3:7.5$ compounds. Alternatively, the borates can be formed according to the equation (134):

$$4NH_4OH + 2ZnSO_4 + 6H_3BO_3 \rightarrow 2ZnO \cdot 3B_2O_3 \cdot nH_2O + 2(NH_4)_2SO_4 + mH_2O$$

A crystal structure determination of the $2:3:7$ compound shows it to have the structural formula $Zn[B_3O_3(OH)_5] \cdot H_2O$ (324).

A borate of composition $3:5:14$ has been synthesized from zinc acetate and boric acid solutions (336), and the borates $1:5:4.5$, $2:3:3$, and $6:5:3$ under hydrothermal conditions (263).

7. Cadmium Borates

Only a few borates of cadmium are known. Cadmium monoborate monohydrate $CdO \cdot B_2O_3 \cdot H_2O$ is formed in cadmium nitrate/borax solutions. The $1:2:5$ borate precipitates from hot boric acid solutions containing cadmium acetate (206). Its dehydration occurs in three stages, with one water molecule lost at 100°C, three at 100–150°C, and the last at over 300°C.

Amorphous products were obtained in the reaction between alkali metal borates and cadmium chloride, analyzed as the $2:3:7$ compound (337). More recently, a patent claiming the formation of $6CdO \cdot 7B_2O_3 \cdot (3-3.8)H_2O$ has appeared (264).

8. Mercury Borates

The mercury compounds $(PhHg)_2HBO_3$ (296), $Hg_3B_2O_6$ (76), and $Hg_4O(BO_2)_6$ (75) have been reported, but there is not conclusive evidence for hydrated borates.

9. Double Metal Borates

Table IX lists many of the known, well characterized hydrated double borates of general formula $a M_2^I O \cdot b M^{II}O \cdot cB_2O_3 \cdot dH_2O$. The structures of kaliborite, hydroboracite, inderborite, probertite, ulexite, $(NH_4)_2O \cdot CaO \cdot 4B_2O_3 \cdot 12H_2O$, and $Rb_2O \cdot SrO \cdot 4B_2O_3 \cdot 12H_2O$ are

TABLE IX

Typical Double-Metal Borates

$(NH_4)_2O \cdot MgO \cdot 3B_2O_3 \cdot x\,H_2O$	$Na_2O \cdot 2CaO \cdot 5B_2O_3 \cdot 10H_2O$, probertite
$3Na_2O \cdot MgO \cdot 12B_2O_3 \cdot 22H_2O$, rivadavite	$Na_2O \cdot 2CaO \cdot 5B_2O_3 \cdot 16H_2O$, ulexite
$Na_2O \cdot 2MgO \cdot 5B_2O_3 \cdot 30H_2O$	$Na_2O \cdot 2SrO \cdot 5B_2O_3 \cdot 16H_2O$
$K_2O \cdot MgO \cdot 3B_2O_3 \cdot 9H_2O$	$K_2O \cdot SrO \cdot 4B_2O_3 \cdot 14H_2O$
$K_2O \cdot 4MgO \cdot 11B_2O_3 \cdot 18H_2O$, kaliborite	$K_2O \cdot BaO \cdot 4B_2O_3 \cdot 14H_2O$
$K_2O \cdot 4MgO \cdot 11B_2O_3 \cdot 14H_2O$, heintzite	$K_2O \cdot 2ZnO \cdot 5B_2O_3 \cdot 10H_2O$
$K_2O \cdot MgO \cdot 3B_2O_3 \cdot 15H_2O$	$(NH_4)_2O \cdot ZnO \cdot 6B_2O_3 \cdot 10H_2O$
$CaO \cdot MgO \cdot 3B_2O_3 \cdot 6H_2O$, hydroboracite	$(NH_4)_2O \cdot 2ZnO \cdot 5B_2O_3 \cdot 10H_2O$
$CaO \cdot MgO \cdot 3B_2O_3 \cdot 11H_2O$, inderborite	$K_2O \cdot CdO \cdot 6B_2O_3 \cdot 10H_2O$
$CaO \cdot 4SrO \cdot 5B_2O_3 \cdot 20H_2O$ (148)	$M_2O \cdot M'O \cdot 4B_2O_3 \cdot 12H_2O$
	$M = NH_4$, K; $M' = Ca$
	$M = Rb$, Cs; $M' = Ba$ (155), Sr, Ca

given in Table I. All the octaborates $M_2O \cdot M'O \cdot 4B_2O_3 \cdot 12H_2O$ are believed to possess the isolated $B_4O_5(OH)_4^{2-}$ ion on the basis of their IR similarity with borax. The preparation and properties of the tabled borates have been reviewed (60).

C. Group III Metal Borate Hydrates

1. Aluminum Borates

The hydrothermal reaction between aluminum oxide and boric acids yields the borates of empirical composition $Al_2O_3 \cdot 3B_2O_3 \cdot 7H_2O$ and $Al_2O_3 \cdot 2B_2O_3 \cdot (2.5-3)H_2O$ (265). By the "indifferent component method" (Section III,C) aluminum borates $Al_2O_3 \cdot B_2O_3 \cdot xH_2O$ ($x = 2$ to 6) are formed by mixing aluminum nitrate and borax solutions in the presence of sodium or potassium chloride (51). The thermal decomposition of the $1:1:3$ compound has been studied (52). High-frequency conductometric titrations of solutions containing aluminum salts and alkali-metal tetraborates have detected the presence of $Al(OH)_2(BO_2)_nH_2O$ in solution (50).

2. Thallium Borates

The only recent authoratitive work on hydrated thallium borates is that by Touboul, who reviewed the field in 1971 (412). It is clear that insufficient data are available to rationalize the thallium borates, but there is definite evidence for the compounds $1:2:1$, $1:2:3$, $1:2:6$, $1:5:8$, $3:1:x$, and $2:5:5$, which is stable up to 100°C.

D. GROUP IV METAL BORATE HYDRATES

Within this group, only lead and zirconium have borates of definite composition. The $PbO \cdot 2B_2O_3 \cdot 4H_2O$ compound is formed by evaporation of acetic acid from lead acetate–boric acid mixtures (206). It dehydrates in three stages: the first water molecule is lost at 50–95°C, two more between 130 and 170°C, and the remainder at over 300°C. The $PbO-B_2O_3-H_2O$ system has been studied at 50, 75, and 100°C (385, 386), and the compounds detected were $1:2:3$ at 50°C; $4:5:2.5$, $1:2:3$, and $3:10:9$ at 75°C; and $4:5:2.5$, $1:2.5:2$, and $3:10:9$ at 100°C. Postulated structures for some of these borates are outlined in Section III,D.

The zirconium tetraborate $ZrO \cdot B_4O_7 \cdot 4H_2O$ is formed from the reaction between zirconyl chloride and boric acid at a temperature of 30°C at pH 4 (25). Its physicochemical properties have been studied (24).

E. TRANSITION-METAL BORATES

The borate chemistry of these elements is mainly restricted to the anhydrous compounds. Although this is an important area, space does not allow adequate coverage of the vast amount of pertinent literature.

1. Chromium Borates

The reaction products of the $Cr_2O_3-B_2O_3-K/NaNO_3-H_2O$ system are the chromium borates $2Cr_2O_3 \cdot 3B_2O_3 \cdot xH_2O$ ($x = 13$ and 18) (145). A basic borate $Cr(OH)(BO_2)_2 \cdot nH_2O$ has been detected by high-frequency conductometry (50).

2. Manganese Borates

The amorphous borate $2MnO \cdot 3B_2O_3 \cdot xH_2O$, prepared by mixing solutions of borax and manganous chloride, provides a useful starting material for the synthesis of other manganese borates. Treatment of this product with boric acid yields the $1:2:9$ and $1:3:8$ borates; boric acid and potassium or sodium chlorides give the $1:1:3$ borate; and when it is dissolved in MOH (M = K or NH_4), the double borates $M_2O \cdot MnO \cdot 6B_2O_3 \cdot 11H_2O$ are obtained (53, 206).

The thermal dehydration of the $1:1:3$ (52) and, in particular, the hexaborates of general formula $1:3:x$ (396) have been studied intensely. Structurally, the $1:2:9$ compound resembles borax and has been assigned the formula $Mn[B_4O_5(OH)_4] \cdot 7H_2O$ (42). On the basis of IR evidence, the hexaborates are believed to possess both tetrahedral and trigonal boron atoms (402).

3. Iron Borates

Most of the hydrated iron borates are formed under hydrothermal conditions. A recent study (163) of the $FeO-Fe_2O_3-B_2O_3-H_2O$ system over the temperature range 135–700°C detected the species $FeO \cdot 2B_2O_3 \cdot 2.5H_2O$, $FeO \cdot 3B_2O_3 \cdot 4.5H_2O$, $FeO \cdot 6B_2O_3 \cdot 5H_2O$, $FeO \cdot 2B_2O_3 \cdot 4.5H_2O$, $6FeO \cdot 7B_2O_3 \cdot H_2O$, $2FeO \cdot B_2O_3 \cdot H_2O$, and $Fe_2O_3 \cdot B_2O_3 \cdot nH_2O$ hydrates.

The borate $FeO \cdot 3B_2O_3 \cdot 8H_2O$ has been prepared at room temperature in an inert atmosphere from borax, boric acid, and ferrous sulfate (401). The $Fe_2O_3 \cdot B_2O_3 \cdot xH_2O$ borate is formed in the $Fe_2O_3-B_2O_3-$ Na/KCl–H_2O systems (49, 145).

The IR spectra (402) and thermal decomposition (396) of the $FeO \cdot 3B_2O_3 \cdot xH_2O$ borates have been measured. The anhydrous iron borates Fe_3BO_6 and $FeBO_3$ exhibit interesting magnetic and magneto-optical properties. In the last 10 years, well over 50 papers have been written on these borates, a few of which have been summarized by Bowden and Thompson (60).

4. Cobalt Borates

The only definite borate hydrates of cobalt are the $CoO \cdot 3B_2O_3 \cdot 8H_2O$ and $CoO \cdot 3B_2O_3 \cdot 10H_2O$ compounds. The octahydrate is prepared by evaporation of acetic acid from cobalt acetate–boric acid mixtures, or by mixing aqueous solutions of cobalt chloride, borax, and boric acid (206). The $1:3:7.5$ borate can form as a solid solution and, in the presence of 3% boric acid, affords the decahydrate (117). The crystal structure determination of this $1:3:10$ compound shows it to possess the hexaborate ion (380). The IR spectra (402) and thermal decomposition (396) of these compounds have been determined.

5. Nickel Borates

The borate $NiO \cdot B_2O_3 \cdot 8H_2O$ is prepared from nickel salt and borax solutions after digesting with boric acid. The decahydrate is formed from the octahydrate in boric acid at temperatures below 10°C; above 80°C the pentahydrate separates out (206). Both the octahydrate and decahydrate have the hexaborate ion $B_6O_7(OH)_6^{2-}$ (378, 380).

A $1:1:2$ compound is obtained in the $NiCl_2-M_2B_4O_7-MCl-H_2O$ (M = K or Na) system (48).

VI. Concluding Remarks

The structure determination of crystalline borates is well advanced, and it is hoped that the present enthusiasm will continue. There are many candidates for further investigation, particularly where spectroscopic or dehydration characteristics have suggested a particular boron anion. For example, the borate $2ZnO \cdot 3B_2O_3 \cdot 3.5H_2O$ loses its water at temperatures in excess of 260°C, implying that its formula water is in the form of hydroxyl groups.

Knowledge of the stability of various polyborate ions is lacking. It is known from structure determinations that the isolated $B_3O_3(OH)_5^{2-}$ is the most common ion of the triborates, but not why there are so few examples of $B_3O_3(OH)_4^-$ and $B_3O_3(OH)_8^{3-}$. Further, why does borcarite have an eight-membered ring in preference to a tetraborate group?

Probably the most neglected field is that of metal borate complexation in solution. Early proposals of ionic species were based on rather dubious evidence, and a great deal of new experimental work is required. With more sophisticated spectroscopic instruments becoming available, both this phenomenon and the related topic of polyborate ions in solution will be easier to observe.

The lack of data on certain metal borate systems has been implied in Section V. Quite apart from thallium, zinc, and lead compounds, the range of known borates of barium and strontium is somewhat limited. There are no obvious reasons why the latter should not have stability ranges approaching those of calcium and magnesium borates. Clearly, more investigations are necessary in this area.

REFERENCES

1. Abbasov, B. S., Abdullaev, G. K., and Rza-Zade, P. F., *Uch. Zap. Azerb. Gos. Univ., Ser. Khim. Nauk* **3,** 21 (1970).
2. Abdullaev, G. K., *Zh. Strukt. Khim.* **17,** 1128 (1976).
3. Abdullaev, G. K., Abbasov, B. S., and Rza-Zade, P. F., *Issled. Obl. Neorg. Fiz. Khim.* p. 147 (1971).
4. Abdullaev, G. K., Abbasov, B. S., and Rza-Zade, P. F., *Issled. Obl. Neorg. Fiz. Khim.* p. 153 (1971).
5. Abdullaev, G. K., Dzhafarov, G. G., and Mamedov, Kh. S., *Azerb. Khim. Zh.* p. 117 (1976).
6. Abdullaev, G. K., and Mamedov, Kh. S., *Issled. Obl. Neorg. Fiz. Khim. Ikh Rol Khim. Promsti, Mater. Nauchn. Konf., 1967* p. 28 (1969).

7. Abdullaev, G. K., and Mamedov, Kh. S., *Zh. Strukt. Khim.* **15,** 157 (1974).
8. Abdullaev, J. K., and Mamedov, Kh. S., *Kristallografiya* **19,** 165 (1974).
9. Abdullaev, G. K., and Mamedov, Kh. S., *Zh. Strukt. Khim.* **17,** 188 (1976).
10. Abdullaev, G. K., and Mamedov, Kh. S., *Kristallografiya* **22,** 271 (1977).
11. Abdullaev, G. K., and Mamedov, Kh. S., *Kristallografiya* **22,** 389 (1977).
12. Abdullaev, G. K., Mamedov, Kh. S., Amiraslanov, I. R., Dzhafarov, G. G., Aliev, O. A., and Usubaliev, B. T., *Zh. Neorg. Khim.* **23,** 2332 (1978).
13. Abdullaev, G. K., Mamedov, Kh. S., Amiraslanov, I. R., and Magerramov, A. I., *Zh. Strukt. Khim.* **18,** 410 (1977).
14. Abdullaev, G. K., Mamedov, Kh. S., and Amirov, S. T., *Kristallografiya* **18,** 1075 (1973).
15. Abdullaev, G. K., Mamedov, Kh. S., and Dzhafarov, G. G., *Kristallografiya* **19,** 737 (1974).
16. Abdullaev, G. K., Mamedov, Kh. S., and Dzhafarov, G. G., *Zh. Strukt. Khim.* **16,** 71 (1975).
17. Abdullaev, G. K., Mamedov, Kh. S., and Dzhafarov, G. G., *Kristallografiya* **20,** 265 (1975).
18. Abdullaev, G. K., Mamedov, Kh. S., and Dzhafarov, G. G., *Azerb. Khim. Zh.* p. 115 (1977).
19. Abdullaev, G. K., Mamedov, Kh. S., and Dzhafarov, G. G., *Azerb. Khim. Zh.* p. 125 (1978).
20. Abdullaev, G. K., and Samedov, Kh. R., *Issled. Obl. Neorg. Fiz. Khim.* p. 160 (1971).
21. Abduragimova, R. A., Rza-Zade, P. F., and Sedel'nikov, G. S., *Izv. Akad. Nauk Turkm. SSR, Ser. Fiz.-Tekh., Khim. Geol. Nauk* p. 35 (1966).
22. Abrahams, S. C., Bernstein, J. L., and Keve, E. T., *J. Appl. Crystallogr.* **4,** 284 (1971).
23. Adachi, M., Mitamura, H., and Eguchi, W., *Kyoto Daigaku Genshi Enerugi Kenkyusho Iho* **51,** 37 (1977).
24. Afanas'ev, Yu. A., Eremin, V. P., and Ryabinin, A. I., *Zh. Fiz. Khim.* **49,** 3018 (1975).
25. Afanas'ev, Yu. A., Ryabinin A. I., and Eremin, V. P., *Dokl. Akad. Nauk SSSR* **215,** 97 (1974).
26. Aleksandrov, N. M., and Pietrzak, J., *Radiospektrosk. Tverd. Tela, Dokl. Vses. Soveshch., 1964* p. 103 (1967).
27. Anderson, J. L., Eyring, E. M., and Whittaker, M. P., *J. Phys. Chem.* **68,** 1128 (1964).
28. D'Ans, J., and Behrendt, K. H., *Kali Steinsalz* **2,** No. 4S, 109/144, 121 (1957).
29. Ashmore, J. P., and Petch, H. E., *Can. J. Phys.* **48,** 1091 (1970).
30. Barbier, Y., Rosset, R., and Tremillon, B., *Bull. Soc. Chim. Fr.* p. 3352 (1966).
31. Bartl, H., and Schuckmann, W., *Neues Jahrb. Mineral., Monatsh.* p. 253 (1966).
32. Baur, W. H., and Tillmanns, E., *Z. Kristallogr.* **131,** 213 (1970).
33. Belina, B., and Veksli, Z., *Croat. Chem. Acta* **37,** 41 (1965).
34. Belokoneva, E. L., Al'shinskaya, L. I., Simonov, M. A., Leonyuk, N. I., Timchenko, T. I., and Belov, N. V., *Zh. Strukt. Khim.* **19,** 382 (1978).
35. Benhassaine, A., *C. R. Hebd. Seances Acad. Sci., Ser. C* **274,** 1442 (1972).
36. Benhassaine, A., *C. R. Hebd. Seances Acad. Sci., Ser. C* **274,** 1516 (1972).
37. Benhassaine, A., *C. R. Hebd. Seances Acad. Sci., Ser. C* **274,** 1933 (1972).
38. Benhassaine, A., and Toledano, P., *C. R. Hebd. Seances Acad. Sci., Ser. C* **272,** 396 (1971).
39. Berger, S. V., *Acta Chem. Scand.* **3,** 660 (1949).

40. Berger, S. V., *Acta Chem. Scand.* **4,** 1054 (1950).
41. Berzina, I., *Tezisy Dokl.—Konf. Molodykh Nauchn. Rab. Inst. Neorg. Khim., Akad. Nauk Latv. SSR, 6th, 1977* p. 18 (1977).
42. Berzina, I., Ozols, J., and Ievins, A., *Kristallografiya* **20,** 419 (1975).
43. Block, S., Burley, G., Perloff, A., and Mason, R. D., *J. Res. Natl. Bur. Stand.* **62,** 95 (1959).
44. Block, S., and Perloff, A., *Acta Crystallogr.* **16,** 1233 (1963).
45. Block, S., and Perloff, A., *Acta Crystallogr.* **19,** 297 (1965).
46. Block, S., Perloff, A., and Weir, C. E., *Acta Crystallogr.* **17,** 314 (1964).
47. Boehlhoff, R., Bambauer, H. U., and Hoffmann, W., *Z. Kristallogr.* **133,** 386 (1971).
48. Boiko, V. F., and Bratolyubova, T. A., *Obshch. Prikl. Khim.* **6,** 65 (1974).
49. Boiko, V. F., and Chudnovskaya, O. N., *Izv. Vyssh. Uchebn. Zaved., Khim. Khim. Tekhnol.* **17,** 1612 (1974).
50. Boiko, V. F., and Chudnovskaya, O. N., *Izv. Vyssh. Uchebn. Zaved., Khim. Khim. Tekhnol.* **19,** 1799 (1976).
51. Boiko, V. F., and Gladkaya, E. P., *Izv. Vyssh. Uchebn. Zaved., Khim. Khim. Tekhnol.* **14,** 21 (1971).
52. Boiko, V. F., Sharai, V. N., Svirko, L. K., and Gladkaya, E. P., *Latv. PSR Zinat. Akad. Vestis, Kim. Ser.* p. 544 (1972).
53. Boiko, V. F., and Svirko, L. K., *Izv. Vyssh. Uchebn. Zaved., Khim. Khim. Tekhnol.* **16,** 511 (1973).
54. Boitard, E., Pilard, R., and Carpeni, G., *J. Chim. Phys. Phys.-Chim. Biol.* **74,** 890 (1977).
55. Boitard, E., Pilard, R., and Carpeni, G., *Int. Symp. Specific Interact. Mol. Ions,* [*Proc.*], *3rd, 1976* Vol. 1, p. 41 (1976).
56. Bokii, G. B., and Kravchenko, V. B., *Zh. Strukt. Khim.* **7,** 920 (1966).
57. Bondareva, O. S., Egorov-Tismenko, Yu. K., Simonov, M. A., and Belov, N. V., *Dokl. Akad. Nauk SSSR* **241,** 815 (1978).
58. Bondareva, O. S., Simonov, M. A., and Belov, N. V., *Kristallografiya* **23,** 491 (1978).
59. Bondareva, O. S., Simonov, M. A., Egorov-Tismenko, Yu. K., and Belov, N. V., *Kristallografiya* **23,** 487 (1978).
60. Bowden, G. H., and Thompson, R., *in* "Mellor's Comprehensive Treatise of Inorganic and Theoretical Chemistry" (R. Thompson, ed.), Vol. V, Part A, pp. 520–650. Longmans, Green, New York, 1980.
61. Bray, P. J., *Mater. Sci. Res.* **12,** 321 (1978).
62. Bray, P. J., Edwards, J. O., O'Keefe, J. G., Ross, V. F., and Tatsuzaki, I., *J. Chem. Phys.* **35,** 435 (1961).
63. Bril, T. W., *Philips Res. Rep., Suppl.* p. 114 (1976).
64. Bronswijk, J. P., *Proc. Int. Conf. Phys. Non-Cryst. Solids, 4th, 1976* p. 101 (1977).
65. Bronswijk, J. P., and Strijks, E., *J. Non-Cryst. Solids* **24,** 145 (1977).
66. Brovkin, A. A., and Nikishova, L. V., *Kristallografiya* **20,** 415 (1975).
67. Brovkin, A. A., Zayakina, N. V., and Brovkina, V. S., *Kristallografiya* **20,** 911 (1975).
68. Burchinskaya, N. B., *Ukr. Khim. Zh.* **30,** 177 (1964).
69. Byrne, R. H., and Kester, D. R., *J. Mar. Res.* **32,** 119 (1974).
70. Cannillo, E., Dal Negro, A., and Ungaretti, L., *Am. Mineral.* **58,** 110 (1973).
71. Carpeni, G., *Bull. Soc. Chim. Fr.* p. 344 (1949).
72. Carpeni, G., *Ann. Fac. Sci. Marseille* **30,** 3 (1960).
73. Carpeni, G., and Souchay, P., *J. Chim. Phys.* **42,** 149 (1945).

74. Carrondo, M. A. A. F. de C. T., and Skapski, A. C., *Acta Crystallogr., Sect. B* **B34,** 3551 (1978).

75. Chang, C. H., and Margrave, J. L., *Inorg. Chim. Acta* **1,** 378 (1967).

76. Chrétien, A., and Priou, R., *C. R. Hebd. Seances Acad. Sci., Ser. C* **271,** 1310 (1970).

77. Christ, C. L., *Am. Mineral.* **45,** 334 (1960).

78. Christ, C. L., *J. Geol. Educ.* **20,** 235 (1972).

79. Christ, C. L., and Clark, J. R., *Z. Kristallogr.* **114,** 321 (1960).

80. Christ, C. L., and Clark, J. R., *Phys. Chem. Miner.* **2,** 59 (1977).

81. Christ, C. L., Clark, J. R., and Evans, H. T., *Acta Crystallogr.* **11,** 761 (1958).

82. Cialdi, G., Corazza, E., and Sabelli, C., *Atti. Accad. Naz. Lincei, Cl. Sci. Fis., Mat. Nat., Rend.* **42,** 236 (1967).

83. Clark, J. R., *Acta Crystallogr.* **12,** 162 (1959).

84. Clark, J. R., *Am. Mineral.* **49,** 1548 (1964).

85. Clark, J. R., Appleman, D. E., and Christ, C. L., *J. Inorg. Nucl. Chem.* **26,** 73 (1964).

86. Clark, J. R., and Christ, C. L., *Z. Kristallogr.* **112,** 213 (1959).

87. Clark, J. R., and Christ, C. L., *Am. Mineral.* **56,** 1934 (1971).

88. Clark, J. R., Christ, C. L., and Appleman, D. E., *Acta Crystallogr.* **15,** 207 (1962).

89. Collin, R. L., *Acta Crystallogr.* **4,** 204 (1951).

90. Cooper, W. F., Larsen, F. K., Coppens, P., and Giese, R. F., *Am. Mineral.* **58,** 21 (1973).

91. Corazza, E., *Acta Crystallogr., Sect. B* **B30,** 2194 (1974).

92. Corazza, E., *Acta Crystallogr., Sect. B* **B32,** 1329 (1976).

93. Corazza, E., Menchetti, S., and Sabelli, C., *Am. Mineral.* **59,** 1005 (1974).

94. Corazza, E., Menchetti, S., and Sabelli, C., *Acta Crystallogr., Sect. B* **B31,** 1993 (1975).

95. Corazza, E., Menchetti, S., and Sabelli, C., *Acta Crystallogr., Sect. B* **B31,** 2405 (1975).

96. Corazza, E., Menchetti, S., and Sabelli, C., *Neues Jahrb. Mineral., Abh.* **131,** 208 (1977).

97. Corazza, E., and Sabelli, C., *Atti Accad. Naz. Lincei, Cl. Sci. Fis., Mat. Nat., Rend.* **41,** 527 (1966).

98. Coulson, C. A., and Dingle, T. W., *Acta Crystallogr., Sect. B* **B24,** 153 (1968).

99. Covington, A. K., and Newman, K. E., *J. Inorg. Nucl. Chem.* **35,** 3257 (1973).

100. Cuthbert, J. D., MacFarlane, W. T., and Petch, H. E., *J. Chem. Phys.* **43,** 173 (1965).

101. Cuthbert, J. D., and Petch, H. E., *J. Chem. Phys.* **38,** 1912 (1963).

102. Cuthbert, J. D., and Petch, H. E., *J. Chem. Phys.* **39,** 1247 (1963).

103. Dal Negro, A., Kumbasar, I., and Ungaretti, L., *Am. Mineral.* **58,** 1034 (1973).

104. Dal Negro, A., Martin Pozas, J. M., and Ungaretti, L., *Am. Mineral.* **60,** 879 (1975).

105. Dal Negro, A., Sabelli, C., and Ungaretti, L., *Atti Accad. Naz. Lincei, Cl. Sci. Fis., Mat. Nat., Rend.* **47,** 353 (1969).

106. Dal Negro, A., Ungaretti, L., and Basso, R., *Cryst. Struct. Commun.* **5,** 433 (1976).

107. Dal Negro, A., Ungaretti, L., and Giusta, A., *Cryst. Struct. Commun.* **5,** 427 (1976).

108. Dal Negro, A., Ungaretti, L., and Sabelli, C., *Am. Mineral.* **56,** 1553 (1971).

109. Dal Negro, A., Ungaretti, L., and Sabelli, C., *Naturwissenschaften* **60,** 350 (1973).

110. Denning, J. H., and Ross, S. D., *Spectrochim. Acta, Part A* **28A,** 1775 (1972).

111. Dharmatti, S. S., Iyer, S. A., and Vijayaraghavan, R., *J. Phys. Soc. Jpn.* **17,** 1736 (1962).

112. Diehl, R., *Solid State Commun.* **17,** 743 (1975).

113. Doonan, D. J., and Lower, L. D., *in* "Kirk-Othmer Encyclopedia of Chemical Technology" (M. Grayson, ed.), 2nd ed., Vol. IV, pp. 67–110. Wiley (Interscience), New York, 1978.

114. Dryssen, D., and Wedborg, M., *Sea* **5**, 181 (1974).
115. Dzene, A., and Svarcs, E., *Latv. PSR Zinat. Akad. Vestis, Kim. Ser.* p. 167 (1977).
116. Dzene, A., and Svarcs, E., *Latv. PSR Zinat. Akad. Vestis, Kim. Ser.* p. 493 (1977).
117. Dzene, A., Svarcs, E., and Ievins, A., *Latv. PSR Zinat. Akad. Vestis, Kim. Ser.* p. 751 (1969).
118. Ebiko, H., and Suetaka, W., *Corros. Sci.* **10**, 111 (1970).
119. Edwards, J. O., Morrison, G. C., Ross, V. F., and Schultz, J. W., *J. Am. Chem. Soc.* **77**, 266 (1955).
120. Egorov-Tismenko, Yu. K., Baryshnikov, Yu. M., Shashkin, D. P., Simonov, M. A., and Belov, N. V., *Dokl. Akad. Nauk SSSR* **208**, 1082 (1973).
121. Egorov-Tismenko, Yu. K., Simonov, M. A., and Belov, N. V., *Dokl. Akad. Nauk SSSR* **210**, 678 (1973).
122. Egorov-Tismenko, Yu. K., Simonov, M. A., Shashkin, D. P., and Belov, N. V., *Kristallokhim. Strukt. Miner.* p. 12 (1974).
123. Eyubova, N. A., Berg, L. G., and Rza-Zade, P. F., *Dokl. Akad. Nauk Azerb. SSR* **23**, 24 (1967).
124. Eyubova, N. A., Berg, L. G., and Rza-Zade, P. F., *Issled. Obl. Neorg. Fiz. Khim.* p. 91 (1970).
125. Fang, J. H., and Newnham, R. E., *Mineral. Mag.* **35**, 196 (1965).
126. Filippova, Z. O., Zhaimina, R. E., Markina, T. I., and Mun, A. I., *Izv. Akad. Nauk Kaz. SSR, Ser. Khim.* **25**, 5 (1975).
127. Follner, H., *Monatsh. Chem.* **104**, 477 (1973).
128. Fornaseri, M., *Period. Mineral.* **18**, 103 (1949).
129. Fornaseri, M., *Period. Mineral.* **19**, 157 (1950).
130. Fornaseri, M., *Ric. Sci.* **21**, 1192 (1951).
131. Frohnecke, J., Hartl, H., and Heller, G., *Z. Naturforsch., B. Anorg. Chem., Org. Chem.* **32B**, 268 (1977).
132. Frohnecke, J., and Heller, G., *J. Inorg. Nucl. Chem.* **34**, 69 (1972).
133. Frydman, M., Sillen, L. G., Hogfeldt, E., and Ferri, D., *Chem. Scr.* **10**, 152 (1976).
134. Gabova, E. L., Bannykh, Z. S., Sushkova, S. G., Gofman, E. E., and Zhitkova, T. N., *Izv. Akad. Nauk SSSR, Neorg. Mater.* **9**, 979 (1973).
135. Gale, W. A., *in* "Boron, Metallo-Boron Compounds and Boranes" (R. M. Adams, ed.), p. 29. Wiley (Interscience), New York, 1964.
136. Galkin, P. S., and Gulin, A. V., *Izv. Sib. Otd. Akad. Nauk SSSR, Ser. Khim. Nauk* p. 6 (1975).
137. Galkin, P. S., and Gulin, A. V., *Deposited Publ.* **VINITI 5997-73** (1973).
138. Garcia-Blanco, S., and Fayos, J., *Z. Kristallogr.* **127**, 145 (1968).
139. Genkina, E. A., Rumanova, I. M., and Belov, N. V., *Kristallografiya* **21**, 209 (1976).
140. Ghose, S., and Wan, C., *Am. Mineral.* **62**, 979 (1977).
141. Ghose, S., and Wan, C., *Am. Mineral.* **64**, 187 (1979).
142. Ghose, S., Wan, C., and Clark, J. R., *Am. Mineral.* **63**, 160 (1978).
143. Giacovazzo, C., Menchetti, S., and Scordari, F., *Am. Mineral.* **58**, 523 (1973).
144. Giese, R. F., *Science* **154**, 1453 (1966).
145. Gladkaya, E. P., and Chudnovskaya, O. N., *Obshch. Prikl. Khim.* **6**, 84 (1974).
146. Gode, H., and Bumanis, R., *Uch. Zap. Latv. Gos. Univ. im. Petra Stuchki* **177**, 44 (1970).
147. Gode, H., and Galvins, A., *Latv. PSR Zinat. Akad. Vestis, Kim. Ser.* p. 365 (1972).
148. Gode, H., Ivchenko, N. P., and Kurkutova, E. N., *Zh. Neorg. Khim.* **20**, 3136 (1975).
149. Gode, H., Ivchenko, N. P., Rau, V. G., and Kurkutova, E. N., *Strukt. Svoistva Krist.* p. 89 (1975).
150. Gode, H., and Kuka, P., *Latv. PSR Zinat. Akad. Vestis, Kim. Ser.* p. 637 (1969).

230   J. B. FARMER

151. Gode, H., and Kuka, P., *Latv. PSR Zinat. Akad. Vestis, Kim. Ser.* p. 520 (1971).
152. Gode, H., and Kuka, P., *Latv. PSR Zinat. Akad. Vestis, Kim. Ser.* p. 106 (1976).
153. Gode, H., Kuka, P., and Zuika, I., *Latv. PSR Zinat. Akad. Vestis, Kim. Ser.* p. 723 (1970).
154. Gode, H., and Kuzyukevich, A. A., *Latv. PSR Zinat. Akad. Vestis, Kim. Ser.* p. 402 (1972).
155. Gode, H., and Majore, I., *Latv. PSR Zinat. Akad. Vestis, Kim. Ser.* p. 344 (1976).
156. Gode, H., and Veveris, A., *Latv. PSR Zinat. Akad. Vestis, Kim. Ser.* p. 628 (1975).
157. Gode, H., and Zuika, I., *Latv. PSR Zinat. Akad. Vestis, Kim. Ser.* p. 724 (1970).
158. Gode, H., Zuika, I., and Adijane, G., *Latv. PSR Zinat. Akad. Vestis, Kim. Ser.* p. 538 (1971).
159. Golovastikov, N. I., Belova, E. N., and Belov, N. V., *Zap. Vses. Mineral. O-va.* **84**, 405 (1955).
160. Good, C. D., and Ritter, D. M., *J. Am. Chem. Soc.* **84**, 1162 (1962).
161. Goulden, J. D. S., *Spectrochim. Acta* p. 657 (1959).
162. Grigor'ev, A. P., *Tr. Soveshch. Eksp. Tekh. Mineral. Petrogr., 8th, 1968* p. 160 (1971).
163. Grigor'eva, T. A., *Vopr. Rudonosn. Yakutii* p. 168 (1974).
164. Griscom, D. L., *Mater. Sci. Res.* **12**, 11 (1978).
165. Guseinov, M. A., Vol'fkovich, S. I., and Aziev, R. G., *Vestn. Mosk. Univ., Ser. 2: Khim.* **16**, 377 (1975).
166. Hainsworth, F. N., and Petch, H. E., *Solid State Commun.* **3**, 315 (1965).
167. Hanic, F., Lindquist, O., Nyborg, J., and Zedler, A., *Collect. Czech. Chem. Commun.* **36**, 3678 (1971).
168. Hansson, I., *Acta Chem. Scand.* **27**, 924 (1973).
169. Hansson, I., *Deep-Sea Res. Oceanogr. Abstr.* **20**, 461 (1973).
170. Hart, P. B., and Brown, C. S., *J. Inorg. Nucl. Chem.* **24**, 1057 (1962).
171. Hart, P. B., and Smallwood, S. E. F., *J. Inorg. Nucl. Chem.* **24**, 1047 (1962).
172. Heller, G., *J. Inorg. Nucl. Chem.* **29**, 2181 (1967).
173. Heller, G., *Fortschr. Chem. Forsch.* **15**, 206 (1970).
174. Heller, G., and Giebelhausen, A., *J. Inorg. Nucl. Chem.* **35**, 3511 (1973).
175. Henderson, W. G., How, M. J., Kennedy, G. R., and Mooney, E. F., *Carbohydr. Res.* **28**, 1 (1973).
176. Hibben, J. H., *Am. J. Sci.* **35-A**, 113 (1938).
177. Hietanen, S., and Sillen, L. G., *Ark. Kemi* **32**, 111 (1970).
178. Hoehne, E., *Z. Anorg. Allg. Chem.* **342**, 188 (1966).
179. Hong, H. Y. P., and Dwight, K., *Mater. Res. Bull.* **9**, 1661 (1974).
180. How, M. J., Kennedy, G. R., and Mooney, E. F., *Chem. Commun.* p. 267 (1969).
181. Huggins, M. L., *Inorg. Chem.* **7**, 2108 (1968).
182. Huggins, M. L., *Inorg. Chem.* **10**, 791 (1971).
183. Ihara, M., and Krogh-Moe, J., *Acta Crystallogr.* **20**, 132 (1966).
184. Imamutdinova, V. M., *Zh. Prikl. Khim.* **40**, 1826 (1967).
185. Ingri, N., *Acta Chem. Scand.* **16**, 439 (1962).
186. Ingri, N., *Acta Chem. Scand.* **17**, 573 (1963).
187. Ingri, N., *Acta Chem. Scand.* **17**, 581 (1963).
188. Ingri, N., *Sven. Kem. Tidskr.* **75**, 199 (1963).
189. Ingri, N., Langerström, G., Frydman, M., and Sillen, L. G., *Acta Chem. Scand.* **11**, 1034 (1957).
190. Iorysh, Z. I., Rumanova, I. M., and Belov, N. V., *Dokl. Akad. Nauk SSSR* **228**, 1076 (1976).

191. Ivchenko, N. P., and Kurkutova, E. N., *Kristallografiya* **20**, 533 (1975).
192. Jahr, K. F., Wegener, K., Heller, G., and Worm, K., *J. Inorg. Nucl. Chem.* **30**, 1677 (1968).
193. Jain, D. V. S., and Jain, C. M., *Indian J. Chem.* **11**, 1281 (1973).
194. Jellison, G. E., and Bray, P. J., *J. Non-Cryst. Solids* **29**, 187 (1978).
195. Jellison, G. E., and Bray, P. J., *Mater. Sci. Res.* **12**, 353 (1978).
196. Jellison, G. E., *Diss. Abstr. Int. B* **38**, 6018 (1978).
197. Kanteeva, I. A., Leont'eva, I. A., Tkachev, K. V., and Gabova, E. L., *Izv. Akad. Nauk SSSR, Neorg. Mater.* **13**, 1102 (1977).
198. Karazhanov, N. A., and Bochkareva, I. V., *Zh. Fiz. Khim.* **42**, 1504 (1968).
199. Karazhanov, N. A., and Bochkareva, I. V., *Zh. Fiz. Khim.* **44**, 1587 (1970).
200. Karazhanov, N. A., Mardanenko, V. K., Kalacheva, V. G., and Guseva, G. P., *Zh. Fiz. Khim.* **47**, 2148 (1973).
201. Karibyan, A. N., Burnazyan, A. S., Babayan, G. G., and Akopyran, F. V., *Prom-st. Arm.* p. 38 (1974).
202. Karibyan, A. N., Burnazyan, A. S., Babayan, G. G., and Sinanyan, I. M., *Arm. Khim. Zh.* **24**, 22 (1971).
203. Karibyan, A. N., Burnazyan, A. S., Babayan, G. G., and Torosyan, E. E., *Arm. Khim. Zh.* **23**, 684 (1970).
204. Karibyan, A. N., Burnazyan, A. S., Sinanyan, I. M., and Babayan, G. G., *Arm. Khim. Zh.* **22**, 303 (1969).
205. Kazanskaya, E. V., Sandomirskii, P. A., Simonov, M. A., and Belov, N. V., *Dokl. Akad. Nauk SSSR* **238**, 1340 (1978).
206. Kesans, A., "Synthesis of Borates in Aqueous Solution and their Investigation." Akad. Nauk Latv. SSR, Riga, 1955.
207. Kesans, A., Svarcs, E., and Vimba, S., *Latv. PSR Zinat. Akad. Vestis* p. 125 (1955).
208. Kesans, A., Vimba, S., and Svarcs, E., *Latv. PSR Zinat. Akad. Vestis* p. 121 (1955).
209. Kesans, A., Vimba, S., and Svarcs, E., *Latv. PSR Zinat. Akad. Vestis* p. 127 (1955).
210. Kessler, G., *Z. Anorg. Allg. Chem.* **343**, 25 (1966).
211. Kessler, G., and Lehmann, H. A., *Z. Anorg. Allg. Chem.* **338**, 179 (1965).
212. Kessler, G., Roensch, E., and Lehmann, H. A., *Z. Anorg. Allg. Chem.* **365**, 281 (1969).
213. Klee, W. E., *Z. Anorg. Allg. Chem.* **343**, 58 (1966).
214. Kocher, J., *Rev. Chim. Miner.* **3**, 209 (1966).
215. Kocher, J., *Bull. Soc. Chim. Fr.* p. 919 (1968).
216. Kocher, J., and Lahlou, A., *C. R. Hebd. Seances Acad. Sci., Ser. C* **269**, 320 (1969).
217. Kocher, J., and Lahlou, A., *Bull. Soc. Chim. Fr.* p. 2083 (1970).
218. Koenig, H., and Hoppe, R., *Z. Anorg. Allg. Chem.* **439**, 71 (1978).
219. Koenig, H., Hoppe, R., and Jansen, M., *Z. Anorg. Allg. Chem.* **449**, 91 (1979).
220. Kolthoff, I. M., *Recl. Trav. Chim. Pays-Bas* **45**, 501 (1926).
221. Kondrat'eva, V. V., *Zap. Vses. Mineral. O-va.* **106**, 490 (1977).
222. Konijnendijk, W. L., and Stevels, J. M., *Mater. Sci. Res.* **12**, 259 (1978).
223. Konnert, J. A., Clark, J. R., and Christ, C. L., *Am. Mineral.* **55**, 1911 (1970).
224. Konnert, J. A., Clark, J. R., and Christ, C. L., *Z. Kristallogr.* **132**, 241 (1970).
225. Konnert, J. A., Clark, J. R., and Christ, C. L., *Am. Mineral.* **57**, 381 (1972).
226. Kravchenko, V. B., *Zh. Strukt. Khim.* **6**, 88 (1965).
227. Kravchenko, V. B., *Zh. Strukt. Khim.* **6**, 724 (1965).
228. Kravchenko, V. B., *Zh. Strukt. Khim.* **6**, 872 (1965).
229. Kriz, H. M., Bishop, S. G., and Bray, P. J., *J. Chem. Phys* **49**, 557 (1968).
230. Kriz, H. M., and Bray, P. J., *J. Magn. Reson.* **4**, 76 (1971).

231. Krogh-Moe, J., *Phys. Chem. Glasses* **3**, 101 (1962).
232. Krogh-Moe, J., *Acta Chem. Scand.* **18**, 2055 (1964).
233. Krogh-Moe, J., *Acta Crystallogr.* **18**, 77 (1965).
234. Krogh-Moe, J., *Acta Crystallogr.* **18**, 1088 (1965).
235. Krogh-Moe, J., *Acta Crystallogr.* **23**, 500 (1967).
236. Krogh-Moe, J., *Acta Crystallogr., Sect. B* **B24**, 179 (1968).
237. Krogh-Moe, J., *Acta Crystallogr., Sect. B* **B28**, 168 (1972).
238. Krogh-Moe, J., *Acta Crystallogr., Sect. B* **B28**, 1571 (1972).
239. Krogh-Moe, J., *Acta Crystallogr., Sect. B* **B28**, 3089 (1972).
240. Krogh-Moe, J., *Acta Crystallogr., Sect. B* **B30**, 578 (1974).
241. Krogh-Moe, J., *Acta Crystallogr., Sect. B* **B30**, 747 (1974).
242. Krogh-Moe, J., *Acta Crystallogr., Sect. B* **B30**, 1178 (1974).
243. Krogh-Moe, J., *Acta Crystallogr., Sect. B* **B30**, 1827 (1974).
244. Krogh-Moe, J., and Ihara, M., *Acta Crystallogr., Sect. B* **B23**, 427 (1967).
245. Krogh-Moe, J., and Ihara, M., *Acta Crystallogr., Sect. B* **B25**, 2153 (1969).
246. Krogh-Moe, J., and Wold-Hansen, P. S., *Acta Crystallogr., Sect. B* **B29**, 2242 (1973).
247. Kujumzelis, Th. G., *Phys. Z.* **39**, 665 (1938).
248. Kuka, P., and Gode, H., *Latv. PSR Zinat. Akad. Vestis, Kim. Ser.* p. 125 (1969).
249. Kuka, P., and Gode, H., *Latv. PSR Zinat. Akad. Vestis, Kim. Ser.* p. 378 (1969).
250. Kurkutova, E. N., Rumanova, I. M., and Belov, N. V., *Dokl. Akad. Nauk SSSR* **164**, 90 (1965).
251. Kurnakova, A. G., *Izv. Sekt. Fiz.-Khim. Anal., Inst. Obshch. Neorg. Khim., Akad. Nauk SSSR* **15**, 125 (1947).
252. Kurtsinovskaya, R. I., and Alekseeva, I. P., *Zh. Prikl. Spektrosk.* **24**, 476 (1976).
253. Kutschabsky, L., *Acta Crystallogr., Sect. B* **B25**, 1811 (1969).
254. Lal, K. C., and Petch, H. E., *J. Chem. Phys.* **43**, 178 (1965).
255. Lefebvre, J., *C. R. Hebd. Seances Acad. Sci.* **241**, 1295 (1955).
256. Lefebvre, J., *J. Chim. Phys.* **54**, 567 (1957).
257. Lehmann, H. A., and Guenther, I., *Z. Anorg. Allg. Chem.* **320**, 255 (1963).
258. Lehmann, H. A., and Jaeger, H., *Z. Anorg. Allg. Chem.* **326**, 31 (1963).
259. Lehmann, H. A., and Kessler, G., *Z. Anorg. Allg. Chem.* **354**, 30 (1967).
260. Lehmann, H. A., Muehmel, K., and Sun, D.-F., *Z. Anorg. Allg. Chem.* **355**, 238 (1967).
261. Lehmann, H. A., and Rietz, G., *Z. Anorg. Allg. Chem.* **350**, 168 (1967).
262. Lehmann, H. A., and Schubert, E., *Z. Chem.* **8**, 346 (1968).
263. Lehmann, H. A., Sperschneider, K., and Kessler, G., *Z. Anorg. Allg. Chem.* **354**, 37 (1967).
264. Lehmann, H. A., and Sperschneider, M., German (East) Patent 63,769 (1968).
265. Lehmann, H. A., and Teske, K., *Z. Anorg. Allg. Chem.* **400**, 169 (1973).
266. Lehmann, H. A., Zielfelder, A., and Herzog, G., *Z. Anorg. Allg. Chem.* **296**, 199 (1958).
267. Levy, H. A., and Lisensky, G. C., *Acta Crystallogr., Sect. B* **B34**, 3502 (1978).
268. Lourijsen-Teyssedre, M., *Bull. Soc. Chim. Fr.* p. 1111 (1955).
269. Machida, K., Adachi, G., and Shiokawa, J., *Acta Crystallogr., Sect. B* **B35**, 149 (1979).
270. Mader, P., and Kolthoff, I. M., *J. Polarogr. Sci.* **14**, 42 (1968).
271. Maeda, M., *J. Inorg. Nucl. Chem.* **41**, 1217 (1979).
272. Manewa, M., and Fritz, H. P., *Z. Anorg. Allg. Chem.* **397**, 290 (1973).
273. Mardanenko, V. K., Karazhanov, N. A., and Kalacheva, V. G., *Zh. Prikl. Khim.* **47**, 439 (1974).

274. Marezio, M., *Acta Crystallogr., Sect. B* **B25**, 1787 (1969).
275. Marezio, M., Plettinger, H. A., and Zachariasen, W. H., *Acta Crystallogr.* **16**, 390 (1963).
276. Marezio, M., Plettinger, H. A., and Zachariasen, W. H., *Acta Crystallogr.* **16**, 594 (1963).
277. Marezio, M., Plettinger, H. A., and Zachariasen, W. H., *Acta Crystallogr.* **16**, 975 (1963).
278. Marezio, M., and Remeika, J. P., *J. Chem. Phys.* **44**, 3348 (1966).
279. Marezio, M., Remeika, J. P., and Dernier, P. D., *Acta Crystallogr., Sect. B* **B25**, 955 (1969).
280. Marezio, M., Remeika, J. P., and Dernier, P. D., *Acta Crystallogr., Sect. B* **B25**, 965 (1969).
281. Maricic, S., Veksli, Z., and Pintar, M., *Phys. Chem. Solids* **23**, 743 (1962).
282. Mariezcurrena, R. A., and Rasmussen, S. E., *Acta Crystallogr., Sect. B* **B29**, 1035 (1973).
283. Martinez-Ripoll, M., Martinez-Carrera, S., and Garcia-Blanco, S., *Acta Crystallogr., Sect. B* **B27**, 672 (1971).
284. Martinez-Ripoll, M., Martinez-Carrera, S., and Garcia-Blanco, S., *Acta Crystallogr., Sect. B* **B27**, 677 (1971).
285. Maya, L., *Inorg. Chem.* **15**, 2179 (1976).
286. Mehta, S. M., and Desai, S. G., *Curr. Sci.* **15**, 128 (1946).
287. Menchetti, S., and Sabelli, C., *Acta Crystallogr., Sect. B* **B33**, 3730 (1977).
288. Menchetti, S., and Sabelli, C., *Acta Crystallogr., Sect. B* **B34**, 45 (1978).
289. Menchetti, S., and Sabelli, C., *Acta Crystallogr., Sect. B* **B34**, 1080 (1978).
290. Merlino, S., *Atti Accad. Naz. Lincei, Cl. Sci. Fis., Mat. Nat., Rend.* **47**, 85 (1969).
291. Merlino, S., and Sartori, F., *Acta Crystallogr., Sect. B* **B25**, 2264 (1969).
292. Merlino, S., and Sartori, F., *Contrib. Mineral. Petrol.* **27**, 159 (1970).
293. Merlino, S., and Sartori, F., *Science* **171**, 377 (1971).
294. Merlino, S., and Sartori, F., *Acta Crystallogr., Sect. B* **B28**, 3559 (1972).
295. Mesmer, R. E., Baes, C. F., and Sweeton, F. H., *Inorg. Chem.* **11**, 537 (1972).
296. Meyer, R., Fleury, S., and Paris, R., *J. Pharm. Belg.* **26**, 555 (1971).
297. Mighell, A. D., Perloff, A., and Block, S., *Acta Crystallogr.* **20**, 819 (1966).
298. Momii, R. K., and Nachtrieb, N. H., *Inorg. Chem.* **6**, 1189 (1967).
299. Moore, P. B., and Araki, T., *Am. Mineral.* **59**, 60 (1974).
300. Moore, P. B., and Araki, T., *Am. Mineral.* **59**, 985 (1974).
301. Moore, P. B., and Ghose, S., *Am. Mineral.* **56**, 1527 (1971).
302. Morimoto, N., *Mineral. J.* **2**, 1 (1956).
303. Mun, A. I., Zhaimina, R. E., and Filippova, Z. O., *Izv. Akad. Nauk Kaz. SSR, Ser. Khim.* **21**, 67 (1971).
304. Murray, A. F., and Lockwood, D. J., *J. Phys. C* **11**, 387 (1978).
305. Nagai, T., and Ihara, M., *Yogyo Kyokaishi* **80**, 432 (1972).
306. Natarajan, M., Faggiani, R., and Brown, I. D., *Cryst. Struct. Commun.* **8**, 367 (1979).
307. Nies, N. P., *in* "Mellor's Comprehensive Treatise of Inorganic and Theoretical Chemistry" (R. Thompson, ed.), Vol. V, Part A, pp. 343–501. Longmans, Green, New York, 1980.
308. Nies, N. P., and Hulbert, R. W., *J. Chem. Eng. Data* **12**, 303 (1967).
309. Nies, N. P., and Hulbert, R. W., French Patent 1,560,173 (1969).
310. Nies, N. P., and Hulbert, R. W., U.S. Patent 3,549,316 (1970).
311. Nikol'skii, B. A., and Plyshevskii, Yu. S., *Zh. Neorg. Khim.* **20**, 2231 (1975).

312. Novgorodov, N. G., Sukhnev, V. S., and Brovkin, A. A., *Izv. Akad. Nauk SSSR, Neorg. Mater.* **9**, 1778 (1973).
313. Novgorodov, P. G., *Vestn. Mosk. Univ., Ser. 4: Geol.* **24**, 91 (1969).
314. Olcay, A., and Berber, R., *Commun. Fac. Sci. Univ. Ankara, Ser. B* **23**, 1 (1976).
315. Onak, T. P., Landesman, H., Williams, R. E., and Shapiro, I., *J. Phys. Chem.* **63**, 1533 (1959).
316. Ostrovskaya, I. V., *Tr. Mineral. Muz., Akad. Nauk SSSR* **19**, 197 (1969).
317. Osugi, J., Sato, M., and Fujii, T., *Nippon Kagaku Zasshi* **89**, 562 (1968).
318. Ozeki, M., and Ishii, D., *Bunseki Kiki* **14**, 153 (1976).
319. Ozols, J., *Latv. PSR Zinat. Akad. Vestis. Kim. Ser.* p. 356 (1977).
320. Ozols, J., Dzene, A., and Ievins, A., *Latv. PSR Zinat. Akad. Vestis, Kim. Ser.* p. 759 (1964).
321. Ozols, J., and Ievins, A., *Latv. PSR Zinat. Akad. Vestis, Kim. Ser.* p. 137 (1964).
322. Ozols, J., Ievins, A., and Pecs, L., *Latv. PSR Zinat. Akad. Vestis, Kim. Ser.* p. 382 (1967).
323. Ozols, J., Ievins, A., and Vimba, S., *Zh. Neorg. Khim.* **2**, 2423 (1957).
324. Ozols, J., Tetere, I., and Ievins, A., *Latv. PSR Zinat. Akad. Vestis, Kim. Ser.* p. 3 (1973).
325. Palkina, K. K., Kuznetsov, V. G., and Moruga, L. G., *Zh. Strukt. Khim.* **14**, 1053 (1973).
326. Pardo, J., Martinez-Ripoll, M., and Garcia-Blanco, S., *Acta Crystallogr., Sect. B* **B30**, 34 (1974).
327. Park, M. J., *Sae Mulli* **16**, 150 (1976).
328. Paton, F., and McDonald, S. G. G., *Acta Crystallogr.* **10**, 653 (1957).
329. Perloff, A., and Block, S., *Acta Crystallogr.* **20**, 274 (1966).
330. Peters, C. R., and Milberg, M. E., *Acta Crystallogr.* **17**, 229 (1964).
331. Peterson, G. E., Carnevale, A., and Kurkjian, C. R., *Proc. Symp. Struct. Non-Cryst. Mater., 1976* p. 125 (1977).
332. Piffard, Y., Marchand, R., and Tournoux, M., *Rev. Chim. Miner.* **12**, 210 (1975).
333. Povarennykh, A. S., *Idei E. S. Federova Sovrem. Kristallogr. Mineral.* p. 118 (1970).
334. Prewitt, C. T., and Buerger, M. J., *Am. Mineral.* **46**, 1077 (1961).
335. Pushcharovskii, D. Yu., Karpov, O. G., Leonyuk, N. I., and Belov, N. V., *Dokl. Akad. Nauk SSSR* **241**, 91 (1978).
336. Putnins, J., and Ievins, A., *Latv. Valsts Univ., Kim. Fak. Zinat. Raksti* **22**, 69 (1958).
337. Putnins, J., and Ievins, A., *Uch. Zap., Rizh. Politekh. Inst.* **2**, 67 (1959).
338. Pye, L. D., Frechette, V. D., and Kreidl, N. J., eds., "Materials Science Research," Vol. 12. Plenum, New York, 1978.
339. Razmanova, Z. P., Rumanova, I. M., and Belov, N. V., *Dokl. Akad. Nauk SSSR* **189**, 1003 (1969).
340. Reardon, E. J., *Chem. Geol.* **18**, 309 (1976).
341. Reburn, W. T., and Gale, W. A., *J. Phys. Chem.* **59**, 19 (1955).
342. Ross, V., and Edwards, J. O., *Am. Mineral.* **44**, 875 (1959).
343. Rumanova, I. M., and Ashirov, A., *Kristallografiya* **8**, 517 (1963).
344. Rumanova, I. M., and Ashirov, A., *Kristallografiya* **8**, 828 (1963).
345. Rumanova, I. M., and Gandymov, O., *Kristallografiya* **16**, 99 (1971).
346. Rumanova, I. M., Gandymov, O., and Belov, N. V., *Kristallografiya* **16**, 286 (1971).
347. Rumanova, I. M., Kurbanov, Kh. M., and Belov, N. V., *Kristallografiya* **10**, 601 (1965).
348. Rumanova, I. M., Razmonova, Z. P., and Belov, N. V., *Dokl. Akad. Nauk SSSR* **199**, 592 (1971).

349. Rza-Zade, P. F., Abduragimova, R., Sedel'nikov, G. S., and Afuzova, A., *Issled. Obl. Neorg. Fiz. Khim., Akad. Nauk Azerb. SSR, Inst. Neorg. Fiz. Khim.* p. 69 (1966).

350. Rza-Zade, P. F., Afuzova, A. O., and Abduragimova, R. A., *Issled. Obl. Neorg. Fiz. Khim. Ikh. Rol Khim. Prom-sti, Mater. Nauchn. Konf., 1967* p. 6 (1969).

351. Rza-Zade, P. F., and Afuzova, A. O., *Mater. Konf. Molodykh Uch. Inst. Neorg. Fiz. Khim. Akad. Nauk Azerb SSR, 1968* p. 204 (1969).

352. Rza-Zade, P. F., and Ganf, K. L., *Azerb. Khim. Zh.* p. 91 (1964).

353. Rza-Zade, P. F., and Ganf, K. L., *Issled. Obl. Neorg. Fiz. Khim., Akad. Nauk Azerb. SSR, Inst. Neorg. Fiz. Khim.* p. 56 (1966).

354. Sadanaga, R., *X-Sen* **5**, 2 (1948).

355. Sadanaga, R., Nishimura, T., and Watanabe, T., *Mineral. J.* **4**, 380 (1965).

356. Sadeghi, N., *Ann. Chim. (Paris)* [2] **2**, 123 (1967).

357. Scarpone, A. J., *Diss. Abstr. Int. B* **30**, 1027 (1969).

358. Schäfer, H., and Sieverts, A., *Z. Anorg. Allg. Chem.* **246**, 149 (1941).

359. Schlatti, M., *Naturwissenschaften* **54**, 587 (1967).

360. Schlenker, J. L., Griffen, D. T., Phillips, M. W., and Gibbs, G. V., *Contrib. Mineral. Petrol.* **65**, 347 (1978).

361. Schneider, W., and Carpenter, G. B., *Acta Crystallogr., Sect. B* **B26**, 1189 (1970).

362. Schorsch, G., and Ingri, N., *Acta Chem. Scand.* **21**, 2727 (1967).

363. Schröder, A., and Hoffman, W., *Neues Jahrb. Mineral., Monatsh.* **12**, 265 (1956).

364. Schuckmann, W., *Neues Jahrb. Mineral., Monatsh.* p. 80 (1968).

365. Serdyuk, V. V., Sazonov, A. M., and Porai-Koshits, A. B., *Zh. Fiz. Khim.* **52**, 2919 (1978).

366. Shamaev, P. N., and Grigor'ev, A. P., *Zh. Neorg. Khim.* **18**, 523 (1973).

367. Shashkin, D. P., Simonov, M. A., and Belov, N. V., *Dokl. Akad. Nauk SSSR* **189**, 532 (1969).

368. Shashkin, D. P., Simonov, M. A., and Belov, N. V., *Dokl. Akad. Nauk SSSR* **195**, 345 (1970).

369. Shchigol, M. B., *Zh. Neorg. Khim.* **4**, 2014 (1959).

370. Shchigol, M. B., *Zh. Neorg. Khim.* **6**, 337 (1961).

371. Shchigol, M. B., *Zh. Neorg. Khim.* **6**, 2693 (1961).

372. Shchigol, M. B., *Zh. Neorg. Khim.* **8**, 1361 (1963).

373. Shchigol, M. B., and Burchinskaya, N. B., *Zh. Anal. Khim.* **7**, 289 (1952).

374. Shchigol, M. B., and Burchinskaya, N. B., *Zh. Neorg. Khim.* **6**, 2504 (1961).

375. Shishido, S., *Bull. Chem. Soc. Jpn.* **25**, 199 (1952).

376. Shitov, O. P., Ioffe, S. L., Tartakovskii, V. A., and Novikov, S. S., *Usp. Khim.* **39**, 1913 (1970).

377. Shubin, A. S., Niko'skii, B. A., Plyshevskii, Yu. S., and Kaverzin, E. K., in "Borates and Borate Systems" (G. Ya. Slaidin, ed.), p. 123–127. Zinatne, Riga, 1978.

378. Silins, E., and Ievins, A., *Latv. PSR Zinat. Akad. Vestis, Kim. Ser.* p. 747 (1975).

379. Silins, E., Ozols, J., and Ievins, A., *Kristallografiya* **18**, 503 (1973).

380. Silins, E., Ozols, J., and Ievins, A., *Latv. PSR Zinat. Akad. Vestis, Kim. Ser.* p. 115 (1974).

381. Simonov, M. A., *Vestn. Mosk. Univ., Ser. 4: Geol.* **30**, 15 (1975).

382. Simonov, M. A., and Belov, N. V., *Krist. Strukt. Neorg. Soedin.* p. 166 (1974).

383. Simonov, M. A., Egorov-Tismenko, Yu. K., and Belov, N. V., *Kristallografiya* **21**, 592 (1976).

384. Simonov, M. A., Yamnova, N. A., Kazanskaya, E. V., Egorov-Tismenko, Yu. K., and Belov, N. V., *Dokl. Akad. Nauk SSSR* **228**, 1337 (1976).

385. Smirnova, G. M., Zdanovskii, A. B., Sushkova, S. G., Zhitkova, T. N., Leont'eva, I. A., and Shishkina, Z. I., *Zh. Neorg. Khim.* **18**, 518 (1973).
386. Smirnova, G. M., Zhitkova, T. N., Sushkova, S. G., Kalitina, L. N., and Shishkina, Z. I., *Zh. Neorg. Khim.* **17**, 1903 (1972).
387. Smith, H. D., and Wiersema, R. J., *Inorg. Chem.* **11**, 1152 (1972).
388. Sokolova, E. V., Azizov, A. V., Simonov, M. A., Leonyuk, N. I., and Belov, N. V., *Dokl. Akad. Nauk SSSR* **243**, 655 (1978).
389. Souchay, P., *Bull. Soc. Chim. Fr.* p. 395 (1953).
390. Souchay, R., and Teyssedre, M., *C. R. Hebd. Seances Acad. Sci.* **236**, 1965 (1953).
391. Spessard, J. E., *J. Inorg. Nucl. Chem.* **32**, 2607 (1970).
392. Sprague, R. W., *in* "Mellor's Comprehensive Treatise of Inorganic and Theoretical Chemistry" (R. Thompson, ed.), Vol. V, Part A, pp. 224–320. Longmans, Green, New York, 1980.
393. Stewner, F., *Acta Crystallogr., Sect. B* **B27**, 904 (1971).
394. Suknev, V. S., *Zh. Prikl. Spektrosk.* **12**, 491 (1970).
395. Svanson, S. E., *Acta Chem. Scand.* **23**, 2005 (1969).
396. Svarcs, E., and Dzene, A., *Proc. Eur. Symp. Therm. Anal., 1st, 1976* p. 291 (1976).
397. Svarcs, E., Grundsteins, V., and Ievins, A., *Zh. Neorg. Khim.* **12**, 2017 (1967).
398. Svarcs, E., Grundsteins, V., and Ievins, A., *Latv. PSR Zinat. Akad. Vestis, Kim. Ser.* p. 263 (1968).
399. Svarcs, E., and Ievins, A., *Latv. PSR Zinat. Akad. Vestis, Kim. Ser.* p. 135 (1956).
400. Svarcs, E., and Ievins, A., *Zh. Neorg. Khim.* **2**, 439 (1957).
401. Svarcs, E., and Ievins, A., *Zh. Neorg. Khim.* **5**, 1676 (1960).
402. Svarcs, E., Rigerte, I., Dzene, A., and Ievins, A., *Latv. PSR Zinat. Akad. Vestis, Kim. Ser.* p. 658 (1972).
403. Svarcs, E. M., and Vimba, S. G., *Zh. Obshch. Khim.* **27**, 23 (1957).
404. Takeuchi, Y., *Acta Crystallogr.* **3**, 208 (1950).
405. Takeuchi, Y., *Acta Crystallogr.* **5**, 574 (1952).
406. Takeuchi, Y., and Kudoh, Y., *Am. Mineral.* **60**, 273 (1975).
407. Teeple, J. E., "The Industrial Development of Searles Lake Brines." American Chemical Society, Washington, D.C. 1929.
408. Tennyson, C., *Fortschr. Mineral.* **41**, 64 (1963).
409. Toledano, P., and Hebrard-Matringe, M. A., *C. R. Hebd. Seances Acad. Sci., Ser. C* **264**, 305 (1967).
410. Toledano, P., and Hebrard-Matringe, M. A., *C. R. Hebd. Seances Acad. Sci., Ser. C* **266**, 1155 (1968).
411. Toledano, P., and Hebrard-Matringe, M. A., *Bull. Soc. Chim. Fr.* p. 1462 (1969).
412. Touboul, M., *Rev. Chim. Miner.* **8**, 347 (1971).
413. Touboul, M., and Amoussou, D., *Rev. Chim. Miner.* **15**, 223 (1978).
414. Valyashko, M. G., and Gode, G. K., *Zh. Neorg. Khim.* **5**, 1316 (1960).
415. Valyashko, M. G., and Vlasova, E. V., *Geokhimiya* p. 818 (1966).
416. Valyashko, M. G., and Vlasova, E. V., *Vestn. Mosk. Univ., Ser. 4: Geol.* **22**, 75 (1967).
417. Valyashko, M. G., and Vlasova, E. V., *Jena Rev.* **14**, 3 (1969).
418. Vegas, A., Cano, F. H., and Garcia-Blanco, S., *Acta Crystallogr., Sect. B* **B31**, 1416 (1975).
419. Vegas, A., Cano, F. H., and Garcia-Blanco, S., *J. Solid State Chem.* **17**, 151 (1976).
420. Vegas, A., Cano, F. H., and Garcia-Blanco, S., *Acta Crystallogr., Sect. B* **B33**, 3607 (1977).
421. Venkatakrishnan, V., and Buerger, M. J., *Z. Kristallogr.* **135**, 321 (1972).
422. Vinogradov, E. E., Zaitseva, S. N., Samoilov, O. Ya., and Yashkichev, V. I., *Zh. Fiz. Khim.* **40**, 2519 (1966).

423. Vlasova, E. V., and Valyashko, M. G., *Zh. Neorg. Khim.* **11**, 1539 (1966).
424. Wan, C., and Ghose, S., *Am. Mineral.* **62**, 1135 (1977).
425. Wang, N., *Neues Jahrb. Mineral., Monatsh.* p. 315 (1971).
426. Wang, N., *Proc. Geol. Soc. China* **14**, 195 (1971).
427. Ward, G. K., and Millero, F. J., *Geochim. Cosmochim. Acta* **39**, 1595 (1975).
428. Waterman, H. H., and Volkoff, G. M., *Can. J. Phys.* **33**, 156 (1955).
429. Waugh, J. L. T., *Struct. Chem. Mol. Biol.* p. 731 (1968).
430. Wegener, K., *J. Inorg. Nucl. Chem.* **29**, 1847 (1967).
431. Weir, C. E., *J. Res. Natl. Bur. Stand., Sect. A* **70A**, 153 (1966).
432. Weir, C. E., and Schroeder, R. A., *J. Res. Natl. Bur. Stand., Sect. A* **68A**, 465 (1964).
433. Wells, A. F., "Structural Inorganic Chemistry," 4th ed. Oxford Univ. Press, London and New York, 1975.
434. White, W. B., Brawer, S. A., Furukawa, T., and McCarthy, G. J., *Mater. Sci. Res.* **12**, 281 (1978).
435. Wieder, H. H., Clawson, A. R., and Parkerson, C. R., U.S. Patent 3,337,292 (1967).
436. Wieder, H. H., Clawson, A. R., and Parkerson, C. R., U.S. Patent 3,337,293 (1967).
437. Wuhan College of Geology, *K'o Hsueh T'ung Pao* **22**, 364 (1977).
438. Yakubovich, O. V., Simonov, M. A., Belokoneva, E. L., Egorov-Tismenko, Yu. K., and Belov, N. V., *Dokl. Akad. Nauk SSSR* **230**, 837 (1976).
439. Yakubovich, O. V., Simonov, M. A., and Belov, N. V., *Dokl. Akad. Nauk SSSR* **238**, 98 (1978).
440. Yakubovich, O. V., Yamnova, N. A., Shchedrin, B. M., Simonov, M. A., and Belov, N. V., *Dokl. Akad. Nauk SSSR* **228**, 842 (1976).
441. Yamnova, N. A., Egorov-Tismenko, Yu. K., Simonov, M. A., and Belov, N. V., *Dokl. Akad. Nauk SSSR* **216**, 1281 (1974).
442. Yamnova, N. A., Simonov, M. A., and Belov, N. V., *Zh. Strukt. Khim.* **17**, 489 (1976).
443. Yamnova, N. A., Simonov, M. A., Kazanskaya, E. V., and Belov, N. V., *Dokl. Akad. Nauk SSSR* **225**, 823 (1975).
444. Ysker, J. St., and Hoffman, W., *Naturwissenschaften* **57**, 129 (1970).
445. Yun, Y.-H., *Diss. Abstr. Int. B* **39**, 4964 (1979).
446. Zachariasen, W. H., *Acta Crystallogr.* **7**, 305 (1954).
447. Zachariasen, W. H., *Acta Crystallogr.* **16**, 380 (1963).
448. Zachariasen, W. H., *Acta Crystallogr.* **16**, 385 (1963).
449. Zachariasen, W. H., *Acta Crystallogr.* **17**, 749 (1964).
450. Zachariasen, W. H., *Acta Crystallogr.* **23**, 44 (1967).
451. Zachariasen, W. H., and Plettinger, H. A., *Acta Crystallogr.* **16**, 376 (1963).
452. Zachariasen, W. H., Plettinger, H. A., and Marezio, M., *Acta Crystallogr.* **16**, 1144 (1963).
453. Zayakina, N. V., and Brovkin, A. A., *Kristallografiya* **21**, 502 (1976).
454. Zayakina, N. V., and Brovkin, A. A., *Kristallografiya* **22**, 275 (1977).
455. Zayakina, N. V., and Brovkin, A. A., *Kristallografiya* **23**, 1167 (1978).
456. Zdanovskii, A. B., and Lyakhovskaya, E. I., *Tr. Vses., Nauchno-Issled. Inst. Galurgii* p. 96 (1959).
457. Zeigan, D., *Z. Chem.* **7**, 241 (1967).
458. Zviedre, I., and Ievins, A., *Latv. PSR Zinat. Akad. Vestis, Kim. Ser.* p. 395 (1974).
459. Zviedre, I., and Ievins, A., *Latv. PSR Zinat. Akad. Vestis, Kim. Ser.* p. 401 (1974).
460. Zviedre, I., Ozols, J., and Ievins, A., *Zh. Neorg. Khim.* **13**, 2613 (1968).
461. Zviedre, I., Ozols, J., and Ievins, A., *Zh. Neorg. Khim.* **14**, 601 (1969).
462. Zyubova, N. A., Berg, L. G., and Rza-Zade, P. F., *Issled. Obl. Neorg. Fiz. Khim.* p. 94 (1970).

ADVANCES IN INORGANIC CHEMISTRY AND RADIOCHEMISTRY, VOL. 25

COMPOUNDS OF GOLD IN UNUSUAL OXIDATION STATES

HUBERT SCHMIDBAUR and KAILASH C. DASH*

Anorganisch-Chemisches Institut der Technischen Universität München
Garching, Federal Republic of Germany

I. Introduction 239
II. Compounds of Gold in Oxidation State −1 240
III. Compounds of Gold in Fractional Oxidation States:
    Gold Clusters 243
    A. A Cluster with an Au_5 Core 243
    B. A Cluster with an Au_6 Core 244
    C. Clusters with an Au_8 Core 245
    D. The Two Types of Au_9 Clusters 246
    E. Clusters with an Au_{11} Core 248
IV. Compounds of Gold in Oxidation State +2 249
    A. Introduction 249
    B. Pseudo-Gold(II) Compounds 249
    C. Gold(II) Species as Reaction Intermediates 251
    D. Mononuclear Gold(II) Compounds 252
    E. Binuclear Gold(II) Compounds Containing a
    Metal–Metal Bond 254
V. Compounds of Gold in Oxidation State +5 257
    A. Hexafluoroaurates(V) 258
    B. Fluorokryptyl(II) Hexafluoroaurate(V) 260
    C. Gold Pentafluoride 260
    References 262

I. Introduction

Compared to the rich chemistry of the neighboring element platinum, the chemistry of gold has developed very slowly. With the advent of a new interest in the role of noble metals in catalysis, in microelectronics, and in pharmacology, gold and its compounds have become the focus of active research in a large number of laboratories throughout the world. New books (77, 84) and review articles (85) witness this recent development and will stimulate further investigations.

* Permanent address: Department of Chemistry, Utkal University, Vani Vihar, Bhubaneswar 751004, India.

Some of the most exciting developments in gold chemistry were initiated by the discovery of unusual oxidation states. This new domain ranges from the rare -1 state in certain alloys of gold, through the gold clusters with nonintegral values between 0 and $+1$, through $+2$ in both mononuclear Au^{2+} and binuclear $[Au_2]^{4+}$ species, to $+5$ in binary or complex fluorides. Thus within only a very few years, there has emerged a much broader scope of the chemistry of gold, which was formerly represented only by Au metal, gold(I), and gold(III).

In this account, the new oxidation states are introduced in the format of a short, but it is hoped complete, summary covering the literature until late 1980. Some of the topics were reviewed previously, but the articles soon became out-dated because of the rapid growth of the fields. Best known are the discussion of gold(II) compounds (13) and the early presentation of gold(V) complexes (5). Gold clusters have been included in the leading reference texts on gold chemistry (77, 85), and the -1 state appeared in a recent summary on liquid metals and liquid semiconductors (39).

II. Compounds of Gold in Oxidation State -1

As a general rule, the chemistry of metal ions in solution is characterized by the ability to form *positive* ions. The stable *negative* metal ions are well characterized only in the gas phase (43). Mercury forms a monoatomic vapor and behaves like an inert gas in this respect. The gold atom, on accepting an electron, fills its 6s shell and is expected to be converted to the auride ion Au^-, whose electronic configuration is the same as that of Hg, $5d^{10}6s^2$. Further, the relatively high electron affinity of Au (2.4 eV) also suggests that it could behave like halogen atoms in forming negatively charged ions. This prediction has been realized, and a few alkali metal *aurides,* $[M]^+[Au]^-$, have been characterized.

As early as 1943, Sommer (101) reported the existence of a *stoichiometric compound* CsAu, exhibiting *nonmetallic* properties. Later reports (53, 102, 103, 123) confirmed its existence and described the crystal structure, as well as the electrical and optical properties of this compound. The lattice constant of its CsCl-type structure is reported (103) to be 4.263 ± 0.001 Å. Band structure calculations are consistent with observed experimental results that the material is a semiconductor with a band gap of 2.6 eV (102). The phase diagram of the Cs–Au system shows the existence of a discrete CsAu phase (32) of melting point 590°C and a very narrow range of homogeneity (42).

The conductivity measurements of *molten* CsAu (42) confirm the nonmetallic electronic structure. The dramatic drop of the electrical conductivity σ in the 50/50 region is shown in Fig. 1. The measurements of electrical transport properties, conductivity, and thermoelectric power of the *liquid Cs–Au alloy system* as a function of temperature and composition, as well as the electromigration studies, show that this liquid alloy consists essentially of Cs^+ and Au^- *ions (38, 56, 94)*.

Further support for this ionicity comes from the recent Mössbauer (*122*) and ESCA (*55, 122*) spectroscopic measurements *of the solid.* The ESCA measurements of a series of gold compounds revealed (*55, 79, 89*) a monotonic relationship between binding energies of the Au ($4f\frac{7}{2}$) levels and formal oxidation states, and provided a reliable probe for the oxidation states for Au in its compounds. The photoelectron spectra of Au^-, Au^0 (metal), Au^+, Au^{2+}, and Au^{3+} compounds showed that the $4f\frac{7}{2}$ electron levels progressively shift to higher energy in the sequence Au^0, Au^+, Au^{2+}, and Au^{3+}, indicating the presence of Au^- in CsAu or RbAu (Fig. 2) (*55*). Both Cs and Au are classical metals with an electronegativity difference, according to the Pauling scale (*72*), that is characteristic of typical salts like CsI (*94*). Thus the Cs atoms appear to donate just one valence electron to each Au atom to fill its 6s shell, turning it into the Au^- ion. Similar results are obtained from a correlation of Mössbauer isomeric shifts and the electronegativity difference in a series of metal aurides (*122*).

Ultraviolet photoemission spectra have also been used to investigate the electronic band structure of *solid CsAu*. The apparent spin–orbit

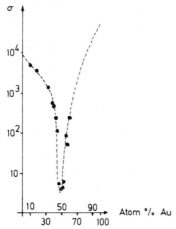

FIG. 1. Electrical conductivity σ [ohm^{-1} cm^{-1}] of liquid CsAu mixtures as a function of the gold content at 600°C (*39*).

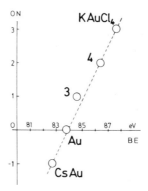

FIG. 2. Binding energy (BE) of the Au(4f $\frac{7}{2}$) level in compounds of different oxydation numbers (ON), from ESCA measurements (*55, 89*). (For compounds 3 and 4, see Section IV,5).

splitting of the Au 5d orbitals is found to be 2.1 eV, i.e., 40% larger than in the Au atom. The results were correlated with theoretical calculations using the various relativistic methods (*37, 62, 70*).

Evaluation of intensities of X-ray diffraction patterns indicates that the CsCl structure is well ordered in crystalline CsAu. *Excess Cs,* which is the primary source for conduction electrons, is dispersed in the lattice with increase of the cell parameter (*105*). Cesium-133 NMR measurements (line shapes, Knight shifts, and relaxation times) confirm this result (*105*). The interpretation of the data for RbAu is less straightforward, however.

The existence of other alkali metal–gold compounds of the same stoichiometry, MAu, has also been reported. Spicer, *et al.* (*103*) investigated RbAu, KAu, and NaAu systems, and Kienast and Verma (*53*) showed the existence of stable phases at the RbAu and KAu stoichiometries. Other alkali metal–gold compounds of different stoichiometries, such as Li_3Au, Li_4Au_5, Na_2Au, $NaAu_2$, NaAu, K_2Au, KAu_2, KAu, KAu_5, RbAu, $RbAu_2$, and $RbAu_5$ also exist (*53, 78*).

Recent mass spectral studies confirm the presence of CsAu molecules in the *gas phase.* From the appearance potentials and the slope of the ionization curve, a dissociation energy of 460 kJ mol^{-1} was deduced, which agrees well with predicted values for a largely ionic bond. It is also very similar to the value arrived at for CsCl, 444 kJ mol^{-1} (*19a*).

More recently, Peer and Lagowski (*75*) presented spectroscopic and electrochemical evidence for the existence of Au^- ions *in liquid ammonia.* This is the only example of a transition metal *anion* in any solvent. In a simple yet elegant technique, the solutions of auride ion, Au^-, could be prepared by the dissolution of metallic gold in ammonia solutions of Cs, Rb, or K, but not in those of Na, Li, Sr, or Ba. The Au^- was shown to be stable even in the absence of any solvated electron. This work also is enlightening with respect to early unexplained results by Zintl *et al.* (*124*), who showed that *two* equivalents of sodium are required in a titration of AuI in liquid ammonia!

III. Compounds of Gold in Fractional Oxidation States: Gold Clusters

Reports of a strong intermetallic bond in the Au_2 molecule in the gas phase (*65*) and the stability of R_3PAuX compounds, prompted Malatesta (*65*) to attempt the synthesis of compounds of the type R_3P—Au—Au—PR_3 by reduction of R_3PAuX with suitable agents, e.g., $NaBH_4$ in EtOH, KOH in MeOH, or $LiAlH_4$ in THF. However, instead of the expected compounds of gold(0), novel clusters of gold were obtained containing Au_6, Au_9, and Au_{11} units, depending on the nature of X and R_3P. These compounds appear to have the metal formally in a nonintegral oxidation state between 0 and +1, and X-ray analysis shows their structures to be closely related to that of the boranes and carboranes (*68*). Most of the cluster molecules or ions are stable to air and water and are easily characterized by modern physical techniques, although yields of most preparations are low.

Application of other synthetic techniques, predominantly metal vapor methods, have added other cluster stoichiometries to this list, which presently includes Au_5, Au_6, Au_8, Au_9, and Au_{11}. The few examples for Au_2 (*3, 69*), Au_3 (*30*), and Au_4 (*36, 76*) will not be considered here, since they do not involve gold in unusual oxidation states. The sections are ordered according to the cluster size, although, historically, the Au_{11} moiety was the first to be detected (*65*).

A. A CLUSTER WITH AN Au_5 CORE

A cluster containing five gold atoms has been prepared by the evaporation of metallic gold into an ethanolic solution of bis(diphenylphosphino)methane (dppmH) and NH_4NO_3. Different mole ratios were used, all of which resulted in the formation of the cation $[Au_5(dppmH)_3(dppm)]^{2+}$. The nitrate forms red crystals, which were shown by X-ray diffraction studies to be composed of NO_3^- anions and cations of the structure shown in Fig. 3 (*108*).

Although the crystals are sensitive to X-ray irradiation, the structure could be refined to satisfactory R values. The Au_5 skeleton can be described as a tetrahedron of four Au atoms with the fifth Au atom

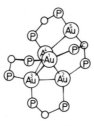

FIG. 3. Cluster structure of the cation in [Au₅(dppmH)₃(dppm)]²⁺(NO₃⁻)₂, as determined by X-ray diffraction. Hydrogen atoms and phenyl groups are omitted (*108*).

being attached to Au(4). A dppmH and a dppm ligand bridge Au(2) and Au(4) with Au(5). The dppm ligand has lost a proton from the CH₂ bridge and is attached to Au(5) through its carbon atom. A similar metal–ligand interaction of dppm is known from only very few other examples (*91*). The compound with its formal oxidation state +0.60 for Au is diamagnetic. Mössbauer spectra and ³¹P NMR are also available but as yet unpublished (*108*). The structure has not yet been related to existing theoretical models.

B. A CLUSTER WITH AN Au₆ CORE

Reduction of (4-CH₃—C₆H₄)₃PAuNO₃ by NaBH₄ in ethanol (molar ratio 4 : 1) yields a red solution from which a tetraphenylborate salt of the cation {[(4-CH₃—C₆H₄)₃PAu]₆}²⁺ can be precipitated in the form of yellow crystals as a small by-product. [The main product appears in the form of red crystals, originally not further specified (*10, 12*).] The X-ray diffraction analysis gave the cation structure shown in Fig. 4.

The core of the cation is an octahedron of gold atoms (formal oxidation state +0.33) with a crystallographic center of symmetry. The six independent edges of the polyhedron can be divided into two classes: two opposite faces, related by the inversion center, have long edges, while the remaining edges are shorter (3.073 and 2.965 Å, respectively). Each of the Au atoms bears one phosphine ligand (*10, 12*). It is

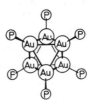

FIG. 4. Cluster structure of the cation in [Au₆(P(C₆H₄-CH₃)₃)₆]²⁺[B(C₆H₅)₄⁻]₂, as determined by X-ray diffraction; p-tolyl groups are omitted (*10*).

important to point out that the presence of hydrido ligands at or even in the cluster cannot be ruled out completely. However, this is rather improbable in view of the crystal stability: the samples were completely stable in air. The only other coinage metal cluster of six atoms has the composition $[HCuP(C_6H_5)_3]_6$ (17).

Au_6 clusters with chelating ligands were also synthesized, but are not yet structurally characterized (68a,b).

A molecular orbital scheme has been proposed for the $[L_6Au_6]^{2+}$ clusters, which shows that the gold 6s orbitals make the dominant contribution to the bonding. Coordination of ligands to the metal clusters encourages a more favorable hydridization of the metal orbitals and results in stronger radial metal–metal bonding (8, 68). The distortion of the octahedron is believed to remove orbital degeneracy in the highest occupied molecular orbital (HOMO) and to give rise to a singlet ground state and diamagnetic properties. Unfortunately, the magnetic properties have not yet been studied, because of insufficient sample quantities.

C. CLUSTERS WITH AN Au_8 CORE

A complex of the type $[(C_6H_5)_3PAu]_8^{2+} \cdot 2NO_3^-$ is prepared on treatment of $[[(C_6H_5)_3P]_8Au_9](NO_3)_3$ (see Section III,D) with an excess of $P(C_6H_5)_3$ in methanol. The octahydrate can be precipitated in 80% yield by addition of diethyl ether. The red product is recrystallized from CH_2Cl_2 and diethyl ether, and the corresponding picrate, alizarinsulfonate (aliz), hexafluorophosphate, perchlorate, and tetraphenylborate were obtained by metathetical exchange. The compounds were formulated as $[L_8Au_8]Y_2$ on the basis of analysis, conductivity measurements, ^{1}H-NMR spectroscopy, and crystallographic determination of molecular structure (66).

In an independent and probably simultaneous study, the above reaction of excess ligand with the Au_9 cluster was carried out in CH_2Cl_2 and followed by ^{31}P-NMR spectroscopy. The equation

$$[L_8Au_9]^{3+} + 2L \rightarrow [L_8Au_8]^{2+} + [LAuL]^+ \tag{1}$$

was found to account for both product distribution and yields. The product composition $[L_{10}Au_9]^{3+}$ suggested earlier (117) was thus ruled out and discarded (66, 107).

The crystal structures of $\{[(C_6H_5)_3P]_8Au_8\}(aliz)_2$ (66) and of $\{[(C_6H_5)_3P]_8Au_8\}(PF_6)_2 \cdot 2CH_2Cl_2$ (118) were determined and the results are in excellent agreement. The structures of the cations are shown in Figs. 5 and 6. The core is a gold-centered cluster with one ligand bonded

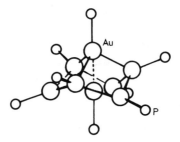

FIG. 5. Cluster structure of the cation $[Au_8(PPh_3)_8]^{2+}$ in the alizarinsulfonate, as determined by X-ray diffraction. Phenyl groups are omitted (66).

to each metal atom. It is related to the Au_{11} cluster by the removal of the three gold atoms of the basal triangle (see Section III,E).

D. THE TWO TYPES OF Au_9 CLUSTERS

A first set of $[(R_3P)_8Au_9]X_3$ cluster compounds, with R = C_6H_5 and 4-CH_3—C_6H_5 and X = NO_3, PF_6, BF_4, or picrate, was synthesized by $NaBH_4$ reduction of R_3PAuNO_3 complexes in ethanol, with metathesis to introduce other anions where necessary. The compounds are green, diamagnetic, crystalline solids, and are best obtained when X has very poor coordinating properties, if any (9, 23, 65). Seven members of this series have been prepared, and the species with R = 4-CH_3—C_6H_4 and X = PF_6 has been investigated by X-ray diffraction. The structure of the cation is shown in Fig. 7. In this cluster, a central gold atom is bound to eight peripheral gold atoms, each attached to a phosphine ligand. The center–periphery distances fall in the ranges 2.689(3) and 2.729(3) Å; peripheral distances are around 2.752(3) and 2.868(3) Å, all being shorter than in gold metal (2.884 Å). The geometry may be con-

FIG. 6. Cluster structure of the cation in $[Au_8(PPh_3)_8]^{2+}(PF_6^-)_2 \cdot 2CH_2Cl_2$, as determined by X-ray diffraction. Phenyl groups are omitted (118).

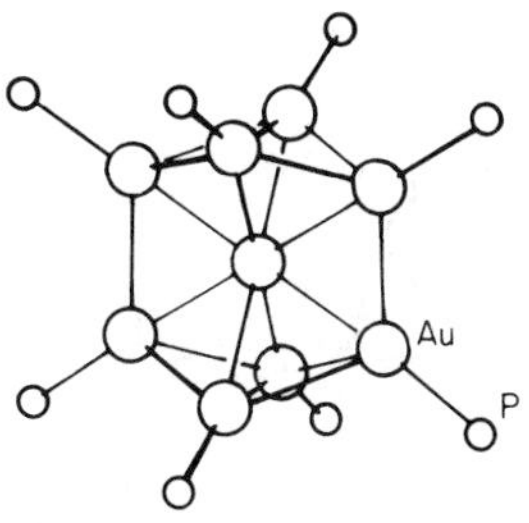

FIG. 7. Cluster structure of the cation in $\{Au_9[P(C_6H_4\text{—}CH_3)_3]_8\}^{3+}(PF_6^-)_3$, as determined by X-ray diffraction (9).

sidered as that of a centered icosahedron from which an equatorial rectangle (two pairs of adjacent vertices from opposite sides) has been removed (65).

Extended Hückel molecular orbital calculations give a diamagnetic closed-shell electronic ground state (^1A_g) with a HOMO/LUMO (lowest unoccupied molecular orbital) gap of 1.7 eV. Computer limitations prevented a full calculation, but the dominant role of the Au 6s orbitals for cluster bonding suggests a simpler and computationally less demanding model (68). In X-ray photoelectron (8) and Mössbauer spectra (116), there is only limited evidence for gold in different coordination sites. $\{[(C_6H_5)_3P]_8Au_9\}(NO_3)_3$ reacts with Cl^-, Br^-, I^-, SCN^-, and CN^- in methanol at $-50°C$ to form the metathesis products, which were characterized by ^{31}P-NMR and Mössbauer spectroscopy (115, 117). Surprisingly, no separate Mössbauer line is observed for the central gold atom. The isomer shift and quadrupole splitting for the peripheral atoms are similar to those in Au_{11} clusters (see Section III,E).

Solutions of the halides and pseudohalides in CH_2Cl_2 are unstable, and the complexes are transformed into Au_{11} clusters or—in the presence of excess ligand—to Au_8 clusters (see Section III,C) (66, 107, 117).

The first *neutral* Au_9 cluster was prepared by reduction of tri(cyclohexyl)phosphinogold(I) thiocyanate, $(c\text{-}C_6H_{11})_3PAuSCN$, with $NaBH_4$, and crystallization from heptane/CH_2Cl_2 as red crystals, dec. 129°C. Its ^{31}P-NMR spectrum shows only one signal at $\delta = 73$ ppm. The structure was determined by X-ray diffraction, and the result is shown in Fig. 8 (26). The formula is thus $[(c\text{-}C_6H_{11})_3P]_5Au_9(SCN)_3$. Gold atoms Au(2), Au(6), and Au(8) are attached to sulfur; Au(1), Au(3), Au(4), Au(5), Au(8), and Au(9) to phosphorus; and the central atom Au(7) bears no ligand. The structure can be derived from the icosahedral framework by removing a triangular set of vertices and one of the opposite gold atoms.

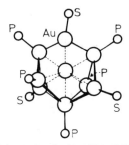

FIG. 8. Cluster structure of the molecule $\{Au_9[P(c\text{-}C_6H_{11})_3]_5\}(SCN)_3$, as determined by X-ray diffraction. Cyclohexyl and CN groups are omitted (26).

E. CLUSTERS WITH AN Au_{11} CORE

Compounds of the composition $L_7Au_{11}X_3$ were the first gold clusters to be discovered (in 1966) (65). They are obtained by reduction of LAuX complexes, where X = I, CN, or SCN. Seven members of the series with L = $P(C_6H_5)_3$, $P(C_6H_4\text{-}4\text{-}CH_3)_3$, $P(C_6H_4\text{-}4\text{-}Cl)_3$, or $P(C_6H_4\text{-}4\text{-}F)_3$, and X = SCN or CN, were fully characterized, and the structures of $\{[(C_6H_5)_3P]_7\text{-}Au_{11}\}(SCN)_3$ and $[(4\text{-}Cl\text{—}C_6H_4)_3P]_7Au_7I_3$ have been determined. The common feature of the two structures, a gold-centered Au_{11} core, is shown in Fig. 9 (2, 11, 22, 67). The center–periphery distances are 2.68 Å and the peripheral distances 2.98 Å. Each peripheral Au atom bears one L or X ligand. The cluster geometry is derived from an icosahedron through replacement of three atoms at the vertices of a triangular face by one atom at the center of the missing face. Space group and unit cell data are available for three other crystalline compounds (2).

Cluster compounds with an Au_{11} core can also be synthesized by the metal vaporization technique. Ethanol films containing L and LAuX at

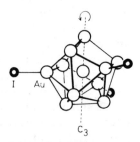

FIG. 9. Cluster structure of the molecule $\{Au_{11}[P(C_6H_4\text{—}Cl)_3]_7I_3\}$, as determined by X-ray diffraction. Phosphine ligands are omitted (2).

$-100°C$ give good yields of the products when exposed to Au vapor at 10^{-2} torr (116). Mössbauer spectra of the Au_{11} clusters exhibit a set of resonance doublets in agreement with the above structure, though assignments are not straightforward (116). X-Ray photoelectron spectroscopy failed to give direct evidence for metal atoms in different coordination sites, however.

The $Au(4f_{\frac{7}{2}})$ binding energies of ~ 83.7 eV may be correlated with the low oxidation state of $+0.27$ (8). The values for the $L_8Au_9X_3$ compounds are similar (84.4 eV).

Molecular orbital calculations (68) give a closed-shell electronic ground-state configuration ($^1A_1$) for an idealized C_{3v} geometry. Other geometrical approaches give slightly modified results. The calculated charge densities show significant differences for the five different sites of the cluster.

IV. Compounds of Gold in Oxidation State +2

A. Introduction

A critical examination of the ionization potentials of the members of group IB—i.e., Cu, Ag, Au—shows that, to attain tervalency, less energy is required for Au than for Cu and Ag. However, more energy is required to reach the Au^{2+} ion starting from the uncharged atom than for either Cu or Ag. In addition, the ligand field factors are significant. In d^9 metal complexes with either a tetragonally distorted octahedral or square planar structure, the odd electron is in the $d_{x^2-y^2}$ orbital, and because of greater ligand field splitting ($\sim 80\%$ greater) in the case of Au as compared to Cu, this orbital has a much higher energy so that the odd electron in Au^{2+} can be easily ionized. It is not surprising, therefore, to observe that divalent Au compounds readily disproportionate to species in monovalent and trivalent states.

B. Pseudo-Gold(II) Compounds

A large number of complexes originally believed to contain Au(II) were later proved to be mixed-valence systems containing equal proportions of linear two-coordinate Au(I) and square-planar tetracoordinate Au(III) species. Thus, $CsAuCl_3$ (33), $(PhCH_2)_2SAuCl_2$ (18, 25, 40), $AuCl(dmg)$ (dmg = dimethylglyoxime) (81), and $AuBr(n\text{-}Bu_2NCS_2)$ (16) are all mixed-valence gold compounds. Also, typically, $Ph_3P\cdot Au(S_2C_2(CN)_2)$ is an Au(I)/Au(III) salt (1) with a linear Au(I) cation and a square-planar complex Au(III) anion (14). The situation is analogous with the related "mnt" complex:

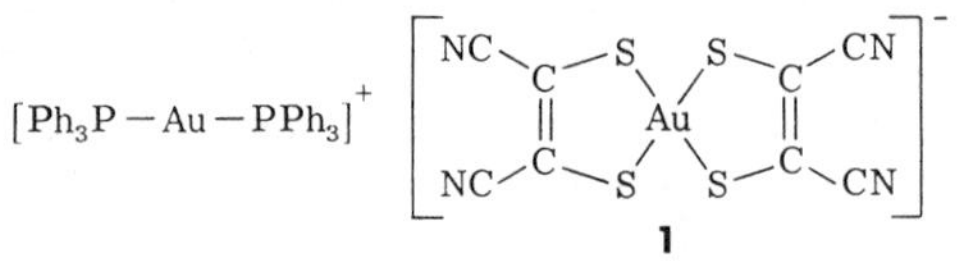

$[(Ph_3P)_2Au]^+[(mnt)_2Au]^-$, mnt = maleonitriledithiolate)

A similar situation probably exists in the following cases: $AuSO_4$ (95), AuO (57), AuS (13), $(PhCH_2)_2SAuX_2$ (X = Br, I) (99), $(nb)_x \cdot Au_2Cl_4$ (nb = norbornadiene; x = 1, 2, or 3) (44). See the accompanying tabulation.

Stoichiometry	Mixed-valence formula
AuO	$Au(I)/Au(III)O_2$
AuS	$Au(I)/Au(III)S_2$
$AuSO_4$	$Au(I)/Au(III)(SO_4)_2$
$CsAuX_3$ (X = Cl, Br, I)	$Cs_2(AuX_2)(AuX_4)$
$(PhCH_2)_2SAuX_2$ (X = Cl, Br)	$(PhCH_2)_2SAuX/(PhCH_2)_2SAuX_3$
$(n\text{-}Bu_2NCS_2)AuCl$	$[(n\text{-}Bu_2NCS_2)_2Au]^+[AuCl_2]^-$
$(dmg)AuCl$	$[(dmg)_2Au]^+[AuCl_2]^-$

Much debate has centered around a compound $AuCl_2$ reported by Thomson (104) and disputed by Krüss and Schmidt (58) in the late 19th century. Though Brewer (19) cited evidence for $AuCl_2(g)$ at high temperatures, the phase diagram and powder-pattern studies (27) of the solid did not support this. Finally the reaction of $AuCl_3$ and $Au(CO)Cl$ in an inert atmosphere was reported (28) to form black crystalline $AuCl_2$, which was diamagnetic and had a powder pattern different from both AuCl and $AuCl_3$ (48). The X-ray structure analysis $(21, 29)$ of this compound shows it to be a Au(I)–Au(III) mixed-valence compound with the tetranuclear formula Au_4Cl_8 and adopting a chair-like arrangement, in which the Au(I) is almost linearly coordinated and the Au(III) has a slightly distorted planar four-coordinate structure (Fig. 10). The distance between two Au(I) centers is 3.095 Å, as compared to 2.884 Å in the metal (74) and to the Au–Au mean distances (2.98 and 3.02 Å) between peripherial gold atoms in some clusters $(11, 12)$.

The electrical conductivity and conductivity activation energy was

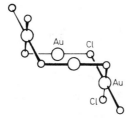

FIG. 10. Molecular structure of Au_4Cl_8, as determined by X-ray diffraction (29).

measured for the mixed-valence compounds $(PhCH_2)_2SAuCl_2$ and $(PhCH_2)_2SAuBr_2$ at pressures up to 360 kbar. The results indicate an enhancement of the conductivity in the range of 100–250 kbar attributed to the compression along the metal–halogen chains leading to nearly equivalent coordination environments for the metal atoms. Similar observations have previously been made with $Cs_2Au_2Cl_6$ (45, 52).

C. GOLD(II) SPECIES AS REACTION INTERMEDIATES

Evidence for the existence of a true mononuclear Au(II) complex was first provided in 1954 by Rich and Taube (79), who, on the basis of kinetic evidence, proved the transient existence of $[AuCl_4]^{2-}$ in aqueous solution, as a reaction intermediate in the Fe(II)-catalyzed exchange of a radioactive Cl^- with $[AuCl_4]^-$ [Eq. (2)].

$$Fe^{2+} + [AuCl_4]^- \rightarrow Fe^{3+} + [AuCl_4]^{2-} \qquad (2)$$

The $[AuCl_4]^{2-}$ is destroyed by disproportionation. It should be noted that kinetic data do not establish the formula of the unstable species; however, $[AuCl_3]^-$ is a possible alternative.

Some reactions involving oxidation of Au(I) to Au(III) complexes are thought to occur by free-radical mechanisms, and Au(II) intermediates are proposed. Thus, oxidation of $MeAu(PMe_3)$ (97) by CF_3I to give $MeAuI(CF_3)(PMe_3)$ proceeds through the formation of $MeAu(CF_3)(PMe_3)$ (49), and the reaction of $MeAuPMe_3$ with PhSH perhaps involves the Au(II) intermediate $MeAu(SPh)(PMe_3)$ (50). In all these cases, Au(II) species are not sufficiently long-lived to be detected directly. However, the oxidation of $[Au(S_2CNEt_2]_2$ by $(Et_2NCS_2)_2$ to form $Au(S_2CNEt_2)_3$ involves the Au(II) species $Au(S_2CNEt_2)_2$, which is sufficiently stable to be detected by ESR methods (106) (Section IV,4).

D. Mononuclear Gold(II) Compounds

1. Phthalocyanine Complex of Au(II)

Phthalocyanine is a divalent ligand with a geometry appropriate for forming a four-coordinated square-planar complex and was thought ideal for generating an Au(II) species. The action of AuBr on molten 1,3-diiminoisoindoline in the absence of solvent indeed yields a neutral gold phthalocyanine (63). The EPR spectra of the complex in 1-chloronaphthalene at 77 K clearly showed the presence of Au(II), a d^9 ion. The g value is 2.065, comparable to 2.042 for copper and 2.093 for silver phthalocyanine (64, 80).

2. Carborane Complex of Au(II)

A sandwich-bonded complex of Au(II) using the (3)-1,2-dicarbollide ligand, $B_9C_2H_{11}^{2-}$, has been prepared (119) by the sodium amalgam reduction of $(Et_4N)\{[\pi$-(3)-$1,2$-$B_9C_2H_{11}]_2Au\}$ in THF. The isolated complex $(Et_4N)_2\{[\pi$-(3)-$1,2$-$B_9C_2H_{11}]_2Au\}$ is deep blue-green and paramagnetic ($\mu_{eff} = 1.79\ \mu_B$) consistent with a d^9 ($s = \frac{1}{2}$) formulation. Its structure was proposed as in Fig. 11, in analogy to that of the Cu(II) complex, for which an X-ray analysis is available.

3. Dithiolate and Related Complexes of Au(II)

The reaction between the gold(I) N,N-dialkyldithiocarbamates and the corresponding thiuram disulfides [Eq. (3)] is a slow process (1).

$$[Au(S_2CNR_2)]_2 + 2(R_2NCS_2)_2 \rightarrow 2\ Au(S_2CNR_2)_3 \tag{3}$$

However, ESR experiments showed that, while the above reactants dissolved separately in C_6H_6 gave no signals, on mixing them four lines

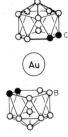

Fig. 11. Proposed structure of the anion in $[(C_2H_5)_4N]_2[Au(B_9C_2H_{11})_2]$. Hydrogen atoms are omitted (119).

of equal intensity appeared (*106*). The observed g value of 2.040 is close to those of Cu(II) and Ag(II) compounds, providing strong evidence for the existence of a Au(II) complex. The hyperfine structure is consistent with the nuclear spin of ^{197}Au ($I = \frac{3}{2}$). The same spectrum is also obtained by dissolving Au(S$_2$CNR$_2$)$_3$, due to the equilibrium

$$Au(S_2CNR_2)_3 \rightleftarrows Au(S_2CNR_2)_2 + 0.5(S_2CNR_2)_2 \qquad R = C_2H_5, \, i\text{-}C_3H_7 \qquad (4)$$

A similar reaction involving the n-butyldithiocarbamate ligand, n-Bu$_2$NCS$_2$ (*15*), was studied, and a four-line ESR spectrum with a g value of 2.039 was again obtained, indicating the presence of Au(II) in the reaction mixture. The existence of a corresponding dialkyldiseleno carbamate derivative Au(Se$_2$CNEt$_2$)$_2$, as a product of disproportionation of Au(Se$_2$CNEt$_2$)$_3$, has been confirmed by ESR spectroscopic methods (*54*):

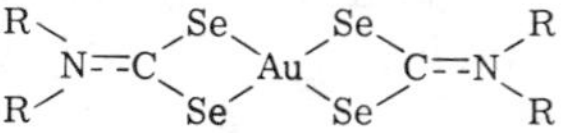

Waters and Gray (*120*) prepared a monomeric and paramagnetic Au(II) complex, (n-Bu$_4$N)$_2$[Au(mnt)$_2$] (**2**) by the borohydride reduction of the corresponding Au(III) complex.

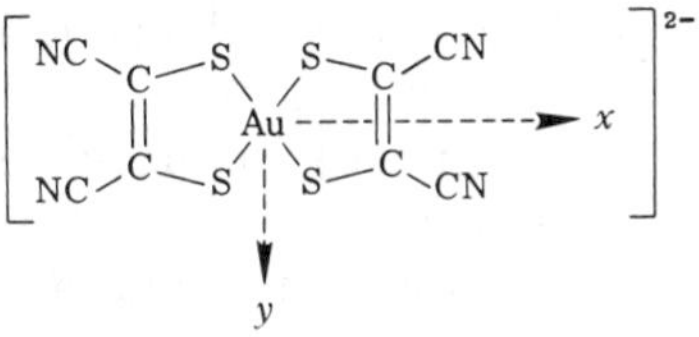

2

The solid is not affected in the presence of air for short periods, but the solution is easily oxidized. However, in the absence of air, the solution is stable indefinitely. The observed magnetic moment ($\mu_{eff} = 1.85$ μ_B) and four equally spaced, equally intense lines in the ESR spectrum clearly establish the divalent state for Au. The conductivity, electronic spectra, and cyclovoltammetry results corroborate this conclusion. The structure is suggested to be square planar (**2**). The same compound is also formed (*121*) by reacting equimolar amounts of Au(I) and Au(III) with the ligand (mnt)$^{2-}$ [Eq. (5)].

$$Au(I) + Au(III) \rightarrow 2 \, Au(II) \qquad (5)$$

As reactants, the Au(I)–Au(III) mixed-valence compounds (PhCH$_2$)$_2$SAuX · (PhCH$_2$)$_2$SAuX$_3$ (X = Cl, Br), n-Bu$_4$NBr, and Li$_2$(mnt) were employed. This reaction, however, does not take place with a variety of other sulfur ligands and appears to be typical for the (mnt)$^{2-}$ ligand.

Some Au(II) complexes of dithiocarbamate ligands are easily cocrystallized with the stable, diamagnetic nickel(II) complexes [Ni(S$_2$CNR$_2$)$_2$], and the EPR spectra of an oriented magnetically dilute sample can be obtained (109–111). In some cases, the spectra showed the anomalous feature of having weak satellites outside the main four-line pattern, which was unequally spaced. However, the results of intensive ESR spectral investigations of the square-planar complex [Au(mnt)$_2$]$^{2-}$ in magnetically dilute single crystals of [Ni(mnt)$_2$][n-Bu$_4$N]$_2$ · 2CH$_3$CN and [M(mnt)$_2$][n-Bu$_4$N]$_2$ (M = Ni, Pd, Pt) over a range of temperature (82) are consistent with a ground-state hole configuration (B$_{1g}$)^2(A$_{1g}$)^1 in D_{2h} symmetry, where A$_{1g}$ is primarily a ligand-based orbital with 15% 6s and smaller 5d admixtures, and B$_{1g}$ is the normally half-filled antibonding orbital in square-planar d^9 complexes. Thus, this complex is best described as an Au(III) complex with a radical-anion ligand. Possibly, the inability of most ligands to form such a structure explains the rarity of formal Au(II) complexes.

In addition to the [Au(S$_2$CNR$_2$)$_2$] and [Au(mnt)$_2$]$^{2-}$ complexes discussed above, good chemical and structural evidence is also provided for the formation of the mixed complex [(S$_2$CNEt$_2$)Au(mnt)]$^-$ in solution, by electrochemical reduction of [Au(S$_2$CNEt$_2$)(mnt)] (107). The pseudo-Au(II) complexes [(Ph$_3$E)Au(mnt)](E = P, As) have been reported by Bergendahl and Waters (14), who pointed out their relationship to the authentic [Au(mnt)$_2$]$^{2-}$ system. They have been, however, reassigned as [(Ph$_3$P)$_2$Au]$^+$[Au(mnt)$_2$]$^-$, a mixed Au(I)–Au(III) complex (1, see Section IV,B). The magnetic susceptibility and ESR measurements on the π-donor–acceptor compound (TTF) · Au[S$_2$C$_2$(CF$_3$)$_2$]$_2$ (TTF = tetrathiafulvalene) exhibit (46, 47) spin Peierls transitions (which is a progressive spin–lattice dimerization occurring below a transition temperature in a system of one-dimensional antiferromagnetic Heisenberg chains) in agreement with a mean-field theory.

In certain gold-containing arsenopyrite ores, FeAsS, a narrow isotropic singlet with a g value of 2.001 has been observed in the ESR spectrum (114). The signal was reduced on lowering the temperature and was absent in samples not containing Au. The reason for the appearance of the EPR signal seems to be the isomorphic admixture of Au(0) and ions of Au(II) in the lattice. Thus, it appears that Au(II) is stabilized by S-donor ligands.

E. BINUCLEAR GOLD(II) COMPOUNDS CONTAINING A METAL–METAL BOND

In addition to the complexes containing a single Au(II) center, a number of binuclear gold(II) complexes have been synthesized by

Schmidbaur and co-workers (*83–90a, 92, 93*). In cyclic gold–ylide complexes (**3**), formed by a transylidation reaction (*87*), the two Au atoms have a parallel two-coordinate linear arrangement of ligands [Eq. (6)].

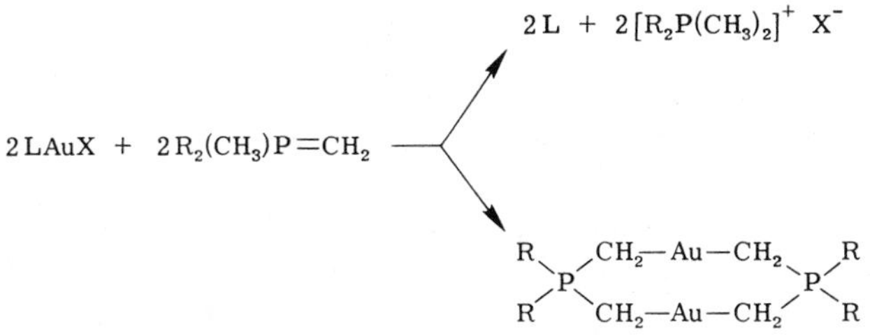

(R = alkyl; X = halogen;
L = phosphine)   **3**   (6)

On reaction with halogens, compound (**3**) affords bicyclic derivatives (**4**) having a *transannular Au–Au bond* (*87*). On treatment with CH₃I, oxidative addition proceeds to give the product (**6**), which on methylation yields (**7**) (*71, 88*).

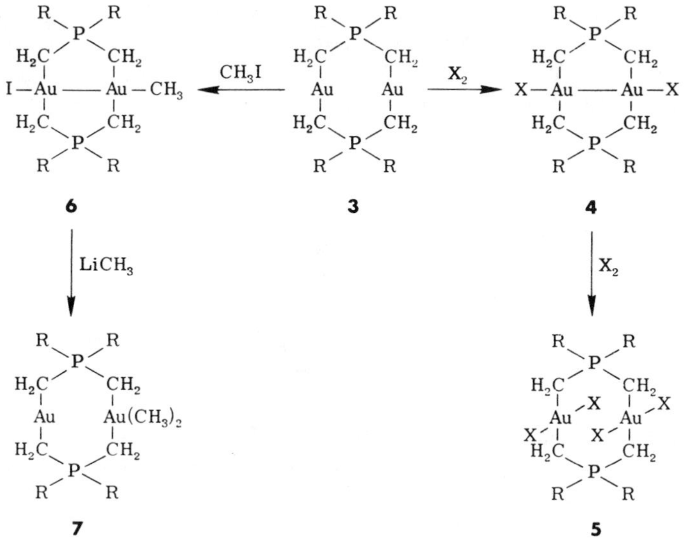

(X = Cl, Br, I;
R = Me, Et, *t*-Bu)   (7)

Further halogenation of the dihalides (**4**) finally results in the rupture of the Au–Au bond to give the Au(III) metallocycles (**5**).

The Mössbauer and ESCA spectra (*89*) confirm the different valence states of the gold atoms in the products (**4**) and (**5**) (see also *8, 24, 34, 35, 116*). The PMR as well as the mass spectra also favor the structure

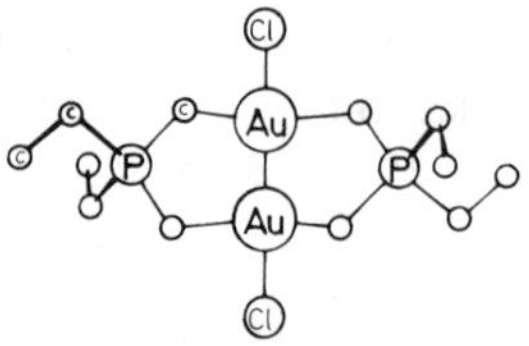

FIG. 12. Molecular structure of $(C_2H_5)_2P[CH_2Au(Cl)CH_2]_2P(C_2H_5)_2$, by X-ray diffraction; see (**4**) in Eq. (7). Hydrogen atoms are omitted (*90*).

(**4**). Four X-ray crystal structure analyses (*71, 90, 90a, 91a*) afforded final proof for the proposed formulas. In (**4**), the two metals of formal oxidation state +2 are separated by a very short distance of 259.7 pm for X = Cl and R = C_2H_5, or 2.67 pm for X = I and R = CH_3 (Figs. 12 and 13). A related sulfur compound is obtained from **3** and thiuramdisulfide (**4**, X = S_2CNR_2) (*88a*).

Another example of an eight-membered metallocycle containing Au(II) centers is the dithiocarbamate complex (**9**) (*20*), obtained from the gold(I) dithiocarbamate complex (**8**) with halogens or pseudohalogens at $-78°$ in CS_2. However, these compounds disproportionate to form mixed-valence salts (**10**). It was shown that the ν_{CN} band of the coordinated dithiocarbamate is sensitive to the oxidation state of the gold, increasing in the order 1495, 1523, and 1547 cm^{-1} for the Au(I), Au(II), and Au(III) complexes, respectively (*20*).

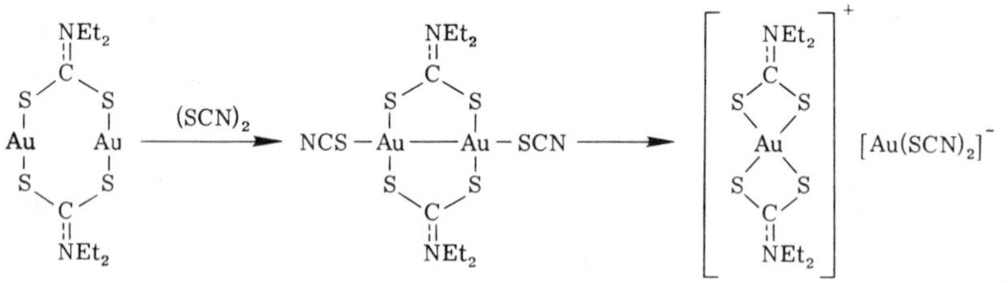

Oxidative addition of halogens to the 1:2 complexes of bis(diphenylphosphino)methane or -propane with gold(I) chloride (**11**) forms the Au(II) complexes (**12**) (*93*).

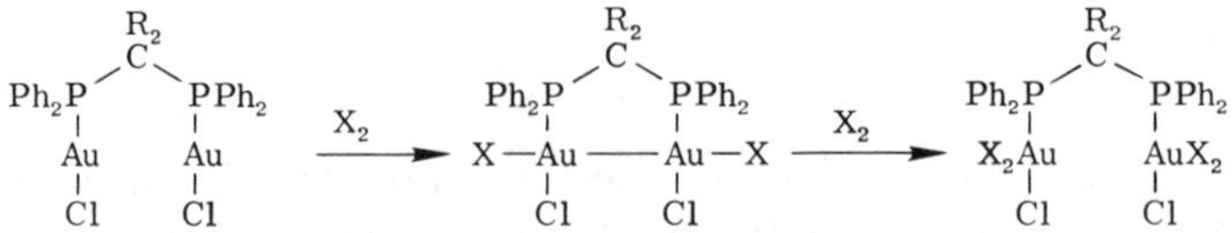

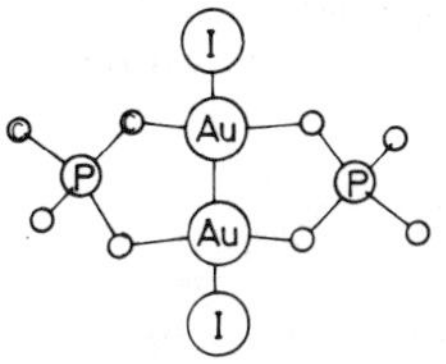

FIG. 13. Molecular structure, by X-ray diffraction, of $(CH_3)_2P[CH_2Au(I)CH_2]_2P(CH_3)_2$; see (4) in Eq. (7). Hydrogen atoms are omitted (71, 84).

A compound of bis(diphenylphosphino)methylamine (15) (92) is prepared similarly. The compound (15) adds one more mole of Cl_2 and is oxidized to the Au(III) complex (16) (92). The ^{197}Au Mössbauer spectra and ESCA measurements indicate that the complexes (12) and (15) are not mixed Au(I)–Au(III) complexes. The Au(II) appears to be present again as the Au_2^{4+} moiety. Because of the insolubility of the compounds (12) and (15), no molecular mass could be determined. Oligomeric formulas are therefore possible (92, 93).

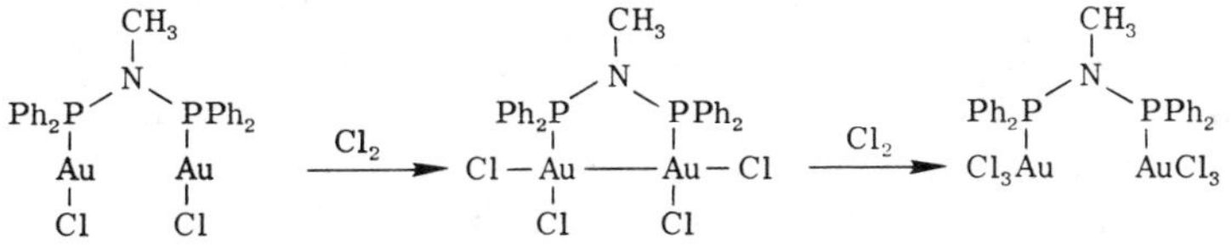

$$(10)$$

V. Compounds of Gold in Oxidation State +5

The chemistry of gold in this oxidation state is limited to the binary fluoride AuF_5 and salts or complexes of the hexafluoroaurate(V) anion AuF_6^-. Until the end of the 1960s, the highest established oxidation state of gold was the trivalent state. The existence of PtF_6 and salts of PtF_6^- with the usual predictable periodic trends did suggest, however, the possibility of the synthesis of gold(V) compounds (4, 5). Attempts to prepare AuF_5 or AuF_6 directly from the elements failed (5, 6), but these efforts resulted in the isolation of a material, the analysis of which indicated at least an oxidation state higher than gold(III).

It is generally true that a given high oxidation state is thermodynamically more stable in an anionic species than in a binary compound (4). Many of the experimental observations are compatible with the electronegativity of the metal atom in the complex anion being appreciably lower than in the corresponding binary compound (73). The first successful approach to the synthesis of gold compounds of higher oxidation state by Bartlett (*5, 6, 59*) therefore emphasized the generation of gold-containing fluoroanions.

A. HEXAFLUOROAURATES(V)

In a pioneering experiment (*59*), AuF_3 was shown to react with XeF_2 at 400°C in an atmosphere of gaseous fluorine at 1000 p.s.i. to form pale yellow crystals of the empirical formula $AuXe_2F_{17}$, later identified as $[Xe_2F_{11}]^+[AuF_6]^-$. XeF_6 is probably formed as an intermediate, and is known to offer excellent conditions for the generation and complexation of metals in high oxidation states (*7, 59, 96, 98*).

$$2XeF_2 + 5F_2 + AuF_3 \rightarrow Xe_2F_{11}^+AuF_6^- \tag{11}$$

The Raman spectrum of the product is similar to that of $[Xe_2F_{11}]^+[AsF_6]^-$, implying the existence of an $[AuF_6]^-$ salt. The crystal structure of the compound was determined by X-ray diffraction, and the result is shown in Fig. 14. The analysis clearly defines an AuF_6 and a Xe_2F_{11} group, the latter consisting of two XeF_5 moieties linked by an additional fluorine atom. All three components possess mirror symmetry. The AuF_6 group is approximately octahedral, with only one Au—F distance (of 1.90 Å) departing significantly from the average value of 1.86(1) Å. The F—Au—F angles are close to 90° (*60, 61*).

The pentafluoroxenyl(VI) salt $XeF_5^+AuF_6^-$ is obtained from the $Xe_2F_{11}^+AuF_6^-$ compound on heating to 110°C under vacuum. It is easily purified by sublimation, yielding a pale yellow-green solid, m.p. 190–192°C, reacting explosively with water. The crystals are monoclinic, space group $P\,2_1/c$ (*5*). The ^{19}F NMR spectra of $XeF_5^+AuF_6^-$ in HF and of $Xe_2F_{11}^+AuF_6^-$ in BrF_5 establish a chemical shift of the anion at $\delta = 109$ ppm (rel. $CFCl_3$). No coupling to the ^{197}Au nucleus is observed (*5, 41*). The spectra also prove the diamagnetism of the material, in harmony with a low-spin d^6 (t_{2g}^6) configuration for Au(V), well separated energetically from any paramagnetic excited state.

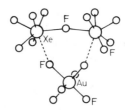

FIG. 14. Molecular structure of $Xe_2F_{11}^+AuF_6^-$, as determined by X-ray diffraction (*61*).

A cesium salt $Cs[AuF_6]$ is obtained if a slight deficiency of CsF is mixed with the above $Xe_2F_{11}^+$ salt and heated to 110°C, when XeF_6 is evolved.

$$[Xe_2F_{11}]^+[AuF_6]^- + CsF \rightarrow Cs[AuF_6] + 2XeF_6 \qquad (12)$$

The resulting pale-yellow product has an X-ray powder pattern virtually identical with those of other complex fluorometallates $Cs[MF_6]$ (59). Its rhombohedral cell (a = 5.24 Å, α = 96.5°) is isomorphous with that of noble metal congeners. The same material was later also prepared from $Cs[AuF_4]$ and elemental fluorine at 250–350°C under 1000 p.s.i. pressure (5, 61). The Raman spectra have been assigned on the basis of an octahedral structure of the anion (5). The ^{197}Au Mössbauer spectra of the cesium salt, of $Xe_2F_{11}^+AuF_6^-$, and of $XeF_5^+AuF_6^-$ show only a singlet absorption without quadrupole splitting as expected for an octahedral environment of the metal. The isomeric shift of about 2.30 mm sec^{-1} suggests a high oxidation state (51).

The related potassium salt, $K[AuF_6]$, was prepared similarly from $K[AuF_4]$ and has very similar X-ray and Raman characteristics (5). A sodium salt is mentioned in a paper on AuF_5. Addition of NaF to this binary fluoride yields $Na[AuF_6]$, which was characterized by its vibrational frequencies (100). CsF gives the same reaction.

Salts of the nitrosyl cation NO^+ and the dioxygenyl cation O_2^+ are also accessible through fluorination of $NO[AuF_4]$, or from AuF_3, oxygen, and fluorine, under pressure at 350 and 500°C, respectively (5). The nitrosyl salt decomposes at 400°C to yield the starting materials, $NO^+[AuF_4]^-$ and F_2. The powerful oxidizing properties of the O_2^+ salt are shown in the reaction with IF_5, which leads to the new hexafluoroaurate(V) $IF_6^+AuF_6^-$, with AuF_3 and molecular oxygen. With the fluorobase IF_7, the same salt is generated, with O_2F as the byproduct (5):

$$3IF_5 + 4O_2^+AuF_6^- \rightarrow 3IF_6^+AuF_6^- + 4O_2 + AuF_3 \qquad (13)$$

$$IF_7 + O_2^+AuF_6^- \rightarrow IF_6^+AuF_6^- + O_2F \qquad (14)$$

The magnetic moment of $O_2^+AuF_6^-$, μ = 1.66 μ_B, indicates the presence of one unpaired electron. The susceptibility data depart only slightly from the Curie–Weiss relationship. X-Ray powder and Raman data are also available (5), and molecular-beam mass spectra have been recorded (31).

$NO^+AuF_6^-$ is also the product of the addition of NOF to the binary AuF_5 to be described in Section V,C (41). The O_2^+ salt is obtained from $KrF^+AuF_6^-$ (see Section V,B) and molecular oxygen (41). A series of other complex fluoroaurates(V) of Kr, Xe, and Br is again derived from AuF_5.

B. FLUOROKRYPTYL(II) HEXAFLUOROAURATE(V)

The compound $FKr—AuF_6$ is treated separately because it appears to be a hexafluoroaurate less ionic in nature than the salts considered in the preceding section. It also plays a key role in the synthesis of many other Au(V) species, in particular the pure binary AuF_5. The compound was prepared by three independent groups in the U.S.A., the United Kingdom, and the U.S.S.R.

On the basis of theoretical considerations, Bartlett postulated the existence of $FKr^+AuF_6^-$ and even of $FAr^+AuF_6^-$, for which an electron transfer from anion to cation should be an energetically unfavorable process (5). Holloway and Schrobilgen carried out the reaction of KrF_2 with gold powder in liquid HF at 20°C and obtained a light yellow solid after removal of excess KrF_2 and HF $in\ vacuo$. The Raman spectrum of the product (under HF at $-80°C$) is consistent with a formulation in which a KrF group is fluorine bridged to an AuF_6 group (of approximate C_{4v} symmetry).

$$7KrF_2 + 2Au \rightarrow 2F\text{-}Kr\cdot\cdot\cdot\cdot FAuF_5 + 5\ Kr \qquad (15)$$

The compound reacts rapidly with molecular oxygen to yield $O_2^+AuF_6^-$. Gaseous xenon is converted into $XeF_5^+AuF_6^-$ and an unidentified red-orange product. Pyrolysis at 60–65°C yields pure AuF_5 (41).

Sokolov $et\ al.$ (100) treated gold foil with KrF_2 dissolved in anhydrous HF at 0–20°C. Again, bright yellow crystals of $KrF_2 \cdot AuF_5$ were obtained, stable to 50°C. At 65–67°C, rapid decomposition is observed into gaseous fluorine and krypton, and solid AuF_5. With a large excess of KrF_2 and at low temperature, a compound of the composition $2\ KrF_2 \cdot AuF_5$ is observed, which is unstable even at 0°C, decomposing rapidly into $KrF_2 \cdot AuF_5$ (100).

Infrared absorptions for $KrF_2 \cdot AuF_5$ and $2KrF_2 \cdot AuF_5$ have been reported. They suggest reduction of the octahedral symmetry of the AuF_6 group to C_{4v} of C_{2v}, probably caused by fluorine bridging. The oxidation state $+5$ of gold was confirmed by Mössbauer spectra (100). $FXe\cdot\cdot\cdot\cdot FAuF_5$ and $Xe_2F_3^+AuF_6^-$ are obtained from AuF_5 and XeF_2.

C. GOLD PENTAFLUORIDE

AuF_5 is displaced from $Xe_2F_{11}^+AuF_6^-$ by SbF_5 (in liquid SbF_5). A mixture of $XeF_5^+Sb_2F_{11}^-$ and $XeF_5^+AuSbF_{11}^-$ is obtained, but no separation is possible (5). (SbF$_5$ and AuF$_5$ appear to undergo codistillation.) Pure AuF$_5$ is easily obtained, however, from the pyrolysis of KrF$_2$ · AuF$_5$ at 60–65°C as a dark red-brown solid (orange when powdered). Its X-ray powder patterns show that no AuF$_3$ is present, and suggest similarities to the structure of the (tetrameric) RuF$_5$. Three bands are observed in the room-temperature Raman spectrum (658, 598, and 228 cm^{-1}). The compound reacts with excess XeF$_2$ in HF or BrF$_5$ to give Xe$_2$F$_{11}^+$AuF$_6^-$. Addition of NOF yields NO$^+$AuF$_6^-$ (41).

Sokolov et al. (100), who also obtained pure AuF$_5$ from the pyrolysis of KrF$_2$ · AuF$_5$ at 65–67°C, describe the product as a red-brownish crystalline material, vigorously hydrolyzed in moist air. It is reported to sublime at 80°C, to undergo a polymorphic transformation at 120–125°C, and to decompose quantitatively above 200°C into AuF$_3$ and F$_2$ without melting. Organic compounds are fluorinated explosively (100). The infrared spectrum was presented in a figure, showing bands at positions similar to those given above. A structure resembling that of α-UF$_5$ was deduced. Nonequivalence of the fluorine atoms is confirmed by the presence of quadrupole splitting in the Mössbauer spectrum (100).

AuF$_5$ forms a complex acid HAuF$_6$ of m.p. 88°C. The species KrF$_2$ · AuF$_5$, 2KrF$_2$ · AuF$_5$, XeF$_2$ · AuF$_5$, 2XeF$_2$ · AuF$_5$, and XeF$_6$ · AuF$_5$ are obtained with KrF$_2$, XeF$_2$, and XeF$_6$, respectively. Br$_2$ yields AuF$_3$ · BrF$_3$, and BrF$_5$ yields BrF$_5$ · AuF$_5$, from which BrF$_7$ · AuF$_5$ is generated with KrF$_2$ in BrF$_5$ (100).

In a third independent study (112, 113), amorphous AuF$_5$ was synthesized by thermal decomposition of O$_2^+$AuF$_6^-$ in vacuo at 160–200°C. A dark red sublimate, formed at a cold finger, melts over the range 75–78°C. (Note the difference from the above reports.) The magnetic susceptibility of $-73 \pm 15 \times 10^{-6}$ emu mol^{-1} agrees well with the value expected for the diamagnetic species. Raman lines were observed at 654, 595, and 219 cm^{-1}. Mass spectra of AuF$_5$ show Au/F clusters with up to three Au atoms, but no molecular ion AuF$_5^+$ is detected. The material used in this study also reacts with XeF$_2$ in liquid HF to form Xe$_2$F$_3^+$AuF$_6^-$, isomorphous with Xe$_2$F$_3^+$IrF$_6^-$ according to X-ray powder photographs (112), and probably resembling the above-mentioned 2XeF$_2$ · AuF$_5$ (100).

Note added in proof. Since the submission of this review, the chemistry of gold clusters has developed very rapidly. A new type of hexatomic gold cluster with a structure consisting of two tetrahedra sharing a common edge was discovered: Au$_6$(P(C$_6$H$_5$)$_3$)$_4$ · (Co(CO)$_4$)$_2$ (107b). An octatomic gold cluster derived from an icosahedron by selecting a set of apices different from those found in previous examples was detected in the cation $Au_8(P(C_6H_5)_3)_7^{2+}$ (*107a*). Degradation reactions of the $Au_9L_8^{3+}$ clusters by other ligands L, SCN^-, or CN^- to form $Au_8L_8^{2+}$, $Au_{11}L_8$-$(SCN)_2^+$, or $Au_{11}L_8Cl_2^+$ cations have been studied in some detail. An intermediate $Au_8L_7^{2+}$ is postulated (*117a*). Most interesting, however, is the synthesis and structural characterization of examples with an Au_{13} cluster in the form of a gold-centered closed icosahedron of gold atoms. The individual species are $Au_{13}(P(CH_3)_2C_6H_5)_{10}Cl_2^{3+}$ (*19b*) and $Au_{13}(P(C_6H_5)_2CH_2P(C_6H_5)_2)_6^{4+}$ (*108a*). An even larger section of the gold metal lattice is preserved in the present record cluster of the formula $Au_{55}(P(C_6H_5)_3)_{12}Cl_6$ (*82a*). Gold clusters were also found in the ternary phase $Rb_4Au_7Sn_2$ (*97a*). Theoretical work on metal clusters including gold appears in two recent reviews (*35a, 67a*).

REFERENCES

1. Åkerström, S., *Ark. Kem.* **14**, 387 (1959).
2. Albano, V. G., Bellon, P. L., and Manassero, M., *J. Chem. Soc., Chem. Commun.* p. 1210 (1970).
3. Andrianov, V. G., Struchnov, Y. T., and Rosinskaya, E. R., *J. Chem. Soc., Chem. Commun.* p. 338 (1973).
4. Bartlett, N., *Angew. Chem., Int. Ed. Engl.* **7**, 433 (1968).
5. Bartlett, N., and Leary, K., *Rev. Chim. Miner.* **13**, 82 (1976).
6. Bartlett, N., and Rao, P. R., *Abstr., 154th Meet. Am. Chem. Soc.* K 15 (1967).
7. Bartlett, N., and Sladky, F. O., *Compr. Inorg. Chem.* **1**, 213–330 (1973).
8. Battistoni, C., Mattogno, G., Cariati, F., Naldini, L., and Sgamellotti, A., *Inorg. Chim. Acta* **24**, 207 (1977).
9. Bellon, P. L., Cariati, F., Manassero, M., Naldini, L., and Sansoni, M., *J. Chem. Soc., Chem. Commun.* p. 1423 (1971).
10. Bellon, P. L., Manassero, M., Naldini, L., and Sansoni, M., *J. Chem. Soc., Chem. Commun.* p. 1035 (1972).
11. Bellon, P. L., Manassero, M., and Sansoni, M., *J. Chem. Soc., Dalton Trans.* p. 1481 (1972).
12. Bellon, P. L., Manassero, M., and Sansoni, M., *J. Chem. Soc., Dalton Trans.* p. 2423 (1973).
13. Bergendahl, T. J., *J. Chem. Educ.* **52**, 731 (1975).
14. Bergendahl, T. J., and Waters, J. H., *Inorg. Chem.* **14**, 2556 (1975).
15. Bergendahl, T. J., and Bergendahl, E. M., *Inorg. Chem.* **11**, 638 (1972).
16. Beurskens, P. T., Blaauw, H. J. A., Cras, J. A., and Steggerda, J. J., *Inorg. Chem.* **7**, 805 (1968).
17. Bezman, S. A., Churchill, M. R., Osborn, J. A., and Wormald, J., *J. Am. Chem. Soc.* **93**, 2063 (1971).
18. Brain, F. H., Gibson, C. S., Jarvis, J. A. J., Phillips, R. F., Powell, H. M., and Tyabji, A., *J. Chem. Soc.* p. 3686 (1952).
19. Brewer, L., *et al.*, *in* "The Chemistry and Metallurgy of Miscellaneous Materials: Thermodynamics" (L. L. Quill, ed.), NNES, Plutonium Proj. Rec., Div. IV, Vol. 19B, p. 239. McGraw-Hill, New York, 1950.

19a. Busse, B., and Weil, K. G., *Angew. Chem., Int. Ed. Engl.* **18**, 629 (1979).
19b. Briant, C. E., Theobald, B. R. C., White, J. W., Bell, L. K., Mingos, D. M. P., and Welch, A. J., *J. Chem. Soc., Chem. Commun.* p. 102 (1981).
20. Calabro, D. C., M. S. Thesis, University of Delaware, (1978).
21. Calderazzo, F., *Pure Appl. Chem.* **50**, 49 (1978).
22. Cariati, F., and Naldini, L., *Inorg. Chim. Acta* **5**, 172 (1971).
23. Cariati, F., and Naldini, L., *J. Chem. Soc., Dalton Trans.* p. 2286 (1972).
24. Charlton, J. S., and Nichols, D. I., *J. Chem. Soc. A* p. 1484 (1970).
25. Coletta, F., Ettore, R., and Gambaro, A., *Inorg. Nucl. Chem. Lett.* **8**, 667 (1972).
26. Cooper, M. K., Dennis, G. R., Henrick, K., and McPartlin, M., *Inorg. Chim. Acta* **45**, L 151 (1980).
27. Corbett, J. D., and Druding, L. F., *J. Inorg. Nucl. Chem.* **11**, 20 (1959).
28. Dell'Amico, D. B., Calderazzo, F., and Marchetti, F., *J. Chem. Soc., Dalton Trans.* p. 1829 (1976).
29. Dell'Amico, D. B., Calderazzo, F., Marchetti, F., Merlino, S., and Perego, G., *J. Chem. Soc., Chem. Commun.* p. 31 (1977).
30. Dimroth, K., and Kanter, H., *Tetrahedron Lett.* p. 545 (1975).
31. Edwards, A. J., Falconer, W. E., Griffiths, J. E., Sunder, W. A., and Vasile, M. J., *J. Chem. Soc., Dalton Trans.* p. 1129 (1974).
32. Elliot, R. P., "Constitution of Binary Alloys," 1st Suppl. McGraw-Hill, New York, 1965.
33. Elliott, N., and Pauling, L., *J. Am. Chem. Soc.* **60**, 1846 (1938).
34. Faltens, M. O., Ph.D. Thesis, Radiat. Lab. UCRL 18706. University of California, Lawrence (1969).
35. Faltens, M. O., and Shirley, D. A., *J. Chem. Phys.* **53**, 4249 (1970).
35a. Gillespie, R. J., *Chem. Soc. Rev.* **8**, 315 (1979).
36. Guy, J. J., Jones, P. G., Mays, M. J., and Sheldrick, G. M., *J. Chem. Soc., Dalton Trans.* p. 8 (1977).
37. Hasegawa, A., and Watanabe, M., *J. Phys. F* **7**, 75 (1977).
38. Hensel, F., *Can. J. Chem.* **55**, 2255 (1977).
39. Hensel, F., *Angew. Chem., Int. Ed. Engl.* **19**, 593 (1980).
40. Hermann, F., *Ber. Dtsch. Chem. Ges.* **38**, 2813 (1905).
41. Holloway, J. H., and Schrobilgen, G. J., *J. Chem. Soc., Chem. Commun.* p. 623 (1975).
42. Hoshino, H., Schmutzler, R. W., and Hensel, F., *Phys. Lett. A* **51A**, 7 (1975).
43. Hotop, H., and Lineberger, W. C., *J. Phys. Chem. Ref. Data* **4**, 539 (1975).
44. Hüttel, R., Reinheimer, H., and Nowak, K., *Tetrahedron Lett.* **11**, 1019 (1967).
45. Interrante, L. V., and Bundy, F. P., *J. Inorg. Nucl. Chem.* **39**, 1333 (1977).
46. Jacobs, I. S., Bray, J. W., Hart, H. R., Jr., Interrante, L. V., Kasper, J. S., and Watkins, G. D., *Phys. Rev. B: Solid State* [3] **14**, 3036 (1976).
47. Jacobs, I. S., Hart, H. R., Jr., Interrante, L. V., Bray, J. W., Kasper, J. S., Watkins, G. D., Prober, D. E., Wolf, W. E., and Bonner, J. C., *Physica (Amsterdam)* **B 86–88**, 655 (1977).
48. Janssen, E. M. W., Folmer, J. C. W., and Wiegers, G. A., *J. Less-Common Metals* **38**, 71 (1974).
49. Johnson, A., and Puddephatt, R. J., *J. Chem. Soc., Dalton Trans.* p. 1360 (1976).
50. Johnson, A., and Puddephatt, R. J., *J. Chem. Soc., Dalton Trans.* p. 115 (1975).
51. Kaindl, G., Leary, K., and Bartlett, N., *J. Chem. Phys.* **59**, 5050 (1973).
52. Keller, R., Fenner, J., and Holzapfel, W. B., *Mater. Res. Bull.* **9**, 1363 (1974).
53. Kienast, G., and Verma, J., *Z. Anorg. Allg. Chem.* **310**, 143 (1961).

54. Kirmse, R., Lorenz, B., Windsch, W., and Hoyer, E., Z. Anorg. Allg. Chem. **384,** 160 (1971).
55. Knecht, J., Fischer, R., Overhof, H., and Hensel, F., J. Chem. Soc., Chem. Commun. p. 905 (1978).
56. Krüger, K. D., and Schmutzler, R. W., Ber. Bunsenges. Phys. Chem. **80,** 816 (1976).
57. Krüss, G., Justus Liebigs Ann. Chem. **237,** 296 (1887).
58. Krüss, G., and Schmidt, F., Ber. Dtsch. Chem. Ges. **20,** 2634 (1887); J. Prakt. Chem. **38,** 77 (1888).
59. Leary, K., and Bartlett, N., J. Chem. Soc., Chem. Commun. p. 903 (1972).
60. Leary, K., Zalkin, A., and Bartlett, N., J. Chem. Soc., Chem. Commun. p. 131 (1973).
61. Leary, K., Zalkin, A., and Bartlett, N., Inorg. Chem. **13,** 775 (1974).
62. Liu, T. L., and Amar, H., Rev. Mod. Phys. **40,** 782 (1968).
63. MacCragh, A., and Koski, W. S., J. Am. Chem. Soc. **87,** 2496 (1965).
64. MacCragh, A., and Koski, W. S., J. Am. Chem. Soc. **85,** 2375 (1963).
65. Malatesta, L., Gold Bull. **8,** 48 (1975).
66. Manassero, M., Naldini, L., and Sansoni, M., J. Chem. Soc., Chem. Commun. p. 385 (1979).
67. McPartlin, M., Mason, R., and Malatesta, L., J. Chem. Soc., Chem. Commun. p. 334 (1969).
67a. Mingos, D. M. P., Pure Appl. Chem. **52,** 705 (1980).
68. Mingos, D. M. P., J. Chem. Soc., Dalton Trans. p. 1163 (1976).
68a. Naldini, L., Cariati, F., Simonetta, G., and Malatesta, L., J. Chem. Soc., Chem. Commun. p. 647 (1966).
68b. Cariati, F., Naldini, L., Simonetta, G., and Malatesta, L., Inorg. Chim. Acta **1,** 315 (1967).
69. Nesmeyanov, A. N., Perevalova, E. G., Grandberg, K. I., Lemenovskii, D. A., Baukova, T. A., and Afanasova, O. B., J. Organomet. Chem. **65,** 131 (1974).
70. Overhof, H., Fischer, R., Vulli, M., and Hensel, F., Ber. Bunsenges. Phys. Chem. **80,** 871 (1976).
71. Paparizos, C., Diss. Case Western Reserve Univ. pp. 1–186 (1977); Diss. Abst. Int. B **38,** 4227 (1978).
72. Pauling, L., "The Nature of the Chemical Bond." Cornell Univ. Press, Ithaca, New York, 1960.
73. Peacock, R. D., Adv. Fluorine Chem. **7,** 113–118 (1973).
74. Pearson, W. B., "Lattice Spacings and Structures of Metals and Alloys." Pergamon, Oxford, 1957.
75. Peer, W. J., and Lagowski, J. J., J. Am. Chem. Soc. **100,** 6260 (1978).
76. Piovesana, O., and Zanazzi, P. F., Angew. Chem. **92,** 579 (1980); Angew. Chem. Int. Ed. Engl. **19,** 561 (1980).
77. Puddephatt, R. J., "The Chemistry of Gold." Elsevier, Amsterdam, 1978.
78. Raub, J. C., and Compton, V. B., Z. Anorg. Allg. Chem. **332,** 5 (1964).
79. Rich, R. L., and Taube, H., J. Phys. Chem. **58,** 6 (1954).
80. Roberts, E. M., and Koski, W. S., J. Am. Chem. Soc. **83,** 1865 (1961).
81. Rundle, R. E., J. Am. Chem. Soc. **76,** 3101 (1954).
82. Schlupp, R. L., and Maki, A. H., Inorg. Chem. **13,** 44 (1974).
82a. Schmid, G., Pfeil, R., Boese, R., Bandermann, F., Meyer, S., Calis, G. M. M., and van der Velden, J. W. A., Chem. Ber. **114,** 3634 (1981).
83. Schmidbaur, H., Acc. Chem. Res. **8,** 62 (1975).
84. Schmidbaur, H., "Gold-Organic Compounds," Gmelin Handbook. Springer-Verlag, Berlin and New York, 1980.

85. Schmidbaur, H., *Angew. Chem., Int. Ed. Engl.* **15**, 728 (1976).
86. Schmidbaur, H., and Franke, R., *Angew. Chem.* **85**, 449 (1973).
87. Schmidbaur, H., and Franke, R., *Inorg. Chim. Acta* **13**, 79 (1975).
88. Schmidbaur, H., and Franke, R., *Inorg. Chim. Acta* **13**, 85 (1975).
88a. Schmidbaur, H., and Mandl, J. R., *Naturwissensch.* **63**, 585 (1976).
89. Schmidbaur, H., Mandl, J. R., Wagner, F. E., van de Vondel, D. F., and van der Kelen, G. P., *J. Chem. Soc., Chem. Commun.* p. 170 (1976).
90. Schmidbaur, H., Mandl, J. R., Frank, A., and Huttner, G., Bejenke, V., Richter, W., *Chem. Ber.* **109**, 466 (1976).
90a. Schmidbaur, H., Mandl, J. R., Frank, A., and Huttner, G., *Chem. Ber.* **110**, 2236 (1977).
91. Schmidbaur, H., and Mandl, J. R., *Angew. Chem.* **89**, 679 (1977).
91a. Schmidbaur, H., Scherm, H. P., and Schubert, U., *Chem. Ber.* **111**, 764 (1978).
92. Schmidbaur, H., Wagner, F. E., and Wohlleben-Hammer, A., *Chem. Ber.* **112**, 496 (1979).
93. Schmidbaur, H., Wohlleben, A., Wagner, F. E., van de Vondel, D. F., and van der Kelen, G. P., *Chem. Ber.* **110**, 2758 (1977).
94. Schmutzler, R. W., Hoshino, H., Fischer, R., and Hensel, F., *Ber. Bunsenges. Phys. Chem.* **80**, 107 (1976).
95. Schottländer, F., *Justus Liebigs Ann. Chem.* **217**, 337 (1883).
96. Schreiner, F., Osborne, D. W., Malm, J. G., and McDonald, G. N., *J. Chem. Phys.* **51**, 4838 (1969).
97. Shiotani, A., and Schmidbaur, H., *J. Organomet. Chem.* **37**, C25 (1972).
97a. Sinnen, H.-D., and Schuster, H.-U., *Z. Naturforsch.* **36b**, 833 (1981).
98. Sladky, F. O., and Bartlett, N., *J. Am. Chem. Soc.* **90**, 5316 (1968).
99. Smith, G. M., *J. Am. Chem. Soc.* **44**, 1769 (1922).
100. Sokolov, V. B., Prusakov, V. N., Ryzhkov, A. V., Drobyshevskii, Yu. V., and Khoroshev, S. S., *Dokl. Akad. Nauk. SSSR* **229**, 884 (1976).
101. Sommer, A. H., *Nature (London)* **152**, 215 (1943).
102. Spicer, W. E., *Phys. Rev.* **125**, 1297 (1962).
103. Spicer, W. E., Sommer, A. H., and White, J. G., *Phys. Rev.* **115**, 57 (1959).
104. Thomson, S., *J. Prakt. Chem.* **13**, 337 (1876); **37**, 105 (1888).
105. Tinelli, G. A., and Holcomb, D. F., *J. Solid State Chem.* **25**, 157 (1978).
106. Vänngard, T., and Åkerström, S., *Nature (London)* **184**, 183 (1959).
107. van der Linden, J. G. M., and van de Roer, H. G. J., *Inorg. Chim. Acta* **5**, 254 (1971).
107a. van der Velden, J. W. A., Bour, J. J., Bosman, W. P., and Noordik, J. H., *J. Chem. Soc., Chem. Commun.* p. 1218 (1981).
107b. van der Velden, J. W. A., Bour, J. J. Otterloo, B. F., Bosman, W. P., and Noordok, J. H., *J. Chem. Soc., Chem. Commun.* p. 583 (1981).
108. van der Velden, J. W. A., Bour, J. J., Vollenbroek, F. A., Beurskens, P. T., and Smits, J. M. M., *J. Chem. Soc., Chem. Commun.* p. 1162 (1979).
108a. van der Velden, J. W. A., Vollenbroek, F. A., Bour, J. J., Beurskens, P. T., Smits, J. M. M., and Bosman, W. P., *Rec. Trav. Chim. Pays-Bas* **100**, 148 (1981).
109. van Rens, J. G. M., and de Boer, E., *Mol. Phys.* **19**, 745 (1970).
110. van Rens, J. G. M., Viegers, M. P. A., and de Boer, E., *Chem. Phys. Lett.* **28**, 104 (1974).
111. van Willigen, H., and van Rens, J. G. M., *Chem. Phys. Lett.* **2**, 283 (1968).
112. Vasile, M. J., Richardson, T. J., Stevie, F. A., and Falconer, W. E., *J. Chem. Soc., Dalton Trans.* p. 351 (1976).
113. Vasile, M. J., and Falconer, W. E., *J. Chem. Soc., Dalton Trans.* p. 316 (1975).

114. Voitsekhovskii, V. N., Berkovskii, P. B., Yaschurzhinskaya, L. V., Chugaev, L. V., and Nikitin, M. V., *Izv. Vyssh. Uchebn. Zaved., Tsvetn. Metall.* p. 60 (1975); *Chem. Abstr.* **83,** 167183n (1975).

115. Vollenbroek, F. A., van den Berg, J. P., van der Velden, J. W. A., and Bour, J. J., *Inorg. Chem.* **19,** 2685 (1980).

116. Vollenbroek, F. A., Bouten, P. C. P., Trooster, J. M., van den Berg, J. P., and Bour, J. J., *Inorg. Chem.* **17,** 1345 (1978).

117. Vollenbroek, F. A., Bour, J. J., Trooster, J. M., and van der Velden, J. W. A., *J. Chem. Soc., Chem. Commun.* p. 907 (1978).

117a. Vollenbroek, F. A., Bour, J. J., and van der Velden, J. W. A., *Rec. Trav. Chim. Pays-Bas* **99,** 137 (1980).

118. Vollenbroek, F. A., Bosman, W. P., Bour, J. J., Noordik, J. H., and Beurskens, P. T., *J. Chem. Soc., Chem. Commun.* p. 387 (1979).

119. Warren, L. F., Jr., and Hawthorne, M. F., *J. Am. Chem. Soc.* **90,** 4823 (1968).

120. Waters, J. H., and Gray, H. B., *J. Am. Chem. Soc.* **87,** 3534 (1965).

121. Waters, J. H., Bergendahl, T. J., and Lewis, S. R., *J. Chem. Soc., Chem. Commun.* p. 834 (1971).

122. Wertheim, G. K., Cohen, R. L., Crecelius, G., West, K. W., and Wernick, J. H., *Phys. Rev. B: Condens. Mater.* [3] **20,** 860 (1979).

123. Wooten, F., and Condas, G. A., *Phys. Rev.* **131,** 657 (1963).

124. Zintl, E., Goubeau, J., and Dullenkopf, W. Z., *Z. Phys. Chem., Abt. A* **154,** 1 (1931).

HYDRIDE COMPOUNDS OF THE TITANIUM AND VANADIUM GROUP ELEMENTS

G. E. TOOGOOD

Department of Chemistry, University of Waterloo
Waterloo, Ontario, Canada

and
M. G. H. WALLBRIDGE

Department of Chemistry and Molecular Sciences
University of Warwick
Coventry, England

I. Titanium, Zirconium, and Hafnium Hydrides 267
  A. Titanium Hydride Compounds 267
  B. Zirconium Hydride Compounds 284
  C. Hafnium Hydride Compounds 303
II. Vanadium, Niobium, and Tantalum Hydrides 305
  A. Vanadium Hydride Compounds 305
  B. Niobium Hydride Compounds 314
  C. Tantalum Hydride Compounds 325
  References 334

I. Titanium, Zirconium, and Hafnium Hydrides

A. TITANIUM HYDRIDE COMPOUNDS

The current literature reflects the fact that molecular species containing a titanium–hydrogen bond are of much interest. This is, in part, due to their possible involvement in reactions involving unsaturated molecules such as alkenes, alkynes, hydrogen, carbon monoxide, and dinitrogen. However, such species remain elusive in many respects, and are often poorly characterized. Considerable progress has been made since the early 1960s, when reports of such compounds started to appear, but even now very few definitive X-ray (or neutron) diffraction studies are available because of the difficulty in obtaining suitable crystals of the air-reactive materials. Spectroscopic identification of Ti–H bonds is also frequently difficult to obtain. The intensity of absorptions associated with Ti–H stretching frequencies in infrared spectra is often low, and samples are often paramagnetic, being derivatives of Ti(III), so that ESR rather than NMR spectra must be used to infer the presence of such bonds. Relatively little use has been made of Raman spectroscopy, and the newer Fourier transform infrared techniques have not yet had time to make an impact on the problem. Chemical reactions, such as the release of hydrogen on treatment of the suspected hydride with acids and the conversion of CCl_4 into $CHCl_3$, have often been used as the means of detecting the Ti–H bond.

There has been no recent comprehensive review of this area, although a book on the organometallic chemistry of titanium, zirconium, and hafnium deals, in part, with some of the hydride derivatives (1). In the present review, the first part of the discussion reflects the fact that much of the early work on organotitanium hydrides was, often unknowingly at the time, interwoven with attempts to prepare titanocene, Cp_2Ti ($Cp = \eta^5\text{-}C_5H_5$). Subsequent sections deal with similar compounds containing an additional metal (e.g., aluminum), miscellaneous titanium hydride compounds, and a summary of the main properties of the above species.

1. Organo- and Cyclopentadienyl Titanium Hydrides

Four closely related species have featured prominently in the development of cyclopentadienyl titanium hydride chemistry, namely $[(\eta\text{-}C_5H_5)_2Ti]_{1-2}$, the so-called titanocene; the titanocene hydride $[(\eta\text{-}C_5H_5)_2TiH]_x$; the fulvalene-containing hydride $\mu\text{-}(\eta^5:\eta^5\text{-}C_{10}H_8)\text{-}\mu\text{-}(H)_2\text{-}[(\eta\text{-}C_5H_5)Ti]_2$; and the C_5H_4-bridged compound $(\eta\text{-}C_5H_5)_2Ti\text{-}\mu\text{-}(\eta:\eta^5\text{-}C_5H_4)\text{-}Ti(\eta\text{-}C_5H_5)$. The nature of the first two of these compounds remains uncertain even now, and related derivatives have often been prepared to assist in characterizing this type of compound. It now appears likely that several of the early experiments designed to prepare titanocene, Cp_2Ti, did in fact lead to the fulvalene titanium hydride derivative, although it should be stressed that this remains to be proved in several cases. In view of this fact, the early chemistry of titanocene-type compounds is reviewed first, in order to clarify the relationship between the different types of compounds described in the various reports.

An air-sensitive, dark-green, diamagnetic crystalline solid, obtained by treating titanium dichloride with NaC_5H_5 in tetrahydrofuran (THF), was reported to be titanocene, $(C_5H_5)_2Ti$, as early as 1956. The compound was easily oxidized to the cation $[(C_5H_5)_2Ti]^+$ and formed an adduct with THF, $(C_5H_5)_2Ti \cdot THF$, that exists as a green paramagnetic solid at low temperatures, and quickly transforms to a brown diamagnetic solid at 25°C (2). A similar green solid was obtained by

Clauss and Bestian (3) when $Cp_2Ti(CH_3)_2$ in hexane at 20°C was subjected to an atmosphere of hydrogen, but the solid was formulated by these authors as a dimeric species $[(C_5H_5)_2Ti]_2$.

Attempts to repeat these experiments did not immediately clarify the situation, since products were often incompletely characterized. The use of Na/Hg amalgam (4, 5), Na/naphthalene (6, 7), and sodium sand (8) in reactions (1), (2), and (3) also led to volatile green solids, again formulated as dimeric species

$$Cp_2TiCl_2 + 2NaHg \xrightarrow{C_6H_5CH_3} [(C_5H_5)_2Ti] + 2NaCl + 2Hg \qquad (1)$$

$$Cp_2TiCl_2 + 2NaC_{10}H_8 \xrightarrow{THF} [(C_5H_5)_2Ti] + 2C_{10}H_8 + 2NaCl \qquad (2)$$

$$Cp_2TiCl_2 + 2Na \xrightarrow{C_6H_5CH_3} [(C_5H_5)_2Ti] + 2NaCl \qquad (3)$$

$[(C_5H_5)_2Ti]_2$, and the unexpected diamagnetism was confirmed. A further unusual feature, deduced from infrared spectral data, was that in one case the compound was not of the π-sandwich type, but instead the dimer apparently contained both π- and σ-bonded cyclopentadienyl groups (8).

The precise nature of "titanocene" therefore remained uncertain up to the late 1960s, although other reports appeared to be consistent with the presence of the $(C_5H_5)Ti$ unit in some form in the various reaction mixtures. Thus the green solution, obtained as in reaction (2), absorbed carbon monoxide slowly to give the known dicarbonyl, $Cp_2Ti(CO)_2$, and although the green solution did not react with 2,2'-dipyridyl, the monomeric, blue-black $Cp_2Ti(dipy)$ compound was obtained by the action of dilithium dipyridyl on Cp_2TiCl_2. However, even at this stage doubts were expressed as to whether titanocene was a simple ferrocene-like molecule (9). A series of experiments also identified an "active titanocene," prepared as in Eq. (2), as a dinitrogen fixing reagent, with $[(C_5H_5)_2Ti]_2N_2$ as the intermediate (10, 11). Experiments related to those above have also claimed to have trapped Cp_2Ti as the $Cp_2Ti(2\text{-bipyrydyl})$ complex when Cp_2TiMe is reacted with aromatic amines (12), and Cp_2Ti is also stabilised by bridging pyrazolyl groups as in $(Cp_2Ti)_2(\mu\text{-pyrazolyl})_2$ (13).

Meanwhile, other reports indicated the presence of titanium hydride species in many reduced Ti(IV) solutions. As early as 1966, Brintzinger suggested on the basis of ESR measurements, that a hydride was present in the dark brown solutions obtained from the action of EtMgCl on Cp_2TiCl (14) and this was later identified using ESR as the anion $(Cp_2TiH_2)^-$, which may also be obtained using magnesium/THF or isopropyllithium as the reducing mixture (15). A neutral hydride was also postulated as an intermediate, but not isolated, in the reactions (16, 17):

$$Cp_2TiCl_2 + i\text{-PrMgBr} \rightarrow [Cp_2TiH] + C_3H_6 \xrightarrow{\text{diene}} Cp_2Ti \text{ (allyl)}$$

$$\xrightarrow[\substack{C_8H_8 \\ (COT)}]{} Cp_2Ti \text{ } (C_8H_9)$$

A related series of anionic $\{[Cp_2Ti(\mu\text{-H})_2TiCp_2]^-M^+$, where M = Li, Na, (I)$\}$ and neutral $[Cp_2TiH_2M$, where M = MgBr, AlCl_2, AlH_2, (II)] hydrides, first reported by Olivé et al. (18, 19), have also been partially characterized from line-width trends in their ESR spectra (20, 21).

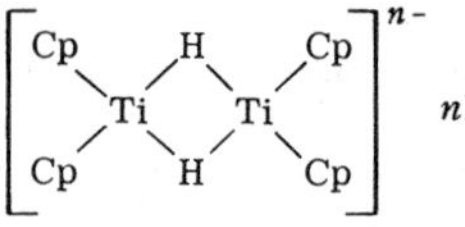

(I) 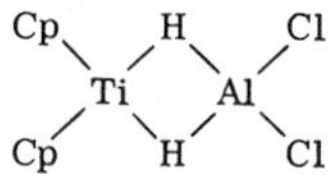   (II)

A series of investigations by Brintzinger and Bercaw in the early 1970s largely clarified the relationship between these titanocene and hydride derivatives. The first isolated hydride, Cp_2TiH, was obtained as a violet solid by treating Cp_2TiMe_2 with hydrogen at low pressure in the *absence* of solvent (22). Under these conditions, only traces of the green "titanocene" referred to be Clauss and Bestian (3) are formed. Since the hydride was diamagnetic a dimeric structure (III) was pro-

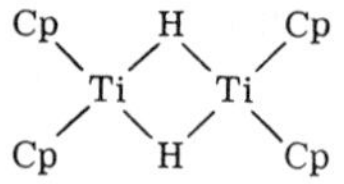

(III) 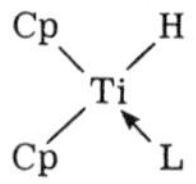   (IV)

posed, and cleavage of the $Ti(\mu\text{-H})_2Ti$ unit by ligands (THF, PPh_3) yields the paramagnetic species (IV). The hydride is apparently only an intermediate, since it decomposes at 150°C in *vacuo* to the green "titanocene" reported earlier.

$$(Cp_2TiH)_2 \xrightarrow[\text{vacuum}]{150°} [Cp_2Ti]_2 \text{ or } [C_{10}H_{10}Ti]_2 + H_2$$
$$\text{green "titanocene"}$$

It also transforms slowly into a gray-green isomer on standing at room temperature (22). This isomer was subsequently suggested to be a polymeric form of the hydride, $(Cp_2TiH)_x$, and is interesting in that on stirring in toluene at 20°C this is also unstable and evolves 0.5 mol of H_2 per mole of Ti with the formation of a dark-colored solution (23).

While the solution species must therefore be formulated as $(Cp_2Ti)_x$, it has been shown to be dimeric in benzene. Although the degree of association in the solid state is uncertain, the infrared spectra of the solid and solution species are similar, but different from that of $(Cp_2TiH)_x$. It is noteworthy that a toluene solution containing $(Cp_2Ti)_2$ is paramagnetic and forms a bright green adduct with THF that decomposes quickly to a brown solid. Such behavior is comparable with that in the initial report claiming the preparation of titanocene (2). The dark-colored solid and solution, initially called "metastable" titanocene, is different from, but isomeric with, the green titanocene. It is a much more reactive compound than the latter, and it is transformed into the green form by heating in toluene at 100°C (23).

The early 1970s also saw the resolution of the problem regarding the precise nature of the green "titanocene" $(C_{10}H_{10}Ti)_2$. While several reports had indicated that treatment of "titanocene" with hydrogen chloride gas did not afford Cp_2TiCl or Cp_2TiCl_2, it was Brintzinger and Bercaw who established that the product of this reaction was in fact $(C_{10}H_9TiCl)_2$ (24). Since the infrared spectrum of "titanocene" contains a strong band at 1230 cm^{-1} that is absent in $(C_{10}H_9TiCl)_2$, Brintzinger and Bercaw suggested that this arises from a Ti(μ-H)$_2$Ti group, and formulated titanocene, $(Cp_2Ti)_2$, as a fulvalene derivative $[Cp(C_5H_4)TiH]_2$ or $(CpTiH)_2(C_{10}H_8)$ (V).

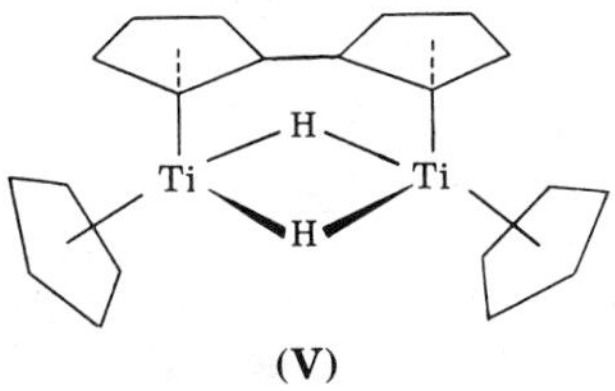

(V)

Although this proposal has not been verified by X-ray crystal data, since suitable crystals have not yet been obtained, there are structural data available on closely related compounds. The hydride (V) is hydrolyzed by water to the bis(hydroxy) derivative (VI), which crystallizes from THF solution as the bis solvate.

$$(CpTiH)_2(C_{10}H_8) + 2H_2O \xrightarrow{\text{THF}} (CpTiOH)_2(C_{10}H_8) \cdot 2C_4H_8O$$
(VI)

The crystal structure of (VI) confirms the presence of a π-bonded fulvalene, with hydroxy groups bridging the two titanium atoms, and the THF hydrogen-bonded to the hydrogens of the hydroxy groups (25).

The Ti–Ti separation (3.195 Å) leaves open the question of how much Ti–Ti interaction is present, and it is significant that the compound is weakly paramagnetic. The presence of a π-bonded fulvalene has also been found in other systems, namely, the mixed aluminotitanium hydrides $(CpTi)_2(H)(H_2AlEt_2)(C_{10}H_8)$ (26) (VII), and $[(C_5H_4)TiHAlEt_2]_2$ $C_{10}H_8$ (26, 27), which are discussed further below, and the niobocene $(CpNbH)_2(C_{10}H_8)$ (28). In addition the ^{13}C-NMR spectrum of titanocene

(VI)   (VII)

confirms the presence of a fulvalene group by comparison with the $(\eta^5:\eta^{5'}$-fulvalene)dicobalt(III,III) dication (29). With the structure of the green "titanocene" therefore established it is very probable that many of the previous formulations as $[(C_5H_5)_2Ti]_2$ require modification (e.g., Refs. 2–8) although some intriguing uncertainties remain. For example, the green titanocene $(Cp_2TiH)_2(C_{10}H_8)$ is relatively unreactive, but the isomeric "metastable" dark form, referred to above as $[(C_5H_5)_2Ti]_x$, reacts rapidly as the dimer in toluene or ether solution with CO and N_2 forming $Cp_2Ti(CO)_2$ and $[(C_5H_5)_2Ti]_2N_2$, respectively (23, 30). A more complex reaction occurs with hydrogen, and the hydride species formed have not been fully identified (23).

Although another active form of titanocene, formulated as $[(C_5H_5)_2Ti]_{1-2}$, has been obtained by van Tamelen et al., it was suggested that this may very well be identical to the metastable form (31). More recently, a black solid active form of titanocene, also formulated as $C_{10}H_{10}Ti$, has been obtained from the diphenyl derivative Cp_2TiPh_2 by photolysis in benzene (32, 33). One report suggests that the titanocene is similar to that obtained earlier from Cp_2TiMe_2 (see Ref. 32), and although the product was not fully characterized it does not show any bands in its infrared spectrum between 1230 and 1220 cm^{-1} from any Ti–H–Ti group vibration (32). An ESR spectrum of another similar product suggests a solution equilibrium as

$$Cp_2Ti(solv)_2 \underset{}{\overset{THF}{\rightleftharpoons}} 2Cp_2Ti(solv)$$

and continuing the photolysis leads to $Cp_2TiH \cdot THF$, as well as $(Cp_2TiH)_x$ (33). An overall reaction scheme has been proposed:

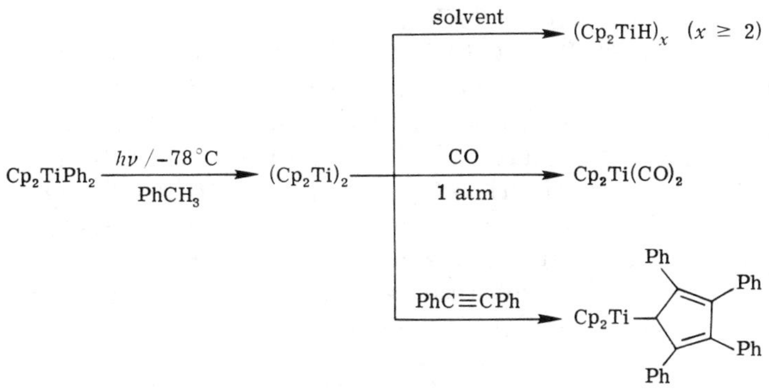

No suggestions have yet been made as to either the precise structures of these active forms or how they differ from the more stable green fulvalene form. Several attempts have been made to prepare similar species by electrochemical reduction of Cp_2TiX_2 and $CpTiX_3$ compounds (e.g., Ref. *30,* and references therein). Blue solutions have been obtained, but the identity of the species at present remains uncertain, although recent studies have indicated that anionic species are the probable products in THF solution according to Refs. *34–36:*

$$Cp_2TiX_2 + e^- \rightleftharpoons [Cp_2TiX_2]^-$$
$$[Cp_2TiX_2]^- + e^- \rightleftharpoons [Cp_2TiX_2]^{2-}$$
$$[Cp_2TiX_2]^{2-} + e^- \rightleftharpoons [Cp_2TiX_2]^{3-}$$
$$[Cp_2TiX_2]^{2-} \rightarrow [Cp_2TiX]^- + X^-$$

where X = Cl, Br, alkyl, aryloxy. However, it has also been reported that controlled reduction of Cp_2TiCl_2 in dimethoxyethane under an atmosphere of CO results in a high yield of $Cp_2Ti(CO)_2$, so that (Cp_2Ti) may be produced at some stage in the reduction *(23).*

While it is not the purpose here to review all Cp–Ti compounds, there is obviously a very close relationship between CpTi moieties and their rearrangement product $(C_5H_4)TiH$ via ring-to-titanium α-hydrogen shift, as has been alluded to above. The nature of the Ti–H species involved in reactions between low-valent titanium compounds and dinitrogen is not always clear. There seems little doubt that some form of bis(cyclopentadienyl) titanium is an active intermediate in dinitrogen coordination reactions *(31, 37–39),* but there is as yet no conclusive evidence that a Ti–H species is involved in the initial reaction *(11),* despite some claims to the contrary *(40).* The existence of different forms of Cp_2Ti is referred to again below.

Further information on the products of reduction of Cp_2TiCl_2 with potassium naphthalene in THF has been elicited by carrying out the reaction at $-80°C$. Removal of the solvent at room temperature followed by extraction with toluene yields $(C_5H_5)_3(C_5H_4)Ti_2$ (**VIII**) as a black powder. Although crystals of this material suitable for X-ray analysis could not be obtained, the structure of the THF adduct, (η-$C_5H_5)_2$-Ti-μ-(η^1: η^5-C_5H_4)-Ti-(η-C_5H_5)(C_4H_8O)-(C_4H_8O) (**IX**), shows the existence of a Ti–Ti bond, with one THF molecule occupying a lattice site. Since (**IX**) is converted to a pure form of (**VIII**) by removal of the THF under vacuum at room temperature, it has been suggested that the structures of (**VIII**) and (**IX**) are similar (41, 42).

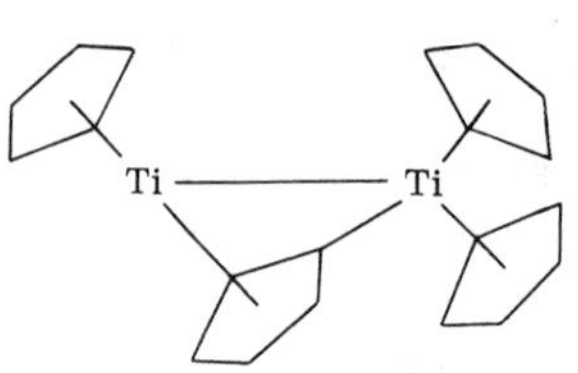

(**VIII**) 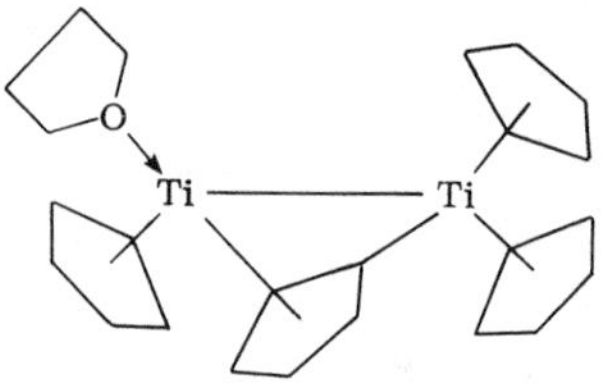   (**IX**)

The black solid (**VIII**) is converted into the green fulvalene titanocene (**V**) at 110°C, and also reacts with H_2, N_2, and alkenes (42), as do the active metastable forms of titanocene (30). With hydrogen, (**VIII**) yields a green-gray precipitate, formulated as $[(C_5H_5)_3(C_5H_4)Ti_2] \cdot H_2$, from toluene solution. The infrared spectrum and deuteration studies show this solid to contain a Ti–H bond, probably with the hydrogen in a bridging position, as either

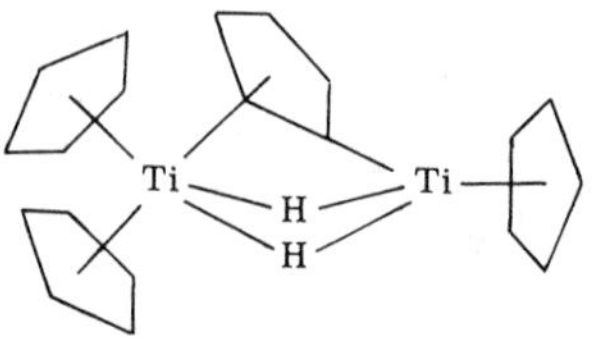

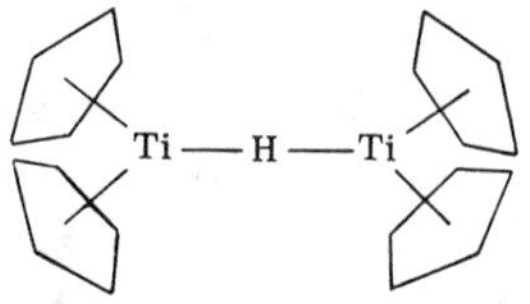

$$(C_5H_5)_3(C_5H_4)Ti_2 + H_2 \underset{\text{vacuum}}{\overset{H_2\ 1-10\ \text{atm}}{\rightleftharpoons}} [(C_5H_5)_3(C_5H_4)Ti_2] \cdot H_2$$

The solid evolves 1 mol of hydrogen per mole of (**VIII**) used, under vacuum, and is therefore similar to the green-gray metastable polymeric form of titancene discussed above and formulated as $(Cp_2TiH)_x$ (23). This solid evolves only 0.5 mol of hydrogen per mole of Ti under vacuum, but both solids show an infrared absorption at 1140 cm^{-1} from the Ti–H bond, as opposed to an absorption at 1450 cm^{-1} from the Ti-μ-$(H)_2$-Ti group in the violet $(Cp_2TiH)_2$ (22). Thus although the two green-gray solids are similar, further properties of each are required before their relationship can be considered as established.

The metallocycle 1,4-tetramethylene bis(cyclopentadienyl)titanium-(IV), obtained by reacting $CpTiCl_2$ with $Li(CH_2)_4Li$, decomposes at $-30°C$, yielding a mixture of 1-butene and ethylene (43). The pathway leading to 1-butene may involve a hydride intermediate, although whether this is formed by α- or β-hydrogen atom elimination is uncertain, e.g.,

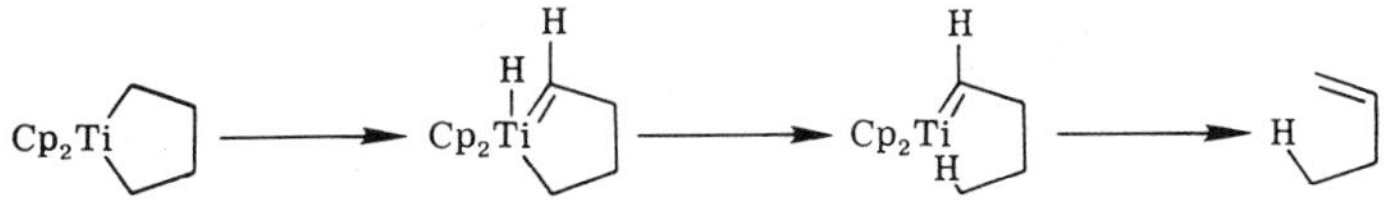

Assuming carbon–carbon bond cleavage occurs in the route leading to ethylene formation, unsuccessful attempts have been made to isolate the bis(ethylene) intermediate by reacting ethylene with titanocene obtained in a way similar to that described above (41). However, carbonylation of the reaction mixture yielded cyclopentanone, inferring that titanocene intermediates are produced (43).

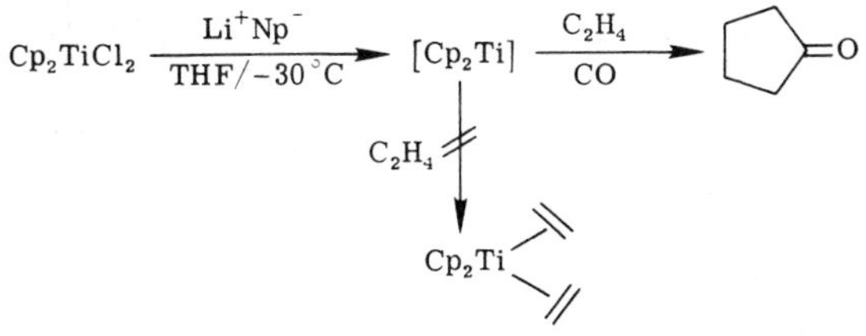

In conclusion, it cannot be assumed at present that all compounds previously formulated as $(C_5H_5)_2Ti$ are in fact (VIII); subtle differences between the various compounds still remain.

A summary of the reactions involving the cyclopentadienyl titanium hydrides is given in Scheme 1 (page 276).

Successful attempts have been made to classify the nature of some of the above species by the use of the pentamethylcyclopentadienyl group, $C_5Me_5(Cp^*)$, which it was supposed avoided the possibility of hydrogen transfer from the Cp^* ring to the metal. Using this modification it has been possible to obtain pure monomeric Cp_2^*Ti as described below; this is now best prepared from its dinitrogen complex, through a series of equilibria (39, 44):

$$2Cp_2^*Ti \underset{\text{vacuum}}{\overset{N_2}{\rightleftharpoons}} [Cp_2^*Ti]_2N_2 \underset{\text{vacuum}}{\overset{2N_2}{\rightleftharpoons}} [Cp_2^*TiN_2]_2 \, N_2$$

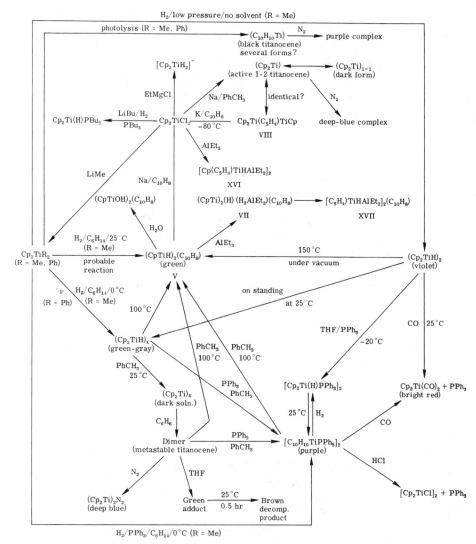

SCHEME 1.

This route is necessary since initial attempts to prepare Cp*₂Ti led to mixtures. Thus while Cp*₂TiMe₂ does not react appreciably with hydrogen (in contrast to Cp₂TiMe₂) when it is refluxed in toluene solution, methane is evolved and a turquoise solid (Cp*)(C₅Me₄CH₂)TiMe, is formed. This solid does react with hydrogen, yielding the diamagnetic dihydride Cp*₂TiH₂, which dissociates in hydrocarbon solvents under vacuum at room temperature to exist in equilibrium with $Cp_2^* Ti$ and $Cp_2^* TiH$, both of which appear to be monomeric in solution (23, 39).

$$(Cp^*)(C_5Me_4CH_2)TiMe + 2H_2 \rightarrow Cp_2^* TiH_2 + CH_4$$

$$Cp_2^* TiH_2 \rightleftharpoons Cp_2^* Ti + H_2$$

$$Cp_2^* TiH_2 + Cp_2^* Ti \rightleftharpoons 2Cp_2^* TiH$$

Passage of dinitrogen into this mixture precipitates the dinitrogen complex as a blue-black microcrystalline solid, and thus provides a synthesis of pure $Cp_2^* Ti$. The yellow-orange $Cp_2^* Ti$ turns yellow-green in solution, and NMR measurements on the paramagnetic Ti(II) species have led to suggestions of the tautomerism:

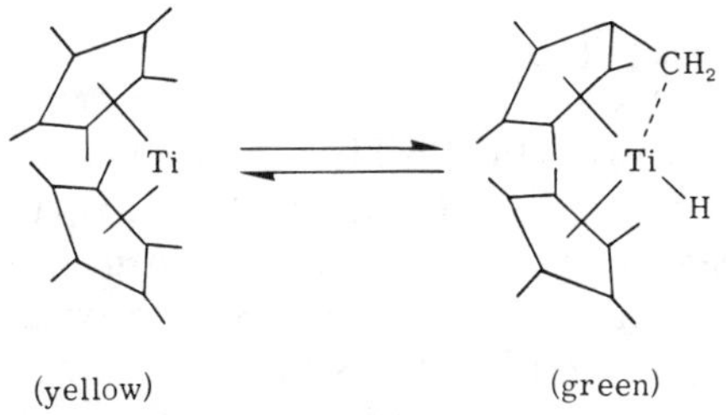

(yellow)   (green)

On prolonged standing at room temperature, $Cp_2^* Ti$ slowly evolves hydrogen yielding the dark violet monomeric $(Cp^*)(C_5Me_4CH_2)Ti$ (39).

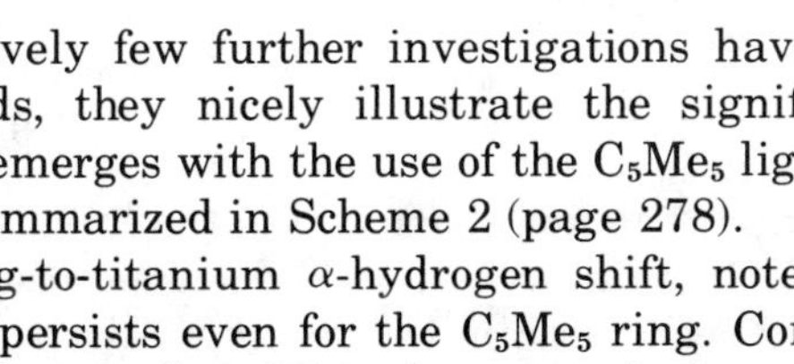

Although relatively few further investigations have been made on these compounds, they nicely illustrate the significantly different chemistry that emerges with the use of the C_5Me_5 ligand. The various reactions are summarized in Scheme 2 (page 278).

Thus the ring-to-titanium α-hydrogen shift, noted above for the C_5H_5Ti moiety, persists even for the C_5Me_5 ring. Confirmation of the fact that the cyclopentadienyl ligand acts as a hydrogen source was also obtained in studies on the dinitrogen-fixing system $Cp_2TiCl_2/Mg/MgI_2$ using deuterated solvents and the C_5D_5 ligand (45).

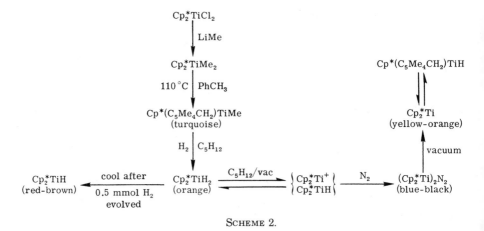

SCHEME 2.

2. *Miscellaneous* Ti–H *Compounds*

The formation of $Cp_2Ti(H)(PPh_3)$ from $(Cp_2TiH)_2$ has been referred to above (Scheme 1). A similar derivative $Cp_2Ti(H)(PBu_3)$ has been detected by ESR measurements when the phosphine is added to a mixture of $Cp_2TiCl_2/LiBu/H_2$ in toluene solution. Such mixtures are hydrogenation catalysts, and the active species has been suggested to be Cp_2TiH existing in a monomer–dimer equilibrium. A complex reaction occurs when Cp_2TiPh_2 is heated with $ZnPh_2$ in ethereal solvents or toluene solution; benzene is evolved and a dark violet diamagnetic solid precipitates. Although neither mass nor NMR spectra could be obtained, other data suggested the solid contained the Ti–H group and a Ti–Ti bond as in (**X**) or (**XI**) (*46*).

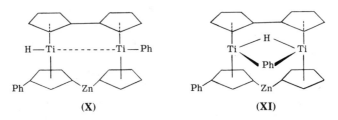

(**X**)   (**XI**)

In contrast to this reaction, the action of the fulvalene titanocene **V** or the complex **VIII** on $ZnPh_2$ results in decomposition of the latter to metallic zinc and benzene. Similar decompositions are observed for $HgPh_2$ (*47*) and $ZnEt_2$ (*48*).

A series of hydrides of empirical formula $Me_2P(CH_2)_2PMe_2MH_3$ (M = Ti, Zr, or Hf) have been prepared from the tetrabenzyl metal compounds by reaction with the phosphine ligand under a high pressure

(200–400 atm) of hydrogen. The compounds appear to be polymeric species containing M–H–M bridge bonds (49). Another interesting series of hydrides has been reported by Wilkinson et al., arising from reduction of [Ti(OPh)$_2$Cl$_2$]$_2$ using either sodium amalgam (or potassium metal) or sodium borohydride (50). With the former, [Ti(OPh)$_2$]$_3$H is formed under argon as an air-sensitive gray powder, and has been assigned the probable structure (XII). If the reaction is carried out under nitrogen, the purple complex [Ti(OPh)$_2$]$_2$(N$_2$)H results, and the structure (XIII) has been suggested. The action of the borohydride in THF solution leads to a yellow solid that also appears to contain a bridging hydride Ti–H–Ti group in [Ti(OPh)$_2$Cl(THF)$_2$]$_2$H (XIV).

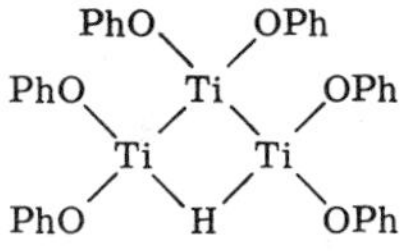

(XII) 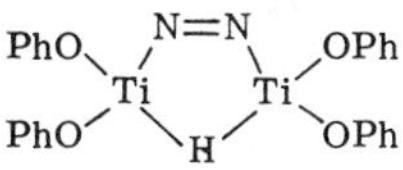   (XIII) 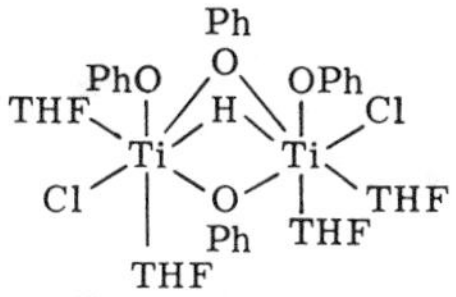   (XIV)

A compound similar to (XIII) has also been prepared from the pyrocatechol compound Ti(O$_2$C$_6$H$_4$)Cl$_2$ by reduction with sodium amalgam in THF, and has been formulated to be similar to (XIII) with the two phenoxide groups at each titanium being replaced by the pyrocatecholate dianion. The ESR signal from both these complexes suggests the presence of one unpaired electron per dimer (50).

3. Hydrides Bridging with Other Elements

Several compounds containing Ti–H–M bonds are well known, in particular those containing a BH$_4$ or AlH$_4$ group. Since the former compounds have been reviewed recently (51), only brief mention is made of them here, but the Ti–H–Al compounds are discussed in more detail. The mode of attachment of the BH$_4$ group to the titanium atom is established, in compounds such as Cp$_2$TiBH$_4$, to be through a double hydrogen bridge

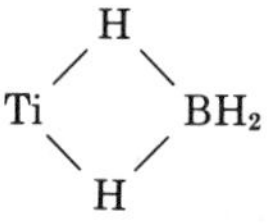

and recent electron diffraction results indicate that a similar model persists in the gas phase (52). An unusual Ti–H–B interaction occurs in the dimeric compound [C$_{16}$H$_{16}$N$_2$O$_2$Ti(BH$_3$)$_2$]$_2$. Here, two BH$_3$ groups are coordinated to nitrogen atoms in the NN'-ethylene-bis(salicylidenaminato) ligand, and also interact, through Ti–H–B bonds, with each metal center, as in (XV), forming a capped octahedral arrangement around each titanium atom (53). The unstable borohydride $Ti(BH_4)_3$ has recently been prepared by grinding solid $TiCl_4$ and $LiBH_4$ together at low temperatures. It decomposes at 20°C, but can be stabilized as the THF adduct $Ti(BH_4)_3 \cdot 2THF$ (54).

$$TiCl_4 + LiBH_4 \rightarrow Ti(BH_4)_3 \rightarrow TiB_2 + 0.5B_2H_6 + 4.5H_2$$

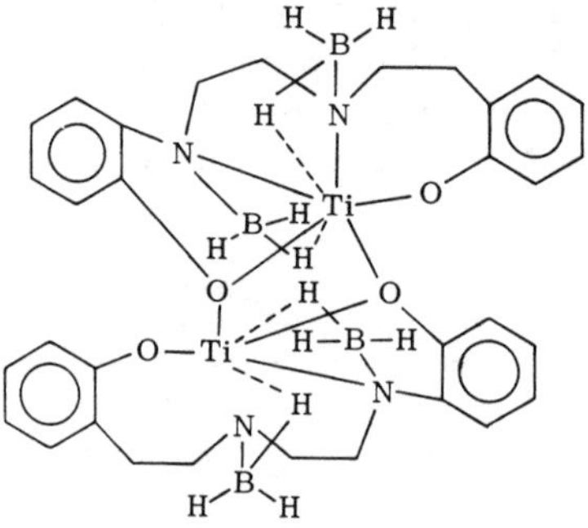

(XV)

Ti–H–Al *compounds*. Some of these have been briefly referred to above—e.g. (VII)—and a novel series of related compounds, (XVI) and (XVII), has been obtained by a series of reactions (26, 27, 55):

$$\left.\begin{matrix} Cp_2TiPh_2 \\ Cp_2TiCl_2 \\ Cp_2TiCl \end{matrix}\right\} \xrightarrow{\text{AlEt}_3} [Cp(C_5H_4)TiHAlEt_2]_2$$

(XVI)

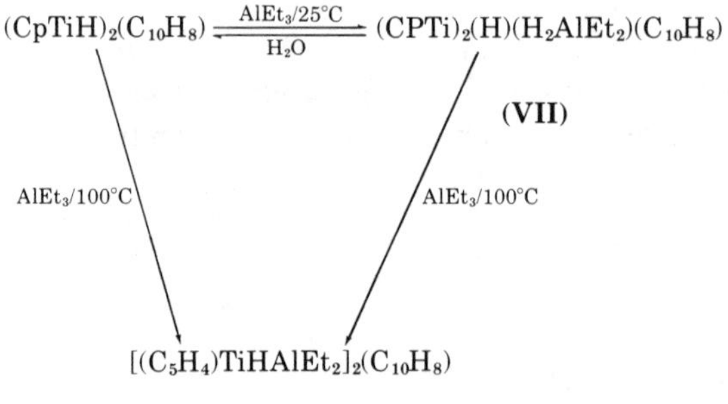

$$(CpTiH)_2(C_{10}H_8) \underset{H_2O}{\overset{\text{AlEt}_3/25°C}{\rightleftharpoons}} (CPTi)_2(H)(H_2AlEt_2)(C_{10}H_8)$$

(VII)

AlEt$_3$/100°C   AlEt$_3$/100°C

$$[(C_5H_4)TiHAlEt_2]_2(C_{10}H_8)$$

(XVII)

The structure of the reddish-purple (VII) has been given (Section I,A,1). Compound (XVII) is similar in that it also contains the fulvalene ligand, but it is unlike (VII) in that it possesses a Ti–Ti bond. On the other hand, both (XVII) and (XVI) are more reminiscent of the compound that is possibly related to the "active" form of titanocene, containing both bridging (C_5H_4) groups and Ti–Ti bonds. However, it is clear that a close structural relationship exists among all three compounds, the two Cp ligands in (XIV) being replaced by the fulvalene

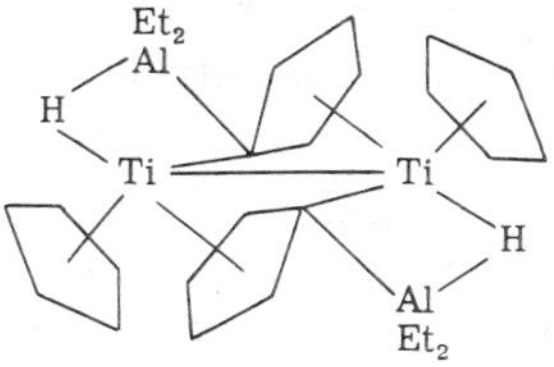

(XVI) 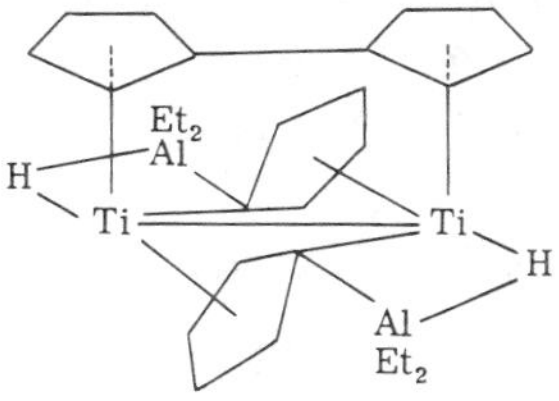   (XVII)

ligand in (XVII). It has not yet been reported whether it is possible to convert (XVI) to (XVII), but such a transformation would appear feasible, especially in view of the relationship among $(Cp_2Ti)_2(H)(H_2AlEt_2)$ $(C_{10}H_8)$, (VII), and (XVII) mentioned above. These reactions are obviously relevant to the nature of the active species involved in the catalytic polymerization of alkenes; the action of $AlEt_3$ on Cp_2TiCl_2 has also been shown to yield $Cp_2TiCl_2AlEt_2$ (56–58), and in the reaction discussed above this has been separated from (XVI) by crystallization (55).

Several studies have been made on reactions involving the interaction of species containing Al–H and Ti–Cl bonds, but so far the precise nature of the products of such reactions has not been defined. The use of ESR spectroscopy to assist in the identification of species such as $Cp_2TiH_2AlH_2$ has already been discussed briefly. A recent investigation, using both calorimetric titrations and ESR methods, has suggested that there are four steps in the $Cp_2TiCl_2/LiAlH_4$ and $CpTiCl_3/LiAlH_4$ reactions. These coincide with Cp_2TiCl_3 or $CpTiCl_3$ to $LiAlH_4$ mole ratios of $2:1$, $1:1$, $1:1.5$, and $1:2$, as summarized in the accompanying tabulation.

Reactants	Mole ratio	Products
$Cp_2TiCl_2/LiAlH_4$	$2:1$	$Cp_2TiH_2AlCl_2 \cdot Cp_2TiCl$
$CpTiCl_3/LiAlH_4$	$2:1$	$Cp(Cl)TiH_2AlCl_2 \cdot CpTiCl_2$
$Cp_2TiCl_2/LiAlH_4$	$1:1$	$[Cp_2TiH_2Al(H)Cl]_2$
$CpTiCl_3/LiAlH_4$	$1:1$	$[Cp(Cl)TiH_2Al(H)Cl]_2$
$Cp_2TiCl_2/LiAlH_4$	$1:1.5$	$[(Cp_2TiH_2)_2Al_2H_3Cl]$
$CpTiCl_3/LiAlH_4$	$1:1.5$	$[(Cp(Cl)TiH_2)_2Al_2H_3Cl]$
$Cp_2TiCl_2/LiAlH_4$	$1:2$	$Cp_2TiH_2AlH_2 \cdot AlH_3$
$CpTiCl_3/LiAlH_4$	$1:2$	$Cp(Cl)TiH_2AlH_2 \cdot AlH_3$

The product formulas are those suggested from both the stoichiometry and the ESR spectra. Thus in the 2 : 1 reaction, the latter technique suggests that both $Cp_2TiH_2AlCl_2$ and Cp_2TiCl are present, presumably in some trinuclear product (59). The presence of Ti–H bonds may be inferred in various solutions of titanium chlorides ($TiCl_4$, Cp_2TiCl_2, etc.) with various metal hydrides, where such solutions are used as catalysts for the hydrogenation (and related reactions) of alkenes (60–63).

4. Catalytic and Other Properties of Ti–H Bonds

The use of both $TiCl_4/LiAlH_4$ and $TiCl_3/LiAlH_4$ mixtures in organic synthesis has been reviewed (64, 65). In addition to the hydrogenation reaction, such mixtures are also active catalysts for the isomerization (66) and hydrometallation (67, 68) of alkenes, although in the latter case aluminum is invariably the metal that becomes attached to the olefinic bond. The relationship between hydrogenation and isomerization, as well as between metathesis and oligomerization leading to polymerization, has been often emphasized. An original proposal, combining both metallocycles and the elimination of α-hydrogen from metal alkyls believed to be involved in metathesis and isomerization reactions respectively, has been put forward to give a mechanism for the homogeneous and heterogeneous transition-metal-catalyzed polymerization of alkenes. In particular for Ziegler–Natta systems, the Ti–alkyl and Ti–H systems are combined as follows (69):

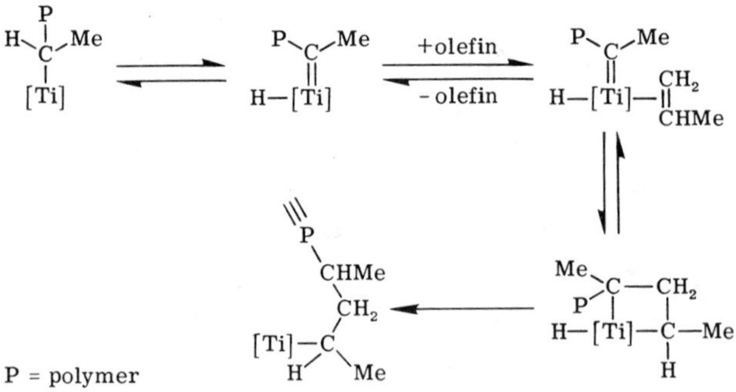

The involvement of the α-elimination reaction in this cycle has been in question following experiments on cyclopentadienyl cobalt complexes, where evidence for olefin insertion for Ziegler–Natta polymerization catalysis has been obtained by labelling experiments, using C_2H_4 with a deuterated cobalt complex (70):

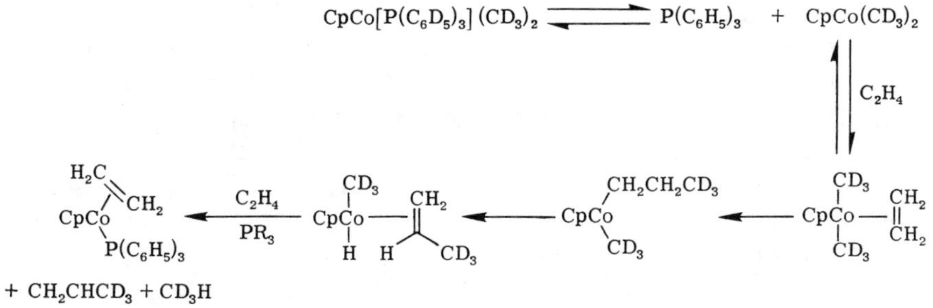

$$CpCo[P(C_6D_5)_3](CD_3)_2 \rightleftharpoons P(C_6H_5)_3 + CpCo(CD_3)_2$$

$$\Big\updownarrow C_2H_4$$

$$+ CH_2CHCD_3 + CD_3H$$

Other studies have shown that both the fulvalene titanocene (**V**), and the cyclopentadiene dititanium compound (**VIII**) act as homogeneous catalysts for the conversion of ethylene into ethane and buta-1,3-diene. In this case, an intermediate containing Ti–H bonds has also been proposed (71).

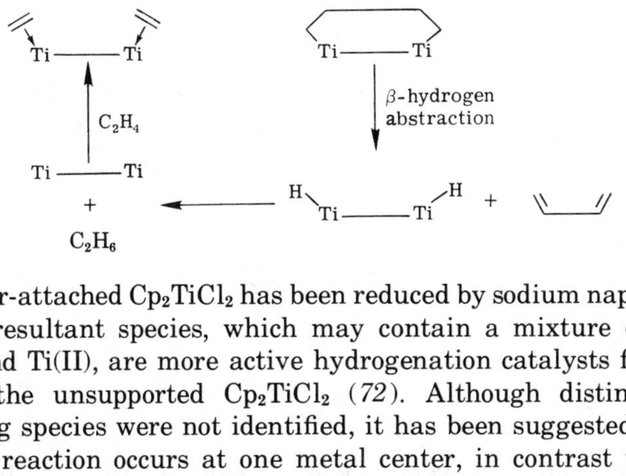

Polymer-attached Cp_2TiCl_2 has been reduced by sodium naphthalide, and the resultant species, which may contain a mixture of Ti(IV), Ti(III), and Ti(II), are more active hydrogenation catalysts for olefins than is the unsupported Cp_2TiCl_2 (72). Although distinct Ti–H-containing species were not identified, it has been suggested that the complete reaction occurs at one metal center, in contrast to earlier suggestions that such reductions involve a bimolecular reaction of an intermediate titanocene alkyl and a titanocene hydride (30).

Various experiments have suggested that, in general, hydride derivatives of the group IVA metals show considerable hydridic character as M^+-H^-. Thus compounds such as $(Cp_2TiH_2)^-$ and Cp_2TiBH_4 react rapidly with acids liberating hydrogen (14, 73, 74), and Cp_2TiBH_4 does not react with amines to liberate BH_3, also consistent with considerable hydridic character of the BH_4 group (73). The reaction of hydride compounds of group IVA, VA, and VIA elements with ketones has been used to assess the relative degree of hydridic character in the M–H bond, and although the nature of the other ligands present is important, there does appear to be a decrease in hydridic character on going from group IVA to VIA (75). Relatively few theoretical studies have been made on Ti–H bonds, but calculations on the unknown species TiF_3H suggest that the hydrogen atom would be hydridic in nature (76). Other calculations have been concerned with the change in stereochemistry through the series TiH_4, TiH_3^+, TiH_3, TiH_3^-, in comparison with their carbon and silicon analogs (77), and have dealt with the transannular ligand superhyperfine splittings in the ESR spectra of species such as $Cl_2TiH_2MX_{0-2}$ (M = Li, Mg, Al; X = H, Cl), where the interaction is primarily through space between the titanium d and transannular atom s orbitals (78). A comprehensive study of bis(cyclopentadienyl) metal compounds has included calculations for the bent systems Cp_2TiH^+, Cp_2TiH_2, and $Cp_2TiH_3^-$ compounds using the models shown below. In each case, the hydrogen atoms lie in the yz plane, which bisects the two Cp rings and contains the twofold rotation axis of the bent Cp_2Ti system. The hydrogen in Cp_2TiH^+ lies off the twofold axis, and the angles between the hydrogen atoms in the bishydride and trishydride compounds are calculated as shown. Although calculations were not carried out for the d^1 complex Cp_2TiH, it was predicted that for the related alkyl compound Cp_2TiR the alkyl group would also lie off the C_2 axis (79).

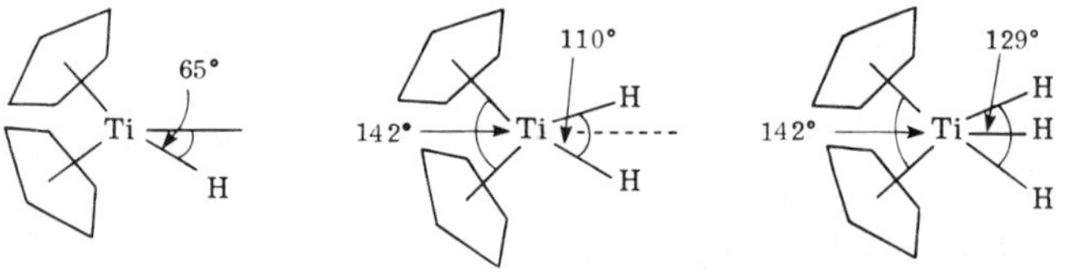

The properties of the various titanium hydride compounds are summarized in Table I.

B. Zirconium Hydride Compounds

The use of zirconium hydride derivatives in organic synthesis (80) and as reagents for coordinating dinitrogen and hydrogenating ligated CO (81, 82) has been reviewed, and are therefore not discussed in detail here. The compounds used in these studies are usually derivatives containing either the (Cp_2Zr) or $[(C_5Me_5)_2Zr]$ group, e.g., $Cp_2Zr(H)Cl$ or $Cp_2^*ZrH_2$, and the properties of these and related compounds are included in the general summary given next. The monohydrides are discussed first, followed by the dihydride compounds.

    a. [HZr(dmpe)$_2$(η^5-C$_6$H$_7$)]. While reduction of ZrCl$_4$(dmpe)$_2$ by Cs/benzene or Na/Hg//THF yields intractable products, in the presence of 1,3-cyclohexadiene a red-brown crystalline hydride is formed in 40% yield. The hydride is stable at ambient temperature, but decomposes evolving hydrogen and benzene above 130°C. The Zr–H group could not be detected spectroscopically (infrared and NMR; in the latter it is assumed that the resonance is obscured by the dmpe or cyclohexadienyl proton resonances between τ 4.5–9.3) (*83, 84*).

$$\text{ZrCl}_4\text{(dmpe)}_2 \ + \ \bigcirc \ \xrightarrow[\text{THF}]{\text{Na/Hg}} \ \text{Zr(H) (dmpe)}_2$$

    b. [Cp$_2$Zr(H)R]$_2$ *and* [Cp$_2^*$Zr(H)R]. Two independent reports (*84, 85*) have shown that the cyclopentadienyl compounds may be prepared from the corresponding chloro derivative:

$$\text{Cp}_2\text{Zr(Cl)R} \ \xrightarrow[\text{or Li[AlH(O-}t\text{Bu)}_3]/\text{DME}]{\text{LiAlH}_4} \ \text{Cp}_2\text{Zr} \overset{\text{H} \quad \text{R}}{\underset{\text{R} \quad \text{H}}{\diagup \diagdown}} \text{ZrCp}_2$$

where R = (cyclohexyl)methyl, cyclohexyl, ethyl, *n*-octyl, neopentyl (*85*), or CH(SiMe$_3$)$_2$ (*86*). These compounds all appear to be essentially dimeric in benzene solution, in contrast to the methyl derivative (R = Me), which is reported to be polymeric (*87, 88*). The silyl group conveys a higher stability on the hydride, since the ordinary alkyl compounds are only stable below −30°C in solution. This general instability of hydridoalkyl compounds compared with the corresponding dihydro or dialkyl compounds has been recognized (*89*), and continues into the pentamethylcyclopentadienyl analogs. Thus Cp$_2^*$Zr(H)(CH$_2$CHMe$_2$), prepared from the action of isobutylene on the hydride Cp$_2^*$ZrH$_2$ (*90*), decomposes slowly at 75°C evolving isobutane, while Cp$_2^*$ZrH$_2$ and Cp$_2^*$ZrMe$_2$ are stable over long periods at 100°C (*91*). The Cp* groups also cause the hydridoalkyl to be monomeric.

    Both the Cp and Cp* hydridoalkyl compounds react with hydrogen, but different mechanisms have been proposed. In both cases there is *no* evidence for the reductive elimination of alkane followed by the oxidative addition of H$_2$ to the Cp$_2$Zr or Cp$_2^*$Zr fragment. In the case of the Cp$_2$Zr(H)R compounds, reaction with D$_2$ leads to incorporation of D into the hydride position, and a mechanism involving deuteride abstraction

TABLE I

TITANIUM HYDRIDE COMPOUNDS AND RELATED DERIVATIVES

Compound	Preparation	Properties	¹H NMR (τ, M–H)	Infrared (cm⁻¹)	Comments	References
Cp_2Ti	$TiCl_2/NaC_5H_5/THF$	Dark-green solid, air-sensitive, diamagnetic. Forms THF adduct, which decomp. to brown solid.	—		Stable in inert atmosphere, no spec. props. available.	2
$(Cp_2Ti)_2$	$Cp_2TiMe_2/H_2/C_6H_{14}$	Dark-green solid, dimeric in soln.	—	No band near 1230 (no Ti–H?)	Probably contains fulvalene–Ti hydride (V).	3
$C_{10}H_{10}Ti$	Photolysis, Cp_2TiMe_2, or Cp_2TiPh_2	Black solid, not completely characterized.	Broad resonances 4–5 (Cp?) 0.5–1.5 (?)	Sharp bands at 790, 1010; no bands at 1220–1230		32, 33
	Redn. Cp_2TiCl_2 with Na/Hg, Na–$C_{10}H_8$, or Na	Green diamagnetic volatile solids, react $CO \rightarrow Cp_2Ti(CO)_2$, $N_2 \rightarrow (Cp_2Ti)_2N_2$	—	1228 (Ti–H), 1262, 1300	From IR presence of (V) in samples indicated.	4–9
	$(Cp_2TiH)_2 \xrightarrow{150°C} (C_{10}H_{10}Ti)_2 + H_2$	Dark-green solid, dimeric in C_6H_6, paramagnetic. Forms green adduct with THF, but turns brown (cf. Ref. 2).	—	790, 1010. Very weak band near 1230 (not assigned in Ref. 23).	Shows similarities to $Cp_3(C_5H_4)Ti_2$ (VIII).	22, 23
	$(Cp_2TiH)_x \rightarrow$ $(Cp_2TiH)_x + H_2$ $\rightarrow (C_{10}H_{10}Ti)_2 + H_2$	Metastable titanocene. Dark-colored soln. converted to green fulvalene hydride (V) on heating to 100°C. Reacts $CO \rightarrow Cp_2Ti(CO)_2$.	—	790, 1010	Also closely similar to (VIII). Dark form of Cp_2Ti not well characterized. Also obtained via $Cp_2TiMe_2/H_2/C_6H_{14}/0°C$ mixture.	22, 23, 30

$(Cp_2Ti)_{1-2}$	Cp_2TiCl_2/Na/toluene under argon	Monomer–dimer equilibrium. Unstable at r.t. Reacts $N_2 \rightarrow Cp_2TiN_2TiCp_2$ as dark-blue complex.	—	790, 1010; no band at 1200–1250. UV max. at 486, 640 nm.	From UV appears very similar to **(VIII)**. Lack of band at 1223 cm^{-1} in infrared indicates $(Cp_2TiH)_2(C_{10}H_8)$ **(V)** not present.	*31*
$(Cp_2TiH_2)^-$	Cp_2TiCl_2 (or Cp_2TiCl)/EtMgCl (or Mg/THF, or Li *i*Pr)	Dark brown in solution, decomp. by H_2O/acids evolving H_2.	ESR shows 1:2:1 triplet (g = 1.992)	—	—	*14, 15, 18*
Cp_2TiH_2M (M^2 = MgBr, AlH_2, $AlCl_2$)	$(Cp_2TiCl)_2/LiAlH_4$/THF	Obtained in soln. only, not isolated.	Identified from ESR data (see also Refs. *20, 21*)	—	—	*18–21*
$(Cp_2TiH)_x$	(i) Cp_2TiMe_2/hexane soln./H_2/0°C (ii) Decompn. $(Cp_2TiH)_2$ at 25°C.	Green-gray isomer of $(Cp_2TiH)_2$. Both hydrides show similar reactions but $(Cp_2TiH)_x$ insol. in ethereal solvents. Unstable in toluene, forms H_2 + metastable titanocene.	—	Broad band at 1140, tentative assignment to Ti–H–Ti vibration in $(-H-Ti-H-Ti-)_x$. Otherwise similar to $(Cp_2TiH)_2$.	Stable when stored as solid under Ar.	*22, 23, 33*
$(Cp_2TiH)_2$	Cp_2TiMe_2/H_2/low pressure/no solvent	Diamagnetic violet solid, decomp. 150°C *in vacuo* to green "titanocene" solid (see above). With 1,3-pentadiene $\rightarrow Cp_2Ti$(allyl), $B_2H_6 \rightarrow Cp_2TiBH_4$, etc.	—	Broad band at 1450 from Ti(μ-H_2)Ti group.	Cleaved by ligands to give paramagnetic comps. Unstable on storage as solid.	*22*
$(CpTiH)_2(C_{10}H_8)$ **(V)** green "titanocene"	Various methods (e.g., see above), typically redn. of Cp_2TiCl_2. Probably present in reaction products reported earlier as "titanocene") (see text).	Diamagnetic green solid, soluble in hydrocarbons. Air- and water-reactive, $H_2O \rightarrow (CpTiOH)_2(C_{10}H_8)$	^{13}C: −105 (Cp), −100 to −122 ppm ($C_{10}H_8$), rel. to TMS. (Ref. *29*).	Band at ~1230 from Ti(μ-H_2)Ti group. Compare with above. Other bands near 800 (790 and 810) and 1010. UV bands at 428, 824 nm.	Structure established by X-ray study on bis(hydroxy) compound (Ref. *25*). Parent ion in mass spectrum. Best prepared by action $Na/C_{10}H_8$ on Cp_2TiCl_2.	*24*

(*table continues*)

TABLE I (Continued)

Compound	Preparation	Properties	¹H NMR (τ, M–H)	Infrared (cm⁻¹)	Comments	References
$Cp_2Ti(C_5H_4)TiCp$ (VIII)	Cp_2TiCl_2/K/naphthalene/THF/–80°C	Black, solid, magnetic props. recorded but not reliable. Green soln. in benzene and THF, latter decomp. to brown soln. (cf. Refs. 2, 22, 23). Reacts H_2, N_2, etc.	Broad resonance at 15 ppm downfield TMS at 32°C—suggests paramagnetic (?). Moves to higher field and broadens on cooling, origin of resonance at $1/T = 0$ is ~7 ppm (cf. Ref. 32 where black titanocene shows Cp protons at 5–6 ppm as broad signal).	Major band at 790, similar to other metallocenes. No splitting of 790 band as there is in (V). Weak band at 1223 cm⁻¹ present due to probable presence small amounts of (V).	No significant band at 824 nm in uv, thus only traces of (V) present. No evidence of M–H bond apart from trace (V).	41, 42
$(CpTi)_2(H)(H_2AlEt_2)(C_{10}H_8)$ (VII)	$(CpTiH_2)(C_{10}H_8)$/Et_3Al/toluene/25°C	Reddish-purple crystals, diamagnetic.	4.74 (C_5H_5); 3.55, 5.39, 6.20 ($C_{10}H_8$); 8.63, 8.79, 9.65, 9.87 (CH_2CH_3); 21.62 (TiH_2AlTi); 31.80 ($TiHTi$).	—	Ti–Ti distance of 3.37Å suggests little bonding, magnetic props. due to exchange in Ti–H–Ti group?	26
$[(C_5H_4)TiHAlEt_2]_2(C_{10}H_8)$ (XVII)	(VII)/Et_3Al/sealed tube/100°C/30 h	Diamagnetic, purple-red solid, m.p. 172–173°C.	1.73, 3.61, 4.16, 5.17, 5.45, 5.75, 5.84 (C_5H_4); 8.47, 8.77, 9.21, 10.00 (CH_2CH_3); 16.59 (TiHAl).	—	Ti–Ti distance 2.91 Å suggests Ti–Ti bond.	26, 27
$[CpC_5H_4)TiHAlEt_2]_2$ (XVI)	Cp_2TiX_2 (X = Ph, Cl, Cl/2)/$AlEt_4$/C_6H_6/70°C/12 h	—	4.92 (Cp); 3.14, 4.17, 5.07, 5.71 (C_5H_4); 8.54, 8.68, 9.36, 9.54 ($AlEt_2$); 17.02 (Ti–H–Al).	—	Separated from $Cp_2TiCl_2AlEt_4$ by crystallization.	55

288

Compound	Preparation	Physical properties	NMR	IR	Remarks	Ref.
$Cp_2Ti(H)PBu_3$	$Cp_2TiCl_2/LiBu/$ $H_2/PBu_3/C_6H_5CH_3$ solution	—	—	—	Hydrogenation catalyst.	47
$(PhC_4H_3ZnC_5H_4)H$ $TiTlPh(C_{10}H_8)$?	$Cp_2TiPh_2/ZnPh_2/$ $Et_2O/60°C/10$ h	Diamagnetic, dark-violet solid. Sensitive to H_2O and O_2.	—	740, 710 (C_6H_5); 820 (C_5H_5); no band at 1230, thus no TiH_2Ti group.	Interaction of two Ti^{3+} centers proposed to rationalize diamagnetism.	46
$[Me_2P(CH_2)_2PMe_2]TiH_3$	$Ti(benzyl)_4/$ ligand/H_2/ 200–400 atm	Polymeric? Black solid.	—	—	—	49
$[Ti(OPh)_2]_3H$	$Ti(OPh)_2Cl_2/2K/Ar$ atmosphere/toluene/25°C. Under N_2 gives purple N_2 complex.	Gray solid, air-reactive. Magnetic props. suggest one unpaired electron per 3 Ti atoms.	Broad ESR signal, no hyperfine splitting.	1258 (σ-bonded) phenoxo group. No band assigned to Ti–H.	Ti–H detected by reaction with $CCl_4 \rightarrow CHCl_3$.	50
$[Ti(OPh)_2Cl(THF)_2]H$	$Ti(OPh)_2Cl_2/NaBH_4/THF$	Yellow solid, only stable in THF. Probably dimeric with one unpaired electron per dimer.	—	440 (terminal Ti–Cl), no band from Ti–H.	Ti–H detected as above.	50
Cp_2^*Ti	$Cp_2^*TiH_2 \rightleftharpoons$ $Cp_2^*Ti + Cp_2^*TiH$ $\left.\begin{array}{l}Cp_2^*Ti \\ Cp_2^*TiH\end{array}\right\} \overset{N_2}{\rightarrow} (Cp_2^*Ti)_2N_2$ $\downarrow$ vac Cp_2^*Ti	Paramagnetic, yellow-orange, monomeric in benzene. Decomp. to violet $(Cp^*)(C_5Me_4CH_2)$ at r.t.	Broad singlet 86.2 ppm downfield from TMS.	Similar to $Cp_2^*TiCl_2$ from 4000–400. No bands from Ti–H bonds.	Tautomerism between yellow Cp_2^*Ti and green (Cp^*) $(C_5Me_4CH_2)TiH$.	39
$Cp_2^*TiH_2$	$(Cp^*)(C_5Me_4CH_2)TiMe$ $+ H_2 \overset{20°C}{\longrightarrow} Cp_2^*TiH_2$	Diamagnetic orange crystals, monomeric in benzene.	Singlets at δ 1.95 (C_5Me_5) and 0.28 (TiH_2) ppm.	Broad band at 1560, from TiH_2 group.	—	23, 39
Cp_2^*TiH	$Cp_2^*TiH_2 \overset{0.5\ H_2}{\longrightarrow} Cp_2^*TiH$	Red-brown, crystalline, monomeric in C_6H_6. Unstable evolving H_2.	Broad singlet at 22.2 ppm (30°C). Shows Curie dependence.	Band at 1575 assigned to Ti–H.	—	39
$Cp^*(C_5Me_4CH_2)TiH$ (green form of Cp_2^*Ti)	Present in soln. of Cp_2^*Ti. Yellow solns. of Cp_2^*Ti at low temps. change to yellow-green on warming.	Exists in equilibrium with Cp_2^*Ti at 25°C.	Shows sharp singlets at 7.82, 8.2, 8.38, 8.64, 8.92. No Ti–H resonance detected.	—	—	39

289

from D_2, followed by electrophilic attack of the residual D^+, was proposed (85):

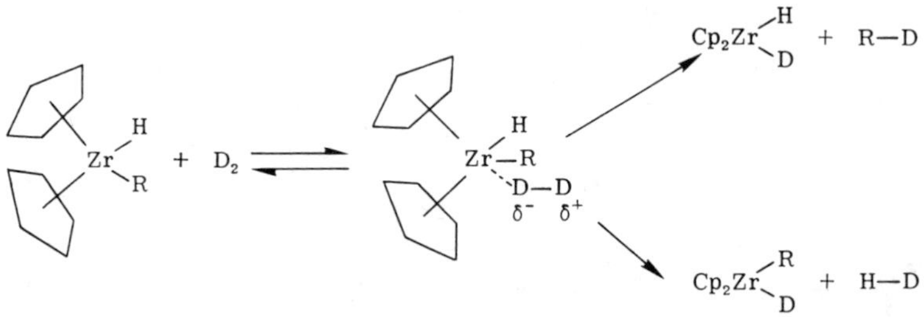

The reverse reaction was shown to be faster than the forward one, leading to the incorporation of deuterium into the alkyl hydride. With $Cp_2^*Zr(H)R$, however, no exchange of deuterium with the hydride position occurred with D_2, and the first step was proposed to be a facile and reversible metal-to-ring hydrogen transfer. This was then followed by oxidative addition of H_2 to the intermediate 14-electron system followed by elimination of the hydrocarbon (91):

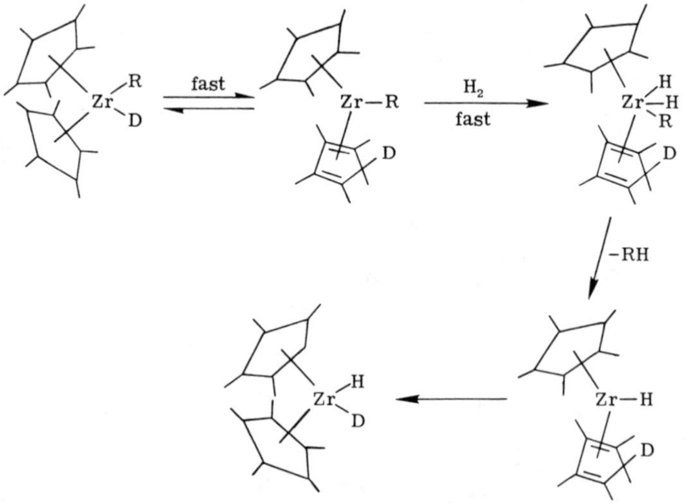

These two pathways have been discussed more recently, and an interesting alternative proposal has been made wherein the formation of an adduct between H_2 and $(C_5R_5)_2ZrR_2^1$ (R,R^1 = H or Me) is proposed. This model utilizes back-bonding from high-energy Zr–R^1 bonding orbitals into the σ^* H_2 orbitals, and is seen as being analogous to the formation of a CO adduct. The model is supported by extended Hückel molecular orbital calculations, and the transition state is suggested to contain an (H--H--H) moiety as a three-center ligand with bonding properties similar to those of an alkyl group. The term *direct hydrogen transfer* has been used to describe this process, and it offers an attractive alternative mechanism for reactions where dihydrogen is activated at a metal center (*92*).

A continuation of the studies on $Cp_2Zr(H)R$ compounds has shown that in addition to hydrogen various tertiary phosphine ligands can also produce reductive elimination of alkane:

$$Cp_2Zr(H)R + 2L \rightarrow Cp_2ZrL_2 + RH$$

where $L = Ph_2PMe$, $PhPMe_2$, $Ph_2P(CH_2)_2PPh_2$, and $Me_2P(CH_2)_2PMe_2$, and $R = $ cyclohexylmethyl. The compound $Cp_2Zr[Me_2P(CH_2)_2PMe_2]$ is stable at room temperature, and is a black-green crystalline solid which is sensitive to both air and moisture. The compounds with Ph_2PMe and $PhPMe_2$ decompose slowly at room temperature in solution, affording deep-red diamagnetic dimeric species, as in (*93*).

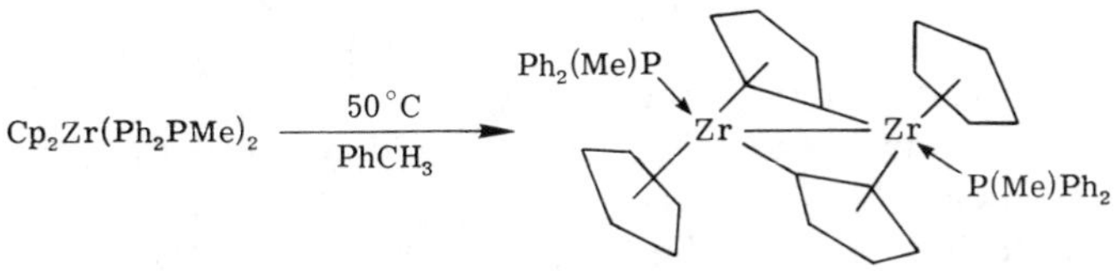

$$Cp_2Zr(Ph_2PMe)_2 \xrightarrow[\text{PhCH}_3]{50\,^\circ C}$$

c. $Cp_2Zr(H)Cl$ and $Cp_2Zr(H)AlH_4$. Addition of $LiAlH_4$, $LiAl(O\text{-}t\text{-}Bu)_3H$ (*87, 88*) or $NaAl(CH_3OCH_2CH_2O)_2H_2$ (*94*) to Cp_2ZrCl_2 in THF precipitates the hydride $Cp_2Zr(H)Cl$ in high yield. Further details of this preparation and that of Cp_2ZrH_2 have been published (*95*). The infrared spectrum of $Cp_2Zr(H)Cl$ indicates the presence of bridging hydrogens in Zr–H–Zr bonds and is consistent with the formulation of the solid as a polymer. Addition of a further mole of $LiAlH_4$ to $Cp_2Zr(H)Cl$ suspended in THF yields a clear solution, from which $Cp_2Zr(H)AlH_4$ may be obtained. This compound is also polymeric and liberates hydrogen on heating in solvents at 60°C, with the formation of a deep-red diamagnetic product that has yet to be fully characterized (*87, 88*). This contrasts with the properties of the related $Cp_2Zr(H)BH_4$, which is volatile at 60°C and reacts further with amine ligand to yield polymeric Cp_2ZrH_2 (*73*).

$$Cp_2Zr(BH_4)_2 \xrightarrow{NMe_3} Cp_2Zr(H)BH_4 \xrightarrow{NMe_3} (Cp_2ZrH_2)_x$$
$$+ BH_3 \cdot NMe_3 \qquad + BH_3 \cdot NMe_3$$

The use of $Cp_2Zr(H)Cl$ in hydrozirconation reactions has been referred to (*80*), and consists of addition to the Zr–H moiety to unsaturated organic molecules, frequently alkenes and alkynes (96). The resulting alkyls, or alkenyls, undergo further reactions—e.g., electrophilic halogenation, oxidation, or insertion reactions (97)—to yield a range of organic products, as in

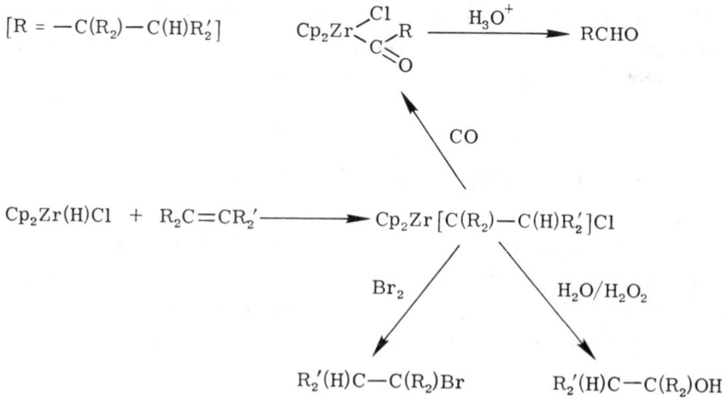

$$[R = -C(R_2)-C(H)R_2'] \qquad Cp_2Zr\overset{Cl}{\underset{C\underset{O}{\overset{R}{<}}}{<}} \xrightarrow{H_3O^+} RCHO$$

$$\uparrow CO$$

$$Cp_2Zr(H)Cl + R_2C=CR_2' \longrightarrow Cp_2Zr[C(R_2)-C(H)R_2']Cl$$

$$\underset{Br_2}{\swarrow} \qquad \underset{H_2O/H_2O_2}{\searrow}$$

$$R_2'(H)C-C(R_2)Br \qquad\qquad R_2'(H)C-C(R_2)OH$$

The use of $Cp_2Zr(Cl)BH_4$ [prepared by the action of $BH_3 \cdot Me_2S$ on $Cp_2Zr(H)Cl$ (98) or by treating Cp_2ZrCl_2 with 1 mole equivalent of $LiBH_4$ (73)] as a selective reducing agent for converting aldehydes or ketones to alcohols has also been recently exploited (98).

d. $Cp_2^*Zr(H)OR$ (R = Me, cyclo-C_5H_7, CH=CHCHCMe$_2$) and $(Cp_2^*ZRH)_2(\mu\text{-OCH}=CHO)$. A series of monohydrides has been obtained by insertion into one of the Zr–H bonds of the pentamethylcyclopentadienyl compound, $Cp_2^*ZrH_2$ (90, 99):

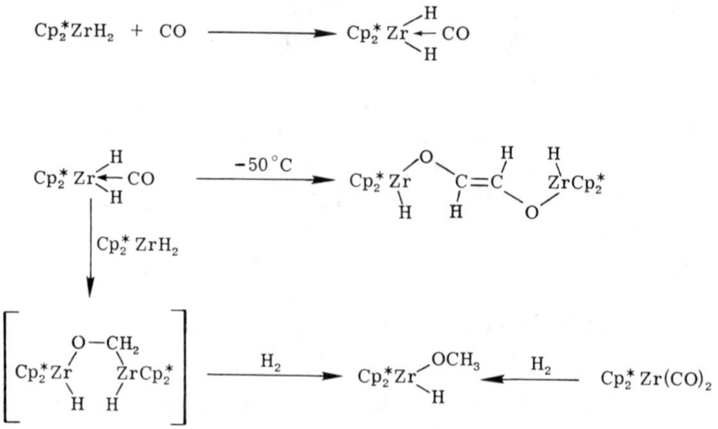

$$Cp_2^*ZrH_2 + CO \longrightarrow Cp_2^*Zr\overset{H}{\underset{H}{\overset{}{\longleftarrow}}}CO$$

$$Cp_2^* Zr\overset{H}{\underset{H}{\overset{}{\longleftarrow}}}CO \xrightarrow{-50^\circ C} Cp_2^*Zr\overset{O}{\underset{H}{<}}C=C\overset{H}{\underset{H}{}} \overset{H}{\underset{O}{}} ZrCp_2^*$$

$$\downarrow Cp_2^* ZrH_2$$

$$\left[Cp_2^*Zr\overset{O-CH_2}{\underset{H\quad H}{<>}}ZrCp_2^* \right] \xrightarrow{H_2} Cp_2^*Zr\overset{OCH_3}{\underset{H}{<}} \xleftarrow{H_2} Cp_2^* Zr(CO)_2$$

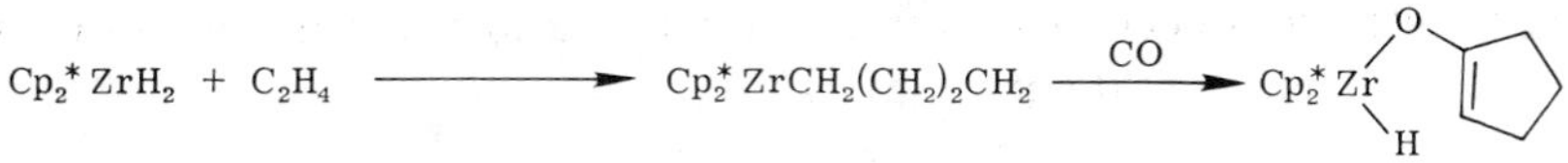

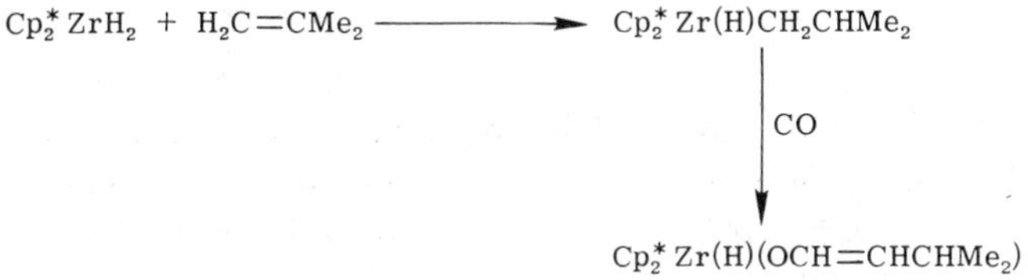

$$Cp_2^* ZrH_2 \; + \; C_2H_4 \longrightarrow Cp_2^* ZrCH_2(CH_2)_2CH_2 \xrightarrow{CO} Cp_2^* \underset{H}{\overset{O}{Zr}}$$

$$Cp_2^* ZrH_2 \; + \; H_2C{=}CMe_2 \longrightarrow Cp_2^* Zr(H)CH_2CHMe_2$$

$$\downarrow CO$$

$$Cp_2^* Zr(H)(OCH{=}CHCHMe_2)$$

If the last reaction of the alkyl hydride with CO is carried out at $-50°C$, an acyl intermediate $Cp_2^* Z(H)(Me_2CHCH_2CO)$ is formed and has been identified from its ^1H-NMR spectrum. Rearrangement to the final product $Cp_2^* Zr(H)(OCH{=}CHCH\,Me_2)$ occurs at $-20°C$.

The question of the mechanism of the insertion of the CO into the Zr–H bond remains to be solved, and different mechanisms may operate. Two routes appear the most likely:

1. Formation of a formyl intermediate by migration of CO into a Zr–H bond and reduction of this by $Cp_2^* ZrH_2$.

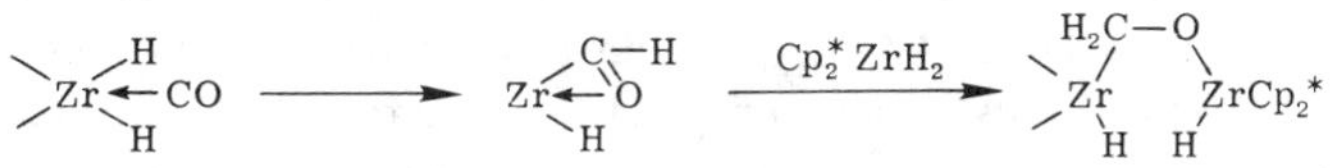

This appears to be favored by the reaction involving isonitriles given in Section I,B,e.

2. Direct reduction, by attack on the coordinated CO.

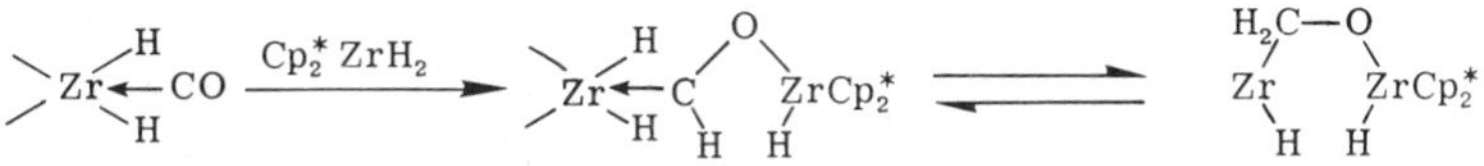

Support for this path comes from the reaction between $Cp_2^* Zr(CO)_2$ and $Cp_2^* ZrH_2$, which forms a cis-enediolate.

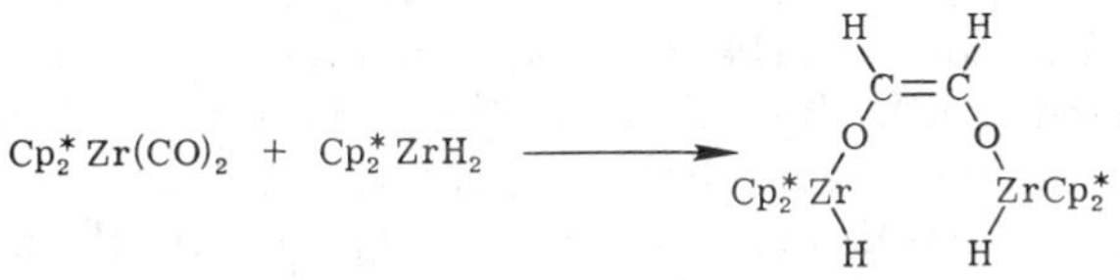

$$Cp_2^* Zr(CO)_2 \; + \; Cp_2^* ZrH_2 \longrightarrow$$

Carbonyl insertion into the Zr–H bond of $Cp_2Zr(H)Cl$ also occurs, yielding the monomeric $[Cp_2Zr(H)Cl]_2CH_2O$, and this has also been suggested to proceed through reduction of a formyl intermediate as in (1) by a further mole of $Cp_2Zr(H)Cl$ (*100*).

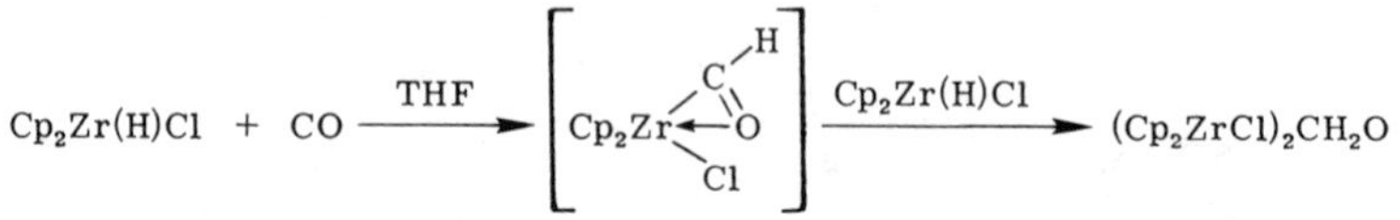

$$Cp_2Zr(H)Cl \; + \; CO \; \xrightarrow{\text{THF}} \; \left[Cp_2Zr \overset{\displaystyle C-H}{\underset{\displaystyle Cl}{\overset{\|}{\longleftarrow} O}} \right] \; \xrightarrow{Cp_2Zr(H)Cl} \; (Cp_2ZrCl)_2CH_2O$$

e. $Cp_2^*Zr(H)CH{=}NMe$ *and* $Cp_2^*Zr(H)NMe_2$. The addition of isonitriles to $Cp_2^*ZrH_2$ affords first an intermediate formimidoyl hydride species—e.g., $Cp_2^*Zr(H)(CH{=}NMe)$—that is generally unstable at room temperature but reacts further with hydrogen forming a stable hydride:

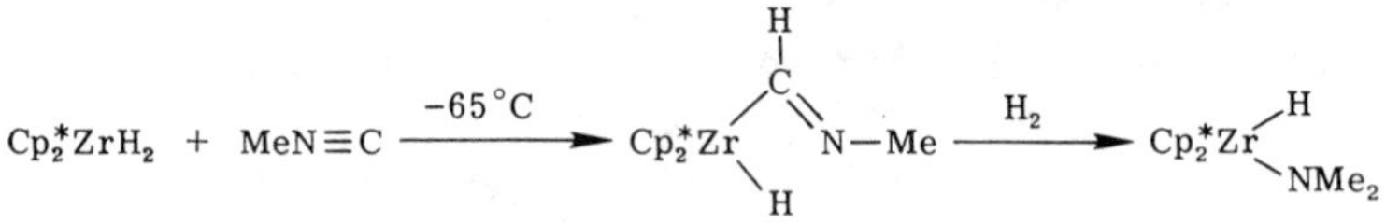

$$Cp_2^*ZrH_2 \; + \; MeN{\equiv}C \; \xrightarrow{-65\,°C} \; Cp_2^*Zr \overset{\displaystyle \overset{H}{\underset{\|}{C}}}{\underset{\displaystyle H}{\diagdown} N{-}Me} \; \xrightarrow{H_2} \; Cp_2^*Zr \overset{\displaystyle H}{\underset{\displaystyle NMe_2}{\diagup}}$$

This sequence provides support for the migratory insertion of the incoming ligand (e.g., MeNC, and by inference, CO) into the Zr—H bond as outlined in mechanism (1) in Section I,B,e above (*101*).

f. $Cp_2^*Zr(H)OCH{=}MCp_2$ (M = Mo, W), $Cp_2^*Zr(H)OCH{=}Nb(H)Cp_2$, *and* $Cp_2^*Zr(H)OCH_2Nb(CO)Cp_2$. The possibility that $Cp_2^*ZrH_2$ might act as a hydride transfer reagent has been realized in reactions where CO is coordinated to another transition metal. Thus, with $Cp_2M(CO)$ (M = Cr, Mo, or W) and $Cp_2Nb(H)CO$, the bimetal species are obtained,

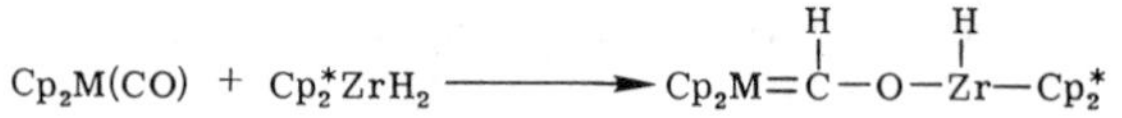

$$Cp_2M(CO) \; + \; Cp_2^*ZrH_2 \; \longrightarrow \; Cp_2M{=}\overset{\displaystyle H}{\underset{}{C}}{-}O{-}\overset{\displaystyle H}{\underset{}{Zr}}{-}Cp_2^*$$

$$Cp_2Nb(H)(CO) \; + \; Cp_2^*ZrH_2 \; \longrightarrow \; Cp_2(H)Nb{=}\overset{\displaystyle H}{\underset{}{C}}{-}O{-}\overset{\displaystyle H}{\underset{}{Zr}}{-}Cp_2^*$$

with the identity of the tungsten derivative being established uniquely by X-ray diffraction, and the W—C bond distance being short, 2.005(13) Å, as expected (*102*). Unpublished observations have shown that several transition-metal carbonyl compounds react with $Cp_2^*ZrH_2$, but only with those containing a single M—CO group are the zirconoxycarbenes, shown above, obtained (*82*).

g. $Cp_2Zr(C_{10}H_7)(H)ZrCp_2$. The isolation of $\mu\text{-}(\eta^1:\eta^5{-}C_5H_4)\eta\text{-}C_5H_5)_3$ Ti_2 (**VII**) from the low-temperature reduction of Cp_2TiCl_2 with potassium naphthalene has been discussed (41); a similar reduction of Cp_2ZrCl_2 has yielded a dark green compound. Although this showed no obvious absorptions due to M–H bonds in its infrared spectrum, a singlet at δ_{SiMe_4} −9.3 in its ^{1}H-NMR spectrum and reactions with HCl and CH_3I indicated that it was a hydride derivative. The structure (**XVIII**) from X-ray diffraction showed a naphthyl ligand bridging two metal atoms, and the hydride, which was not detected from the X-ray data, was also assumed to be in a bridging position (103).

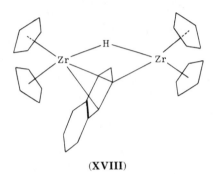

(**XVIII**)

While (**XVIII**) is stable in the solid state, it decomposes slowly in solution, turning blue or purple, to produce other zirconium hydride species that have yet to be characterized.

h. $HZrCH_2CH_2CMe_3$. Reaction of Zr atoms with alkanes (e.g., isobutane, neopentane) condensed into a matrix at 77 K results in oxidative addition of the metal to both C–C and C–H bonds. In the latter case, the presence of a black solid hydride has been inferred from hydrolysis reactions and deuteration studies (e.g., 104).

$$Zr + neopentane \begin{cases} \rightarrow HZrCH_2CMe_3 \\ \\ \rightarrow H_3CZrCMe_3 \end{cases}$$

i. $Cp(C_5H_4)ZrH$. Although this compound has not been isolated, it has been postulated, from ESR spectra, as an intermediate in the photolysis of Cp_2ZrMe_2. The compound is photolabile and degrades further by cleavage of the Zr–H bond to form the Zr(II) derivative (105).

$$Cp_2ZrMe_2 \overset{h\nu}{\rightarrow} Cp_2Zr(III)Me + Me \cdot \rightleftharpoons (Cp_2ZrMe)_2 \underset{h\nu}{\overset{\Delta}{\rightleftharpoons}} Cp_2Zr + Cp_2ZrMe_3$$

$$Cp_2Zr \overset{\Delta}{\rightarrow} Cp(C_5H_4)Zr(III)H \overset{h\nu}{\rightarrow} [Cp(C_5H_4)Zr]_n + H \cdot$$

1. Dihydride Compounds

a. Cp_2ZrH_2 and $(C_5H_4R)_2ZrH_2$. The white polymeric hydride $(Cp_2ZrH_2)_x$ was first reported in 1967, and may be prepared by three methods, two of which are (73, 87, 88)

$$Cp_2Zr(BH_4)_2 \xrightarrow{2NMe_3} (Cp_2ZrH_2)_x + 2BH_3 \cdot NMe_3$$

$$(Cp_2ZrCl)_2O + LiAlH_4 \xrightarrow{THF} (CpZrH_2)_x + LiCl + (AlOCl)_2$$

The $LiAlH_4$ may be replaced by four moles of $LiAl(O\text{-}t\text{-}Bu)_3H$, but when only two moles of the latter are used, the hydride precipitates as $(Cp_2ZrH)_2O(Cp_2ZrH_2)$ (88). The $(Cp_2ZrH_2)_x$ acts as a catalyst for the hydrogenation of alkenes and alkynes (106).

The third preparative route consists of the hydrogenolysis of M–C bonds in the reaction (107)

$$(\eta^5\text{-}C_5H_4R)_2MCl_2 \xrightarrow[-20°C]{MeLi} (\eta^5\text{-}C_5H_4R)M(CH_3)_2 \xrightarrow[80°C]{H_2,\ 60\ atm} (\eta^5\text{-}C_5H_4R)_2MH_2$$

where R = H, Me, Me_2CH, Me_3C, $PhCH_2$, or $Ph(Me)CH$ (two stereoisomers), and M = Zr or Hf. This method has been used to prepare an optically active dimethyl compound $(C_5H_5)[C_5H_4CH(Me)Ph]ZrMe_2$, although when this was used to hydrogenate alkenes under hydrogen pressure the products showed no induced optical activity, probably due to the distance of the chiral center from the catalytic center.

b. $(C_5Me_5)_2ZrH_2$ and the adducts $(C_5Me_5)_2ZrH_2 \cdot L$ (L = CO, PF_3). While $Cp_2^*ZrH_2$ may be obtained from $Cp_2^*ZrCl_2$ by treatment with $Li(BEt_3H)$ or $Na[AlH_2(OCH_2CH_2OCH_3)_2]$, it is better prepared from the nitrogen derivative (81, 82, 90, 99).

$$2Cp_2^*ZrCl_2 \xrightarrow[N_2]{Na/Hg} [Cp_2^*ZrN_2]_2N_2 + 4NaCl$$

$$[Cp_2^*ZrN_2]_2N_2 + 2H_2 \xrightarrow[PhCH_3]{0°C} 2Cp_2^*ZrH_2 + 3N_2$$

Unlike the polymeric $(Cp_2ZrH_2)_x$, this pale yellow compound is soluble in hydrocarbons and is monomeric. Since it is a 16-electron compound, it not unexpectedly forms adducts; however, the somewhat surprising feature is that it forms only $Cp_2^*ZrH_2 \cdot L$ (where L = CO or PF_3; only stable below room temperature), and does not combine with other ligands such as $P(OMe)_3$, PPh_3, or PMe_3. The precise structure of these adducts remains uncertain. Although they are usually depicted as possessing a symmetrical structure (**XIX**), in a recent study where CO coordinates to $Cp_2Zr(C_6H_5)_2$ the resulting η^3-benzoyl compound has its carbonyl oxygen in the lateral position, indicating the initial formation of an unsymmetrical adduct (**XX**) (*108*).

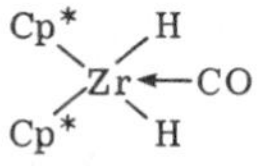

(**XIX**) 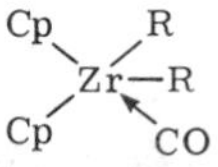   (**XX**)

The hydridic nature of the hydrogen atoms in $Cp_2^*ZrH_2$ has been emphasized (*75*), and the $Zr^+\!-\!H^-$ character of the bond is illustrated by reactions with HCl, CH_3I, and CH_2O (*81*), as in

$$Cp_2^*ZrH_2 + 2HCl \xrightarrow{-80°C} Cp_2^*ZrCl_2 + 2H_2$$

$$Cp_2^*ZrH_2 + 2CH_3I \longrightarrow Cp_2^*ZrI_2 + 2CH_4$$

c. $H_2Zr(tetrahydroindenyl)_2$ and $H_2Zr(COT)$. Hydrogenation (H_2, 80 atm, 140°C/8 h) of $(\pi\text{-indenyl})_2ZrMe_2$ affords the tetrahydroindenyl hydride derivative $(C_9H_{13})_2ZrH_2$, which is soluble in aromatic hydrocarbons and dimeric in benzene. The ^{1}H-NMR spectrum of the dimer indicates that there are two sets of hydride hydrogens that are exchanging at 30°C, but not at -36°C, and that within the two sets, each of the two hydrogens are magnetically equivalent. It has therefore been suggested that at least one set occupies bridging positions between two zirconium atoms (*109*).

The reduction of $Zr(OR)_4$ compounds with Et_2AlH in the presence of cyclooctatetraene (COT) produces the dihydride $(COT)ZrH_2$. This reacts with protonic hydrogens, liberating hydrogen; but since the dihydrides in general act as efficient hydrogenating catalysts for alkenes, the final hydrocarbon product formed is essentially a mixture of cyclooctatriene and cyclooctadiene (*110*), e.g.,

$$(COT)ZrH_2 \xrightarrow{2EtOH} C_8H_{12} + Zr(OEt)_2 \xrightarrow{\frac{1}{2}O_2} OZr(OR)_2$$

$$\Big\downarrow 2EtOH$$

$$H_2 + Zr(OEt)_4$$

Different products may be obtained from the initial reaction; thus when tetraallylzirconium is treated with COT and Et_2AlH, the 1:1 and 1:2 adducts $(COT)_2Zr(HAlEt_2)_{1 \text{ or } 2}$ are formed. In addition, the complex $(COT)_2Zr \cdot (HAlEt_2)_2$ may be formed in the reaction shown below, and it is postulated to ionize in toluene solution to form an anion containing a Zr–H–Al bridge bond (110).

$$Zr(OR)_4 + 4AlEt_3 + 2COT \rightarrow (COT)_2Zr + 4Et_2AlOR + [4C_2H_5]$$

$$(COT)_2Zr + 2Et_2AlH \rightarrow (COT)_2Zr \cdot (HAlEt_2)$$
$$\updownarrow$$
$$[(COT)_2ZrH(Et_2)AlH]^- + [Et_2Al]^+$$

d. H₃Zr(dmpe)₂. This compound has been referred to in the section on titanium, and is prepared by reacting zirconium tetraphenyl with hydrogen under high pressure in the presence of the ligand $Me_2P(CH_2)_2PMe_2$(dmpe) (49).

2. Spectroscopic Properties of Zirconium Hydrides

These are summarized in Table II. The Zr–H stretching frequency now appears to lie in the region 1600–1500 cm^{-1}, rather than 2000–1800 cm^{-1} as observed for the group VI–VIII transition-metal hydrides (81). Frequencies for ν(M–H) have been observed at 1710 cm^{-1} and 1735 cm^{-1} for Cp₂NbH₃ (111) and Cp₂TaH₃ (112), respectively, and main group hydrides (e.g., Ge, Sn, Pb) show similar absorptions near 2000 cm^{-1} (113). Some uncertainty still exists since bridging hydride vibrational modes (e.g., Zr–H–Zr) have not been positively identified. Thus (Cp₂ZrH₂)$_x$ shows broad absorptions associated with Zr–H vibrations at 1540–1520 cm^{-1} and ~1300 cm^{-1} (73, 87); assuming a polymeric structure of the general type

$$\begin{array}{c} \text{H} \\ | \\ \text{Cp}_2\text{Zr—H—ZrCp}_2 \\ | \qquad | \\ \text{H} \qquad \text{H} \\ | \end{array}$$

would suggest that the terminal and bridging vibrations should be assigned to the bands at ~1500 cm^{-1} and ~1300 cm^{-1}, respectively. The polymeric Cp₂Zr(H)Cl shows only one ν(Zr–H) absorption at 1390 cm^{-1}, presumably arising from a bridging Zr–H–Zr bond (87).

In the ¹H-NMR spectra of zirconium hydride compounds, the unusual feature compared with later transition-metal hydrides is the low-field chemical shift of the hydride hydrogen. For Cp$_2^*$ZrH₂, δ_{Zr-H} is 7.46 (90); a weak resonance signal at δ 4.5 has been tentatively assigned to the simple hydride hydrogen in Cp₂Zr(H)BH₄ (73); and similar signals for Cp$_2^*$TiH₂ and Cp$_2^*$HfH₂ have been observed at δ 0.28 and 15.6, respectively (81, 90). A more comprehensive listing of the hydride shifts for many of the compounds discussed above has been given, and all the

TABLE II

ZIRCONIUM HYDRIDE COMPOUNDS

Compound	Preparation	Properties	¹H NMR (τ, M–H)	Infrared (cm⁻¹)	Comments	References
$(\eta^3\text{-}C_6H_7)_2Zr(H)(dmpe)_2$	$ZrCl_4(dmpe)_2$/Na/Hg/THF	Red-brown crystals, decomp. above 130°C evolving H_2. Reacts HCl $\rightarrow H_2$ (from Zr–H).	Zr–H not detected, C_6H_7 ligand from 2.59–5.5	Zr–H not detected (parent ion found in mass spectrum).	Catalyst for disproportionation [diagram]	83, 84
$[Cp_2Zr(H)R]_2$ R = alkyl, cyclohexyl, $CH(SiMe_3)_2]$	$Cp_2Zr(Cl)R$/LiAlH₄ or $Li[AlH(O\text{-}tBu)_3]$/DME	Colorless solids, dimeric in benzene except R = Me (polymeric). Generally unstable at r.t. except silyl derivative.	~12 (Zr–H)	1500 and 1310 in $[Cp_2Zr(H)Me]_n$ from Zr–H–Zr	—	85, 86
$Cp_2Zr(H)Cl$	Cp_2ZrCl_2/LiAlH₄/THF	Polymeric white solid, –Zr–H–Zr–bonds. Reacts alkenes, alkynes, etc.	Insolubility limits investigation (Ref. 88).	1390 (Zr–H–Zr)	Further addn. LiAlH₄ gives $Cp_2Zr(H)AlH_4$, also probably polymeric.	87, 88, 94, 95
$Cp_2Zr(H)BH_4$	$Cp_2Zr(BH_4)_2$/1 mol NMe₃	Volatile white solid, hydrolyzed in air releasing hydrogen.	5.47 (Zr–H?); 4.30 (Cp); 10.20 (BH_4).	1623 (Zr–H); 2400, 1945, 1132 (BH_4).	—	73
$Cp_2Zr(H)(C_{10}H_7)ZrCp_2$ (XVIII)	Cp_2ZrCl_2/K/$C_{10}H_8$/THF/–80°C	Dark-green solid. Decomposes in soln., reacts MeI $\rightarrow CH_4$. No magnetic data.	19.3 (Zr–H); 4.0 (Cp); 2–5 ($C_{10}H_7$).	No Zr–H band detected, strong band at ~800 from Cp groups.	Structure from X-ray data, H in Zr–H bond not located.	103
$ZrH(CH_2CMe_3)$	Zr atoms/isobutane/condense in matrix at 77 K	Black solid, not fully characterized, Zr–H bond deduced from hydrolysis.	—	Unstable on warming from matrix temp. No spectroscopic data.	Oxidative addition of C–H bond to metal atom.	104
$(Cp_2ZrH_2)_x$	$Cp_2Zr(BH_4)_2$/2 mol NMe₃ or $[Cp_2ZrCl]_2O$/LiAlH₄/THF/25°C	White polymeric solid.	—	1520 and 1300 (Zr–H–Zr).	Hydrogenation catalyst.	73, 87, 88, 95

(table continues)

TABLE II (Continued)

Compound	Preparation	Properties	¹H NMR (τ, M–H)	Infrared (cm⁻¹)	Comments	References
[(C₅H₄R)₂ZrH₂]ₓ (R = Me, Me₂CH, etc.)	(C₅H₄R)₂ZrMe₂/H₂/60 atm/80°C	White polymeric solids, except where R = Me₃C or Me₂CH when compounds are dimers. Sensitive to air and moisture.	—	—	Hydrogenation catalysts for alkenes and alkynes.	107
[(C₉H₁₃)₂ZrH₂]₂ (C₉H₁₃ = tetrahydro-indenyl)	(C₉H₁₃)ZrMe₂/H₂/80 atm/140°C/8 h	Sol. hydrocarbons, dimeric in C₆H₆. D₂ affects exchange in ZrH₂ groups.	5.41, 11.56 (Zr–H).	1545, 1285 (Zr–H–Zr).	Two sets of ZrH₂ protons in dimer are inequivalent in H-NMR spectrum.	109
(COT')ZrH₂	Zr(OR)₄/Et₂AlH/COT	Reacts H⁺ → H₂; EtOH → C₈H₁₂ + Zr(OEt)₄ + H₂.	—	1537, 1310 (Zr–H).	Bridged species, e.g., (COT')Zr(HAlEt₂)₁ or ₂, also prepared.	110
(Cp₂ZrH₂·AlMe₃)₂	(Cp₂ZrH₂)ₓ/AlMe₃/C₆H₆	Pale-blue, proposed dimer, unstable r.t. → unidentified crimson comp.	7.08 (Zr–H–Zr); 9.08 (Zr–H–Al); 4.50 (Cp); 10.41 (Al–Me).	1780 (Zr–H–Al); 1350 (Zr–H–Zr).	—	116
Cp₂Zr(μ-H)₃Al(iBu)₂	(Cp₂ZrH₂)ₓ/2(iBu)₂AlH/ C₆H₆/25°C	Deep-blue oil, unstable at r.t.	10.28] Zr–H–Al 12.03] Zr–H–Al 4.43 (Cp)	—	Reacts (iBu)₂AlCl → Cp₂Zr(H)(μ-H)₂ [Al(iBu)₂]Cl. Also AlMe₃ → Cp₂Zr(H) (μ-H)₂[Al(iBu)₂] (AlMe₃).	117
(Cp₂ZrH₂)₂[Al(iBu)₂R]₂ (R = neohexyl)	Cp₂ZrH₂/2(iBu)₃AlH	Reacts with further (iBu)₂AlH → trihydride Cp₂Zr(H) (μ-H)₂Al(iBu)R].	—	—	Indicates interchange of neohexyl/H groups between Zr and Al.	117
[Cp₂Zr(H)]₂OCp₂ZrH₂	Cp₂ZrH₂/2 mol LiAl(O-tBu)₃H/THF	Pale-pink insoluble polymer. Reacts acids, all Zr–H bonds → H₂.	—	1510, 1385, 1240 (Zr–H–Zr)	—	88
(dmpe)₂ZrH₃	Zr(benzyl)₄/H₂/dmpe/ C₆H₆/200 atm	Dark-brown solid, reacts HCl → H₂.	—	1280 (Zr–H–Zr)	—	49
Cp₂*ZrH₂	Cp₂*ZrCl₂/LiBEt₃H or (Cp₂*ZrN₂)₂N₂/2H₂/ PhCH₃/0°C	Monomeric, pale-yellow solid, sol. hydrocarbons. Forms adducts Cp₂*ZrH₂·L (L = CO or PF₃) at low temps.	2.54 (Zr–H); 7.98 (Cp*).	1555 (Zr–H)	—	81, 82, 90, 99

Compound	Preparation	Product description	NMR	IR	Notes	Ref.
$Cp^*_2Zr(H)(CH_2CHMe_2)$	$Cp^*_2Zr(H)/CH_2{=}CMe_2$ at r.t.	Pale-yellow, monomeric in benzene, stable to 70°C when $CH_2{=}CMe_2$ evolved.	3.57 (Zr–H); 8.07 (Cp*).	—	Deuterated species also prepared.	90
$Cp^*_2Zr(H)(OCH_3)$	$CP^*_2Zr(CO)_2/H_2/110°$ or $Cp^*_2Zr(CO)/$ $Cp^*_2ZrH_2/H_2$	White crystals, monomeric, reacts $HCl \rightarrow Cp^*_2ZrCl_2$.	4.3 (Zr–H); 6.13 (OCH$_3$); 8.04 (Cp*).	1590 (Zr–H); 1140 (C–O).	—	90, 99
$(Cp^*_2ZrH)_2(OCH{=}CHO)$	Allow $Cp^*_2ZrH_2(CO)$ to warm to –50°C.	Yellow crystalline solid, reacts MeI → $[Cp^*_2ZrI]_2$ (OCH=CHO).	4.27 (Zr–H); 8.06 (Cp*); 3.45 (OCH=CHO).	1580 (Zr–H); 1205 (C–O).	X-Ray studies on $(Cp^*_2ZrI)_2$ (OCH=CHO) in accord with proposed structures for hydride.	90
$Cp^*_2Zr(H)$ $(\overline{OC{=}CHCH_2CH_2CH_2})$	$\overline{Cp^*_2Zr{-}CH_2CH_2CH_2}$ CO/25°C or $Cp^*_2Zr(CO)_2/C_2H_4/h\nu$	White crystalline solid, reacts MeI → CH$_4$ + $Cp^*_2Zr(I)(\overline{OC{=}CH}$ $\overline{CH_2CH_2CH_2})$.	3.93 (Zr–H); 8.03 (Cp*); 5.48 ⎤ 7.5–8.0 ⎦ (C$_5$H$_7$O).	1538 (Zr–H); 1628 (C=C); 1275 (C–O).	—	90
$Cp^*_2Zr(H)(Me_2CHCH_2CO)$	$Cp^*_2Zr(H)(CH_2CHMe_2)$ CO/1 atm/–50°C	Transient acyl intermediate, rearranges at –20°C to $Cp^*_2Zr(H)$ (OCH=CHCHMe$_2$).	6.34 (Zr–H); 8.18 (Cp*); 7.46 ⎤ 8.80 ⎦ (C$_4$H$_9$O).	—	—	90
$Cp^*_2Zr(H)$ (OCH=CHCHMe$_2$)	Final product of rearrangement at –20°C given above.	Orange oil, but not isolated pure.	3.96 (Zr–H); 8.06 (Cp*); 3.37 ⎤ 5.39 ⎦ (C$_4$H$_9$O).	1550 (Zr–H)	—	90
$Cp^*_2Zr(H)NMe_2$	$Cp^*_2ZrH_2/MeNC/H_2/1$ atm/reaction warmed from –80° to 25°C.	Reacts MeI → $Cp^*_2Zr(I)NMe_2$ + CH$_4$.	3.59 (Zr–H); 8.06 (Cp*); 7.59 (NMe$_2$).	1550 (Zr–H)	Intermediate reaction product is formimidoyl hydride $Cp^*_2Zr(H)$ C(H)=NMe.	101
$Cp^*_2Zr(H)(OCH{=}MCp_2)$ (M = Mo or W)	$Cp^*_2ZrH_2/Cp_2M(CO)/$ $C_6H_5CH_3/{-}80°C$	W comp. brown crystalline solid, thermally stable.	4.28 (W comp.); 4.20 (Mo comp.); 4.06 (Cr comp. at –3°C).	1560 (Zr–H, W comp.); 1543 (Zr–H, Mo comp.)	X-Ray structure on tungsten compound. Cr compound decomp. at 25°C.	102
$Cp^*_2Zr(H)$ $[OCH{=}Nb(H)Cp_2]$	$Cp^*_2ZrH_2/Cp_2NbH(CO)/{-}80°C$	Product in nearly quantitative yield from NMR.	4.30 (Zr–H); 4.98 (Cp); 8.01 (Cp*); 13.14 (Nb–H); ^{13}C also reported.	1567 (Zr–H); 1701 (Nb–H).	—	102
$Cp^*_2Zr(H)$ $[OCH_2Nb(CO)Cp_2]$	$Cp^*_2ZrH_2/Cp_2NbH(H)$ (CO) under CO at 25°C	—	4.77 (Zr–H); 5.28 (Cp); 7.99 (Cp*).	1553 (Zr–H); 1898 (C≡O).	—	102

301

resonances lie in the region δ 0–7.46 (90). Although the metal atom in each of these compounds possesses formally a d^0 electron configuration and may therefore be considered to be more like the main-group elements in some of its properties, the reason for the low-field shift remains in doubt at present.

3. Zirconium Hydrides Containing Bridge Bonds Zr–H–M (M = B or Al)

This section does not review comprehensively all the tetrahydroborate derivatives, since these have been adequately covered elsewhere (50). The aluminohydride compound $Cp_2Zr(H)AlH_4$ has been mentioned, but several similar hydrides have been identified. Such compounds are relevant in particular to known reactions where titanium(IV) or zirconium(IV) compounds catalyze the addition of Al–H bonds to alkenes and alkynes (114, 115).

a. $[Cp_2ZrH_2 \cdot AlMe_3]_2$. The addition of $AlMe_3$ to a suspension of $(Cp_2ZrH_2)_x$ in benzene leads to a pale-violet solution from which a pale-blue compound $Cp_2ZrH_2 \cdot AlMe_3$ may be isolated in high yield. This is postulated to be a dimer on the basis of spectroscopic data and is unstable at room temperature, forming an unidentified crimson compound and evolving methane and hydrogen (116).

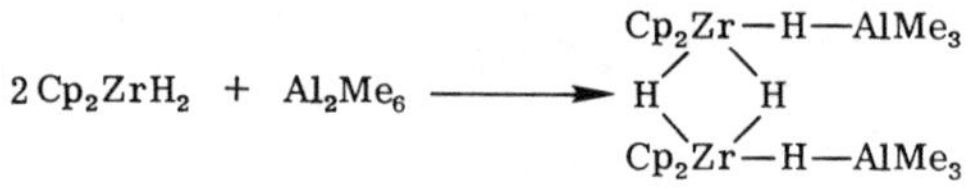

$$2\,Cp_2ZrH_2 \;+\; Al_2Me_6 \;\longrightarrow\; \begin{array}{c} Cp_2Zr-H-AlMe_3 \\ H \diagdown \quad \diagup H \\ Cp_2Zr-H-AlMe_3 \end{array}$$

A similar product has been reported to result from the action of $(i\text{-}Bu)_2$ AlH on $Cp_2Zr(neo\text{-hexyl})_2$, and here alkyl–hydride exchange clearly occurs. Further addition of $(i\text{-}Bu)_2AlH$ displaces $(i\text{-}Bu)_2Al(neo\text{-hexyl})$ and affords the trihydride (117):

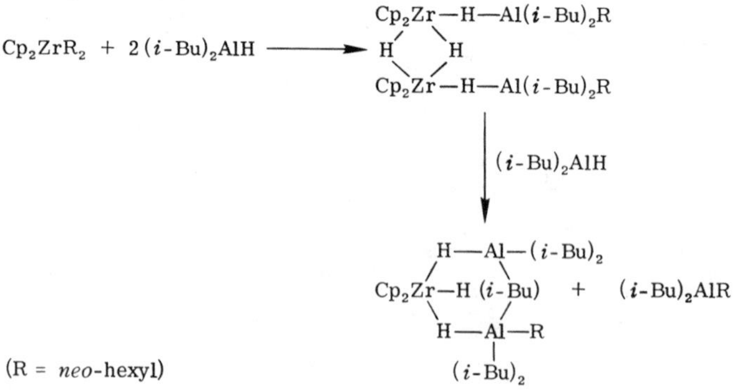

$$Cp_2ZrR_2 \;+\; 2\,(i\text{-}Bu)_2AlH \;\longrightarrow\; \begin{array}{c} Cp_2Zr-H-Al(i\text{-}Bu)_2R \\ H \diagdown \quad \diagup H \\ Cp_2Zr-H-Al(i\text{-}Bu)_2R \end{array}$$

$$\downarrow (i\text{-}Bu)_2AlH$$

$$\begin{array}{c} H-Al-(i\text{-}Bu)_2 \\ Cp_2Zr-H\,(i\text{-}Bu) \quad + \quad (i\text{-}Bu)_2AlR \\ H-Al-R \\ (i\text{-}Bu)_2 \end{array}$$

(R = neo-hexyl)

b. $Cp_2ZrH_3AlR_2$, $Cp_2Zr(H)(\mu\text{-}H_2\text{-}Al\text{-}R_2\text{-}R\text{-}AlR_2')$, *and* $Cp_2Zr(H)[\mu\text{-}H_2\text{-}(AlR_2)\text{-}Cl]$. If Cp_2ZrH_2 reacts with $(i\text{-}Bu)_2$ AlH instead of $AlMe_3$, the deep-blue oily complex (**XXI**) is formed. While this decomposes at room temperature, treatment of (**XXI**) with $(i\text{-}Bu)_2AlCl$ yields the thermally stable compound (**XXII**) as a pale-violet oil. Alternatively, treatment of (**XXI**) with $AlMe_3$ produces the mixed alkyl oily complex (**XXIII**) (*96, 117*).

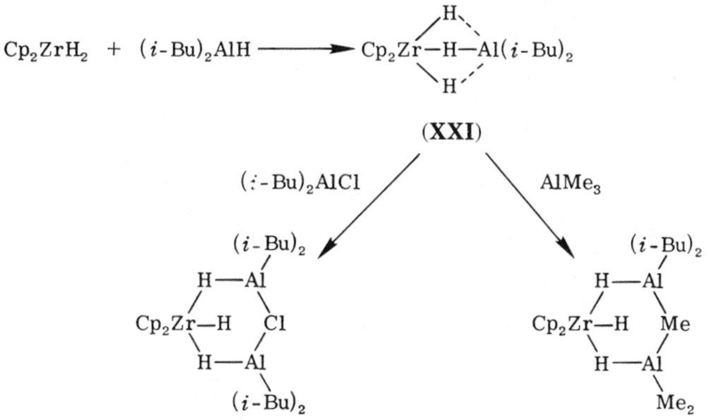

$$Cp_2ZrH_2 \ + \ (i\text{-}Bu)_2AlH \longrightarrow Cp_2Zr\underset{H}{\overset{H}{\diagdown}}H{-}Al(i\text{-}Bu)_2$$

(**XXI**)

$(i\text{-}Bu)_2AlCl$ $AlMe_3$

$$\begin{array}{ccc}
& (i\text{-}Bu)_2 & \\
& H{-}Al & \\
Cp_2Zr{-}H & & Cl \\
& H{-}Al & \\
& (i\text{-}Bu)_2 &
\end{array}$$

$$\begin{array}{ccc}
& (i\text{-}Bu)_2 & \\
& H{-}Al & \\
Cp_2Zr{-}H & & Me \\
& H{-}Al & \\
& Me_2 &
\end{array}$$

(**XXII**)   (**XXIII**)

The compound (**XXII**) is also obtained upon addition of 3 mol $(i\text{-}Bu)_2AlH$ to Cp_2ZrCl_2 in benzene. It is interesting that this compound absorbs 2 mol CO, and the resulting golden-yellow solution on hydrolysis yields a series of alcohols ROH (R = Me, Et, n-Pr, n-Bu). The mechanism has been postulated to involve the initial dissociation of 1 mol $(i\text{-}Bu)_2AlH$, followed by coordination of CO to zirconium; subsequent chain propagation occurs as shown below (*118*).

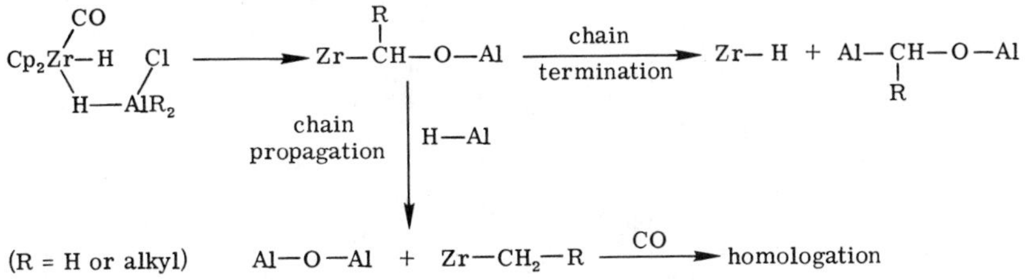

$$\begin{array}{ccc}
CO & & \\
Cp_2Zr{-}H & Cl & \\
& H{-}AlR_2 &
\end{array} \longrightarrow Zr\overset{R}{\overset{|}{-}}CH{-}O{-}Al \xrightarrow[\text{termination}]{\text{chain}} Zr{-}H \ + \ Al\overset{}{-}CH{-}O{-}Al$$

chain propagation $\Big\downarrow$ H—Al

(R = H or alkyl)   Al—O—Al + Zr—CH_2—R $\xrightarrow{\ \ CO\ \ }$ homologation

The properties of the various zirconium hydride compounds are summarized in Table II.

C. HAFNIUM HYDRIDE COMPOUNDS

In contrast to the extensive zirconium series, there have been relatively few hafnium hydride compounds reported (Table III). The

TABLE III

HAFNIUM HYDRIDE COMPOUNDS

Compound	Preparation	Properties	^{1}H NMR (τ, M–H)	Infrared (cm^{-1})	Comments	References
Cp$_2$Hf(H)Cl	Cp$_2$HfCl$_2$/LiAlH$_4$	White solid, reacts alkenes, appears similar to Zr analog.	—	—	—	119
(Cp$_2$HfH$_2$)$_x$	Cp$_2$HfMe$_2$/H$_2$	White insoluble polymeric solid.	—	—	Methyl comp. from Cp$_2$HfCl$_2$/LiMe/−20°C.	107
[(C$_5$H$_4$R)$_2$HfH$_2$]$_x$ (R = Me, PhCH$_2$, PhCHMe, Me$_3$C, Me$_2$CH)	Cp$_2$HfMe$_2$/H$_2$/60 atm/80°C	For R = Me$_3$C, Me$_2$CH, x = 2, sol. hydrocarbons. For R = Me, PhCH$_2$, PhCHMe, polymers, insol. hydrocarbons.	—	—	More stable thermally than Zr analogs.	107
(dmpe)$_2$HfH$_3$	Hf(benzyl)$_4$/dmpe/ C$_6$H$_6$/H$_2$/400 atm/90°C	Tan solid.	—	1300 (Hf–H–Hf?)		49

monohydride chloride $Cp_2Hf(H)Cl$ has been prepared by the action of $LiAlH_4$ on Cp_2HfCl_2 and, as expected, it shows similar reactions to the zirconium analog in reactions with alkenes and alkynes. For example, with 1-heptene it yields $Cp_2HfCl[(CH_2)_6Me]$, which on treatment with bromine liberates the alkyl bromide $Me(CH_2)_6Br$ (119).

Reference has already been made to $Cp_2Zr(H)BH_4$; preliminary results have shown that a similar hafnium compound may be obtained (73). The hafnium dihydride is polymeric like $(Cp_2ZrH_2)_x$, and may also be prepared by hydrogenolysis of Cp_2HfMe_2 (107). The dihydrides $(C_5H_4R)_2HfH_2$ [R = Me, Me_2CH, Me_3C, $C_6H_5CH_2$, $C_6H_5(Me)CH$ (two stereoisomers)] have been obtained by the same reaction used for the zirconium compounds, namely treatment of the corresponding dichloride with methyllithium followed by treatment with hydrogen under pressure (107). The trihydride $(dmpe)_2HfH_3$ has been included in the general patent referred to earlier (49).

II. Vanadium, Niobium, and Tantalum Hydrides

The chemistry of hydrido complexes of group V metals seems to reflect the usual tendency for vanadium to behave differently from the other two elements, although generalizations are probably premature in such a new field. Certainly niobium and tantalum form numerous hydrides of similar composition and properties, appearing to have no vanadium counterparts to date, but there have been no systematic investigations involving all three elements under comparable conditions, as is clear from the following discussions.

A. VANADIUM HYDRIDE COMPOUNDS

Compounds containing vanadium–hydrogen bonds were first reported in the early 1960s, but only recently have any been reasonably well characterized. With one exception, all such compounds involve carbon monoxide (or the equivalent π-acceptor, PF_3) as coligands, and many also contain chelating phosphines or arsines, or the cyclopentadienyl moiety. Those compounds containing terminal vanadium–hydrogen bonds show characteristic ^{1}H-NMR absorption in the region τ 13.5–18.5, except for the trihydrides (τ = 8.86), but no infrared absorptions attributable to $\bar{\nu}_{M-H}$ have been reported. The spectroscopic properties of each of the vanadium hydrides are summarized in Table IV, together with other relevant information.

In the treatment that follows, the monohydrides are considered first,

TABLE IV

VANADIUM HYDRIDE COMPLEXES

Compound	Preparation	Properties	^{1}H NMR (τ, M–H)	Infrared (cm^{-1})	Comments	References
HV(CO)$_6$	V(CO)$_6$ + (1)Na/Hg + (2)H$_3$PO$_4$				Acidity functions measured but only salts are well characterized.	120, 121
HV(CO)$_5$(PPh$_3$)	V(CO)$_6$/PPh$_3$ + (1)Na/Hg + (2)H$_3$PO$_4$					
HV(PF$_3$)$_6$	V(CO)$_6^-$ + PF$_3$ + irradiation + protonation	Pale-yellow crystals, sublimes 60°C (10^{-2} torr), stable to 135°C.	18.4 multiplet [solution in Ni(PF$_3$)$_4$]	$\bar{\nu}_{V-F}$ 846, 911	Diamagnetic. X_{31} -236×10^{-6} cm^3 mol^{-1}; mass spec. shows HV(PF$_3$)$_6^+$ but not V(PF$_3$)$_6^+$.	122
HV(CO)$_4$(dppm)	[Et$_4$N][V(CO)$_4$(dppm)] protonation	Yellow crystals, m.p. 75–82°C (dec), stable under N$_2$ at 10°C, turns brown in days at room temperature.	15.04 triplet (THF-d_8, 20°C); 14.37 triplet (benzene-d_6); $^2J(^1$H–^{31}P) 21.9 Hz.	$\bar{\nu}_{C-O}$ 1989, 1901, 1875, 1857 (0.02 M THF) or 1990, 1885, 1858, 1815 (KBr)	^{51}V NMR reported	123, 124
HV(CO)$_4$(dppe)	[Et$_4$N][V(CO)$_4$(dppe)] protonation	Yellow solid changing (r.t.) to brown (with no detectable spectral changes, 123) or green/black [V(CO)$_4$dppe] (125); solutions deteriorate rapidly.	15.03 triplet (THF-d_8, 20°C); $^2J(^1$H–^{31}P) 26.0 Hz.	$\bar{\nu}_{C-O}$ 1990, 1878 (0.02 M THF) or 1988, 1890, 1867, 1856 (Nujol)	^{51}V NMR reported	123, 125
HV(CO)$_4$(dppp)	[Et$_4$N][V(CO)$_4$(dppp)] protonation	Yellow solid (turning brown at r.t.).	14.91 triplet (THF-d_8, 20°C); $^2J(^1$H–^{31}P) 25.9 Hz.	$\bar{\nu}_{C-O}$ 1990, 1909, 1867, 1831 (0.02 M THF)	^{51}V NMR	123
HV(CO)$_4$(dppb)	[Et$_4$N][V(CO)$_4$(dppb)] protonation	Yellow solid (turning brown at r.t.).	15.15 triplet (THF-d_8, 20°C); $^2J(^1$H–^{31}P) 26.7 Hz.	$\bar{\nu}_{C-O}$ 1988, 1895, 1867, 1839 (THF)	^{51}V NMR	123

Compound	Preparation	Appearance/properties	^{1}H NMR	$\tilde{\nu}_{C-O}$	Other NMR	Ref.
HV(CO)$_4$(dmpe)	[Et$_4$N][V(CO)$_4$(dmpe)] protonation	Pale-yellow crystals stable under N$_2$ (r.t.) (turns brown in air); toluene solutions stable under Ar (0°C); m.p. 123°C (dec).	14.58 triplet (toluene, d_8, 20°C); $^2J(^1$H–^{31}P) 23.4 Hz.	$\tilde{\nu}_{C-O}$ 1981, 1865 (THF) or 1973, 1863, 1842 (Nujol)	NMR shows no temperature dependence between +20 and −65°C.	126
HV(CO)$_4$(arphos)	[Et$_4$N][V(CO)$_4$(arphos)] protonation	Yellow crystals similar to (dppm) complex.	15.08 doublet (THF-d_4, 20°C); 14.53 doublet (benzene-d_6); $^2J(^1$H–^{31}P) 25.9 Hz.	$\tilde{\nu}_{C-O}$ 1990, 1875 (THF) or 1985, 1885, 1868, 1852, 1822 (KBr)	^{51}V NMR	123, 127
HV(CO)$_4$(diars)	[Et$_4$N][V(CO)$_4$(diars)] protonation	Pale-yellow crystals stable under N$_2$ (r.t.). Solutions air sensitive, but toluene solutions under Ar are stable (0°C).	14.48 singlet (toluene, d_8, 20°C).	$\tilde{\nu}_{C-O}$ 1989, 1975 (THF) or 1989, 1900, 1863, 1830 (Nujol)	NMR shows no temperature dependence between +20 and −67°C.	123, 126
HV(CO)$_3$PhP(CH$_2$CH$_2$PPh$_2$)$_2$	As for diphosphines	Yellow crystals	14.54 sextet (THF-d_6, 60°C); $^2J(^1$H–^{31}P) 20 and 39.7 Hz.	$\tilde{\nu}_{C-O}$ 1927, 1920, 1835, 1820 (THF)	^{51}V NMR δ −1640; ^{31}P NMR δ +90	123, 124
HV(CO)$_3$MeC(CH$_2$PPh$_2$)$_3$	As for diphosphines	Yellow	15.06 (broad)	$\tilde{\nu}_{C-O}$ 1908, 1825, 1810	^{51}V NMR δ −1528; ^{31}P NMR δ +27, −28	124
HV(CO)$_4$MeC(CH$_2$PPh$_2$)$_3$	As for diphosphines		15.20 (triplet); $^2J(^1$H–^{31}P) 26.1 Hz.	$\tilde{\nu}_{C-O}$ 1987, 1866	^{51}V NMR δ −1608	124
HV(CO)$_4$P(CH$_2$CH$_2$PPh$_2$)$_3$	As for diphosphines		15.32 (triplet); $^2J(^1$H–^{31}P) 25 Hz.	$\tilde{\nu}_{C-O}$ 1991, 1877	^{31}P NMR δ +87, +76, −15.6	124
HV(CO)$_3$P(CH$_2$CH$_2$PPh$_2$)$_3$	As for diphosphines	Red needles	14.54 (sextet); $^2J(^1$H–^{31}P) 39.7, 20.0 Hz.	$\tilde{\nu}_{C-O}$ 1922, 1911, 1819	^{51}V NMR δ −1690; ^{31}P NMR δ +99, +83, −15.6	124
HV(CO)$_4$(Ph$_2$PCH$_2$CH$_2$CH$_2$PPhCH$_2$)$_2$	As for diphosphines	Yellow powder	15.20 (triplet); $^2J(^1$H–^{31}P) 23.0, 27.0 Hz.	$\tilde{\nu}_{C-O}$ 1990, 1905, 1877, 1845	^{51}V NMR δ −1685; ^{31}P NMR δ +68	124
HV(CO)$_3$(Ph$_2$PCH$_2$CH$_2$CH$_2$PPhCH$_2$)$_2$	As for diphosphines		15.08 (quartet); $^2J(^1$H–^{31}P) 29.2, 21.4, 32.1 Hz.	$\tilde{\nu}_{C-O}$ 1905, 1806	^{51}V NMR δ −1665; ^{31}P NMR δ +96, +84, +80, −15.1, −18.0	124

(*table continues*)

TABLE IV (Continued)

Compound	Preparation	Properties	¹H NMR (τ, M–H)	Infrared (cm⁻¹)	Comments	References
cis-HV(CO)₂P(CH₂CH₂PPh₂)₃	Irradiation of HV(CO)₃-P(CH₂CH₂PPh₂)₃	Red crystalline powder	13.84 (quartet of doublets); $^2J(^1H-^{31}P)$ 39.7, 9.6 Hz.	$\bar{\nu}_{C-O}$ 1829, 1772	^{51}V NMR δ −1541; ^{31}P NMR δ +100	124
trans-HV(CO)₂(Ph₂PCH₂CH₂CH₂PPhCH₂)₂	Irradiation of HV(CO)₃-(Ph₂PCH₂CH₂PPhCH₂)₂	Red needles	13.62 triplet of triplets; $^2J(^1H-^{31}P)$ 113.0, 17.7.	$\bar{\nu}_{C-O}$ 1884, 1764	^{51}V NMR δ −1658; ^{31}P NMR δ +111, +98	124
HV(CO)₃(mesitylene)	[V(CO)₄(mesitylene)]-PF₆ + (1) NaI and (2) NaBH₄ or Na/Hg	Orange solid stable in air (−20°C) but decomposing slowly at +25°C. Dec. under vac 69°C. THF solutions stable (under Ar). Other solutions dec. at 25°C. Toluene stable at −20°C.	15.8	$\bar{\nu}_{C-O}$ 1961, 1895, 1879 (hexane)	NMR line-width is temperature-dependent. Mass spec. shows strong peaks for [HV(CO)ₙ-(mesit)⁺] and [V(CO)ₙ(mesit)⁺] where n = 0, 1, 2, 3.	128
[CpV(H)(CO)₃]⁻	(a) CpV(CO)₄ + (1) Na/Hg + (2) H₂O + (3) pptn. as PPN⁺ salt; or (b) CpV(CO)₄ electrolysis in THF/H₂O; or (c) Na[CpV(CO)₄] + Et₄NCl in CH₃CN at room temperature	PPN⁺ salt is orange-yellow, m.p. (dec) 201°C. Air sensitive as solids or in solution (THF). Et₄N⁺ salt is relatively air stable. Slow decomposition (2 days) in solution (THF or CH₃CN).	16.10 (THF, d_8)	$\bar{\nu}_{C-O}$ 1890, 1780 (THF)	NMR line-width is temperature dependent.	129–131
			16.33(Et₄N⁺ salt)	$\bar{\nu}_{C-O}$ 1889, 1775		124

Compound	Preparation	NMR	IR	Description	Comments	Reference
{[CpV(CO)₃]₂H]⁻	CpV(CO)₄ + [CpV(H)(CO)₃]⁻ + uv irradiation or [CpV(H)(CO)₃]⁻ plus bromoalkanes		$\bar{\nu}_{C-O}$ 1857, 1817	Very reactive materials spectroscopically observed but not isolated.		129, 130
Cp₂V(H)(μ-Cl)₂AlCl₂	Cp₂VCl₂ + Et₂AlCl₂ in CH₂Cl₂/n-heptane			Reactive intermediate observed by ESR.	ESR study	132, 133
CpV(H)₂(CO)₃	[CpV(H)(CO)₃]⁻ + H⁺			Suggested as an intermediate.		130, 134
[H₂V(CO)₄(diars)]⁺	[HV(CO)₄(diars)] + H⁺			Postulated intermediate.		135
H₃V(CO)₃(diars)	[Et₄N][V(CO)₄(diars)] plus excess HX(X = Cl, Br, or I) in THF	8.86 triplet (CH₂Cl₂, D₂, −20°C).	$\bar{\nu}_{C-O}$ 1840, 1779 (THF)	Red-violet crystals stable in air (r.t.) for several weeks. Under Ar, stable to 130°C.	NMR linewidth very temperature dependent.	126, 135
H₃V(CO)₃(dmpe)	[HV(CO)₄(dmpe)] plus t-butyl chloride/H₂O		$\bar{\nu}_{C-O}$ 1827, 1740 (THF)	Unstable violet-red solid.	Spectra similar to H₃V(CO)₃(diars).	135
Cp₂V(μ-H)₂BH₂	Cp₂VCl₂/NaBH₄		$\bar{\nu}_{B-H_t}$ 2442, 2418 $\bar{\nu}_{B-H_b}$ 1745, 1650	Dark-violet pyrophoric solid	Mol. weight (cryoscopy), Raman spectrum. ¹H NMR shows all BH₄ protons equivalent (r.t.) at τ = 19.46.	136, 137
	(a) [PPN⁺][(CpV(H)(CO)₃]⁻ BH₃·THF (b) CpV(CO)₄ + PPN⁺BH₄⁻ (irradiation)	5.86 (Cp) 2.48 (PPN⁺)	$\bar{\nu}_{C-O}$ 1846, 1735 (THF) $\bar{\nu}_{C-O}$ 1835, 1721 (KBr) $\bar{\nu}_{B-H}$ 2360 (KBr)	Green crystals (PPN⁺ salt)	Elemental analysis.	130
[V(BH₄)ₙ] ?						138, 139

followed by the di- and trihydrides, and the section concludes with discussion of compounds containing bridging $V-(H)_n-BH_m$ linkages.

1. Monohydrides of Vanadium

a. $HV(CO)_6$ and $HV(CO)_5(PPh_3)$. Protonation of carbonyl anions is a well-established general route to hydride complexes and has been used for the production of these vanadium compounds: $V(CO)_6^-$ can be made from $V(CO)_6$ by reduction with sodium amalgam, or from VCl_3 by reaction with sodium and carbon monoxide in diglyme at 160°C; $V(CO)_5(PPh_3)^-$ can be prepared by treatment of $V(CO)_6$ with triphenylphosphine (1:3 mole ratio) in hexane, followed by reduction with sodium amalgam. Phosphoric acid has been found to be the most convenient protonating agent for either anion.

Although numerous metal salts of these acids have been made and analyzed, the acids themselves are poorly characterized. However, it is known that they are strongly acidic, as indicated by their pK_a values, which have been determined potentiometrically (120, 121).

b. $HV(PF_3)_6$. In view of the similar ligand properties of CO and PF_3, it is not surprising to find that this complex has been produced in an analogous manner to that used for the carbonyl hydride:

$$[Na(diglyme)_2]^+[V(CO)_6]^- \xrightarrow[\text{UV, diglyme}]{PF_3} [V(PF_3)_6]^- \xrightarrow{H_3PO_4} HV(PF_3)_6$$

The ^1H-NMR spectrum [in $Ni(PF_3)_4$] showed a characteristic hydride resonance at $\tau = 18.4$, and the mass spectrum showed the molecular ion and fragments derived by successive loss of PF_3 from the ions $[HV(PF_3)_5]^+$ and $[V(PF_3)_5]^+$ but gave no indication of $[V(PF_3)_6]^+$. The compound is thermally stable to at least 135°C (122).

c. $[HV(CO)_m(phosphine)]$ complexes and related compounds involving arsenic ligands. Compounds of this nature are known for a variety of "di-," "tri-," and "tetraphosphines" and some arsenic analogs. For the diphosphines and diarsines, only tetracarbonyls (**XXIV – XXX**) have been reported; but when "tri-" or "tetraphosphines" are employed, $m = 2, 3$, or 4 carbonyls (**XXXI – XXXIX**) are known, though in these cases the tetracarbonyls appear to be the least stable.

d. "Diphosphine" compounds: $[HV(CO)_4\widetilde{PP}](\widetilde{PP} = Ph_2P(CH_2)_nPPh_2;$ $n = 1-4;$ **XXIV–XXVII**); $HV(CO)_4\{Me_2P(CH_2)_2PMe_2\}$ (**XXVIII**); $HV(CO)_4\{Ph_2As(CH_2)_2PPh_2\}$, (**XXIX**); and $HV(CO)_4(diars)$, (**XXX**). Various methods of preparation have been reported for these compounds, but all involve protonation of the corresponding "phosphine" carbonylate anions, and they differ only in the manner in which this has been accomplished. Treatment of the anions with silica gel, presumably by virtue of potentially acidic {Si—OH} groups, has proved a convenient method for compounds (**XXIV**)–(**XXIX**) (*123, 124*). Other means have included *t*-butyl chloride and water (for **XXV** and **XXVII**) (*125, 126*), and $Ph_2AsHCl_2 \cdot 2H_2O$ in benzene (for **XXX**) (*123, 126*).

The hydrides are described as microcrystalline yellow powders, which change to ochre or brown upon standing at room temperature. These color changes are not accompanied by any observable changes in the infrared or NMR spectra (*123*) unless the storage is prolonged, in which case more profound changes may occur. For example, (**XXV**) is converted to paramagnetic green-black $V(CO)_4(dppe)$ (*125*). At lower temperatures ($-10°C$), and under N_2, the solid complexes are stable for at least several months; however, solutions of the hydrides deteriorate rapidly unless oxygen is rigorously excluded. Room-temperature ^1H-NMR (THF-d_8) spectra reveal the hydridic resonance in the range $\tau = 14.5$ to $\tau = 15.15$ and have been interpreted for compounds (**XXIV**)–(**XXVIII**) as showing either a fluxional motion of the hydride about the entire seven-vertex polyhedron or that the hydride is fixed so that the two phosphorous atoms are magnetically equivalent with respect to it (*123*). However, the ^1H-NMR spectrum of (**XXX**) shows no evidence of fluxionality, since there is no change in the $\tau = 14.48$ resonance between $+20°C$ and $-67°C$ (*126*). ^{31}P-NMR and ^{51}V-NMR spectra were also obtained for all compounds but add little to the conclusions reached from the proton spectra (*123, 124*).

The hydrides readily lose their hydrogen upon reaction with other materials. In some cases, the hydrogen is transferred to the reactant, which then occupies the coordination position originally held by the hydride ligand (*127*), e.g., $HV(CO)_4\widetilde{PP} + Et_3N \rightarrow HNEt_3V(CO)_4PP$ and $HV(CO)_4(dppm) + isoprene \rightarrow (\eta^3\text{-dimethylallyl})V(CO)_3(dppm)$.

e. "Triphosphine" compounds: $HV(CO)_3PhP(CH_2CH_2PPh_2)_2$, (**XXXI**); $HV(CO)_3MeC(CH_2PPh_2)_3$, (**XXXII**). These compounds have been made by the silica-gel protonation method used for their diphosphine counterparts (*123, 124*), and their ^1H-, ^{31}P- and ^{51}V-NMR spectra have been interpreted in favor of a model involving restricted dynamic behavior based on a hydride-face-capped octahedron similar to that of $HTa(dmpe)_2(CO)_2$ (*124*).

f. "Tetraphosphine" compounds: $HV(CO)_4P(CH_2CH_2PPh_2)_3$, (**XXXIV**); $HV(CO)_3P(CH_2CH_2PPh_2)_3$, (**XXXV**); $HV(CO)_4(Ph_2PCH_2CH_2 PPhCH_2)_2$, (**XXXVI**); $HV(CO)_3(Ph_2PCH_2CH_2PPhCH_2)_2$, (**XXXVII**); *cis*-$HV(CO)_2P(CH_2CH_2PPh_2)_3$, (**XXXVIII**); *and trans*-$HV(CO)_2(Ph_2PCH_2)$-

$CH_2(PPhCH_2)_2$, (**XXXIX**). Compounds (**XXXIV**)–(**XXXVII**) have been made via the silica-gel protonation route just described, and the dicarbonyl complexes (**XXXVIII**) and (**XXXIX**) by UV irradiation of their tricarbonyl analogs (*124*).

   g. $HV(CO)_3(mesitylene)$. This compound is obtained as a bright orange powder (decomp. 69–72°C) from the reaction of $[IV(CO)_3$ (mesitylene)] with $NaBH_4$ in THF at 0°C. It shows a characteristic hydride resonance at $\tau = 15.8$ (toluene), and further evidence of its nature is found in its conversion to $[V(CO)_3(mesitylene)]^-$, isolable as the tetrabutylammonium salt, by either sodium amalgam or sodium hydroxide (*128*).

   h. $[CpV(H)(CO)_3]^-$ *and* $[\{CpV(CO)_3\}_2H]^-$. Treatment of a slurry of $Na_2[CpV(CO)_3]$ in THF with 1 eq H_2O yields solutions containing $[CpV(H)(CO)_3]^-$, which may be isolated as its orange-red $(Ph_3P)_2N^+$ salt. A corresponding deuteride is produced if D_2O is used (*129, 130*). More recently it has been found that a higher yield, in a more air-stable form (as the Et_4N^+ salt), results from the reaction of $Na_2[CpV(CO)_3]$ with Et_4NCl in acetonitrile (*124*). The ion also is produced during the electrolysis in THF/H_2O of either $CpV(CO)_4$, or $Cp_2V(CO)$ under a CO atmosphere (*131*). Irradiation of this species in the presence of CpV-$(CO)_4$ produces $[\{CpV(CO)_3\}_2H]^-$, which is believed to be a V–H–V bridged dimer, although it has not been fully characterized. Due to its extreme sensitivity, it has not been isolated from solution (*130*).

   The ^{1}H-NMR spectrum of $[CpV(H)(CO)_3]^-$ shows the hydride resonance at $\tau = 16.10$, and the nature of the ion has been further demonstrated by its chemical reactions. For example,

$$[CpV(H)(CO)_3]^- + RX \rightarrow RH + \text{organometallics}$$

and

$$[CpV(H)(CO)_3]^- + RCOX \rightarrow RCHO + \text{organometallics}$$

where R is an alkyl group. The hydrogen acquired by the R or RCO group is established as coming from the hydride (not the solvent) by reactions of the deuteride analog, e.g., $[CpV(D)(CO)_3]^- + 1$-bromooctane $\rightarrow C_8H_{17}D$. The organometallic products of these reactions are complex and include the bridged dimer and $CpV(CO)_4$. A detailed kinetic and mechanistic study of these reactions has been carried out (*130*).

   i. $Cp_2V(H)(\mu\text{-}Cl_2)(AlCl_2)$. The reaction between Cp_2VCl_2 (in methylene chloride) and Et_2AlCl (in heptane) yields a paramagnetic product designated as the hydrido complex on the basis of its ESR

n. $Cp_2V(\mu\text{-H})_2BH_2$ *and* $[CpV(CO)_2(\mu\text{-H})_2BH_2]^-$. These well-characterized compounds have recently been reported (*130, 136, 137*). The former is produced when $LiBH_4$ reacts with Cp_2VCl_2 at $-12°C$ in 1,2-dimethoxyethane, and an analogous (BD_4) complex is also known. The violet crystalline materials decompose slowly at room temperature even under nitrogen, but infrared, Raman, and NMR studies have been performed and have revealed bidentate coordination of the BH_4 group. Although all four protons are equivalent at room temperature (τ = 19.46), the barrier to bridge-terminal hydrogen exchange is high enough to permit observation of the instantaneous structure in the low-temperature ($-100°C$) ^{1}H-NMR spectrum. At this temperature, the bridge hydrogens show $\tau \approx 34$ and the terminal ones are calculated to lie at $\tau \approx 5$ (obscured by the solvent). The infrared and Raman spectra indicate that the bridge expansion (involving the VH_2B segment, and having considerable $\bar{\nu}_{M-H}$ character) occurs at 1395 cm^{-1} (1030 cm^{-1} for the deuterated material) (*136, 137*). The second compound results when mixtures of $CpV(CO)_4$ and $[(Ph_3P)_2N^+BH_4^-]$ are irradiated. However, a better method starts with $[CpV(H)(CO)_3]^-$, which is reacted with $THF \cdot BF_3$. The $(Ph_3P)_2N^+$ salt is green and shows spectra similar to $Cp_2V(\mu\text{-H})_2BH_2$; its composition is based upon these similarities, its elemental analysis, and its method of preparation. Reactions leading to each compound are thought to proceed via $[Cp(CO)_3V\text{—}H\text{—}BH_3]^-$ (*130*).

o. $V(BH_4)_n$. Finally, mention should be made of the compound $V(BH_4)_n$, which was originally (*138*) considered to be a V(III) material but is now reported as $V(THF)_x(BH_4)_2$, containing V(II) (*139*).

B. Niobium Hydride Compounds

Niobium forms more than 30 compounds in which either terminal niobium–hydrogen or bridging $Nb-(\mu\text{-H})_n-M$ linkages (M is another metal or boron; n = 1, 2, or 3) are present. Although most contain only one Nb–hydrogen bond, up to five per molecule are known; but even in the latter, some other ligands are always present and the nature of the co-ligands is clearly important in determining the stability of all the hydrido niobium complexes. Cyclopentadienyl groups are present in nearly all cases [the exceptions involve the chelating phosphine $Me_2PCH_2CH_2PMe_2$, (dmpe)], together with additional two-electron donors such as carbon monoxide, alkenes or alkynes, and monophosphines. However, the exact role of these co-ligands is not clear, and the discovery of most compounds owes more to serendipity than to rational synthesis.

The terminal hydrides display a $\bar{\nu}_{Nb-H}$ infrared absorption in the range 1620–1740 cm^{-1} [falling to 1520–1550 for the (dmpe) compounds], and ^{1}H-NMR absorption attributable to the hydride ligands in the τ range 10.8–17.8. The latter rises to 17–22 for the bridged compounds. Specific data for the compounds are found in Table V. The pattern used in the discussion of the vanadium hydrides is continued, so that the treatment begins with monohydrides and ends with the bridged species.

1. Simple Hydrides

a. HNb(dmpe)$_2$(CO)$_2$ and HNb(dmpe)$_2$(C$_2$H$_4$)$_2$. These compounds are produced from Me$_5$Nb(dmpe) according to the equations

$$Me_5Nb(dmpe) + H_2 + dmpe \xrightarrow[60°C]{THF} H_5Nb(dmpe)_2$$

$$H_5Nb(dmpe)_2 + 2L \xrightarrow[25°C]{pentane} HNb(dmpe)_2(L)_2 + 2H_2$$

where L = CO or C$_2$H$_4$; the (C$_2$D$_4$)$_2$ complex can also be made.

The infrared spectra of the compounds show a weak to medium band at 1550 cm^{-1} attributable to the Nb–H stretch. The carbonyl band is strong and occurs at 1700–1750 cm^{-1}; the olefinic ν_{C-H} stretching mode shows up as a shoulder at 3020 cm^{-1}, absent in the (C$_2$D$_4$) spectrum, which shows peaks at 2290, 2220, 2200, and 2170 cm^{-1} due to coordinated olefin. The ^{1}H-NMR spectrum of HNb(dmpe)$_2$(CO)$_2$ in C$_6$D$_6$ at 25°C shows a triplet of triplets at $\tau = 14.46$ for the hydride resonance. This shifts to $\tau = 12.94$ in the olefin complexes. These findings are analogous to those for the corresponding HTa(dmpe)$_2$L$_2$, complexes which are isomorphous (and, presumably, isostructural) with the niobium compounds. ^{31}P-NMR and ^{13}C-NMR studies have also been performed on these materials. It is noteworthy that the exchange of hydrogen between the hydride and the olefin in HNb(dmpe)$_2$(C$_2$H$_4$)$_2$ is much slower than in the isoelectronic [HMo(dppe)$_2$(C$_2$H$_4$)$_2$]$^+$. The hydride τ12.94 resonance of the niobium compound shows no broadening characteristic of the onset of exchange with ethylene protons, in marked contrast to that seen in the molybdenum complex, where four protons on one of the ethylene ligands equilibrate with the hydride ligand [rapidly on the ^{1}H-NMR timescale (100 MHz) at +7°C] by the postulated reversible formation of an ethyl ligand (140, 141).

b. [Cp$_2$Nb(H)L] complexes. Hydrido compounds of this general formula are known where L = CO (142–144), an alkene (various) (145, 146), an alkyne (various) (147), or a phosphine (various) (142, 145, 148). A η^2-COT complex has been postulated as an intermediate in the reaction of $Cp_2Nb(C_8H_8)$ with $NaBH_4$ (*149*). There is some evidence that Cp_2NbH is also produced in this reaction (*170*).

Syntheses.   Most (and, presumably, all) of these compounds can be made from Cp_2NbH_3 according to the reaction

$$Cp_2NbH_3 + L \xrightarrow[\text{benzene}]{\text{reflux}} Cp_2Nb(H)L + H_2$$

However, some care must be taken where $L = C_2H_4$ as the ethyl/ethylene complex is the main product if excess ethylene is available (*142, 145, 147*). The more readily available Cp_2NbCl_2 has been used as a starting material in a general reaction, which produces the alkene complexes from the corresponding alkyl Grignard. To date, n-propyl, i-propyl, and n-butyl Grignards have been successfully used (*146*). Cp_2NbCl_2 was also the starting point for the production of some of the phosphine complexes (*148*), such as

$$Cp_2NbCl_2 \xrightarrow{NaBH_4} Cp_2Nb(BH_4) \xrightarrow{PPh_3} Cp_2Nb(H)(PPh_3)$$

Spectroscopic Properties.   The infrared spectra of these $Cp_2Nb(H)L$ complexes contain a $\bar{\nu}_{Nb-H}$ band in the region 1630 cm^{-1} (for the phosphines) to 1730 cm^{-1} (for the olefin) (*142, 145*). Their ^{1}H-NMR spectra show a band due to the hydride hydrogen in the vicinity of $\tau 17$ (in the phosphines), but considerably lower for the olefins (12.95) and alkynes (10.8). The crystal structure of $Cp_2Nb(H)(CO)$ has been determined but the niobium-bonded hydrogen was not located (*144*). X-ray data are available for $Cp_2Nb(Et)(C_2H_4)$, which, at least in solution, is known to be similar to $Cp_2Nb(H)(C_2H_4)$ from ^{1}H and ^{13}C NMR studies (*153*).

Bonding.   The bonding in $Cp_2M(H)(CO)$ and $Cp_2M(H)(C_2H_4)$ complexes has been discussed from the molecular-orbital viewpoint. It is of interest that such compounds are predicted only for those metals with two or fewer d electrons in the M(III) state (*79*).

Reactions. $Cp_2Nb(H)(CO)$ reacts with Lewis acids to produce compounds, of varying stability, in which the acid coordinates to the hydride ligand. For example,

$$Cp_2Nb(H)(CO) + BF_3 \xrightarrow{\text{toluene}} Cp_2Nb(CO)(HBF_3) \quad \text{(Ref. 150)}$$

$$Cp_2Nb(H)(CO) + AlEt_3 \longrightarrow Cp_2Nb(CO)(HAlEt_3) \quad \text{(Ref. 151)}$$

$$Cp_2Nb(H)(CO) + H^+ \xrightarrow[\text{THF, <0°C}]{\text{HCl}} \{Cp_2Nb(CO)(H_2)Cl\} \quad \text{(Ref. 152)}$$

The latter spontaneously decomposes to $Cp_2Nb(CO)Cl$ and H_2 at room temperature, and the aluminum compound is not much more stable.

However, the boron adduct appears to be indefinitely stable under ambient conditions.

It was reported originally that $Cp_2Nb(H)(CO)$ reacted with H_2 to yield a brown material from which some methanol could be obtained (142); however, it is now known that the latter derived from an impurity present in the carbonyl hydride, which was made using $NaAlH_2$ $(OCH_2CH_2OCH_3)_2$ and Cp_2NbCl_2. $Cp_2Nb(H)(CO)$ made by other methods shows no such impurity and generates no methanol-producing materials when reacted with H_2 (143). $Cp_2Nb(H)(C_2H_4)$ reacts with excess ethylene to yield the ethyl complex referred to above; however, in the presence of H_2, ethylene is catalytically reduced to ethane, probably via an unisolated $[Cp_2Nb(H)_2(Et)]$ intermediate (145, 153). $[Cp_2Nb(H)(alkyne)]$ complexes react nearly quantitatively with acid producing the corresponding cis olefins, possibly via an intermediate of the form $[Cp_2Nb(H)_2(alkyne)]^+$. When methyl salts are used, the product is methane rather than the methylated olefin, since the elimination of CH_4 from the intermediate $[Cp_2Nb(H)(Me)(alkyne)]^+$ occurs more rapidly than the insertion of alkyne into the Nb–H bond. The hydrido complexes are rapidly converted into their iodo analogs by CH_3I (147). In $[Cp_2Nb(H)(phosphine)]$ when the phosphine is Ph_3P or $PhMe_2$, and presumably in other cases too, protonation by dilute aqueous HCl in the presence of a suitable precipitating anion gives rise to salts such as $[Cp_2Nb(H)_2(PPhMe_2)]^+(PF_6^-)$, with $\bar{\nu}_{Nb-H}$ at 1740 cm^{-1} and a doublet at $\tau = 13.96$ in the ^1H-NMR spectrum assignable to the NbH_2 hydrogens (148).

c. $[(HNb(\eta^5-C_5H_5)(\eta^{1,5}-C_5H_4)_2]$.   Thermal decomposition of Cp_2NbH_3 in benzene at 80°C, in the absence of other ligands, yields a yellow solid, to which the name "niobocene" has sometimes been applied since the compound's elemental analysis is consistent with the Cp_2Nb formulation. However, the observed diamagnetism and the ^1H-NMR spectrum rule out this idea. Instead, they suggest a dimer with bridging C_5H_4 groups and Nb–H linkages, and this has been confirmed by an X-ray diffraction study (154, 155). The presence of an M–H bond is also demonstrated by the infrared peak at 1650 cm^{-1} and the $\tau = 12.07$ ^1H-NMR resonance. The structure (idealized C_2 symmetry) is shown below. The Nb–H and Nb–Nb bond lengths are 1.70(3) and 3.105(6) Å,

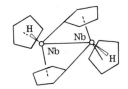

(AlH_4^-) or ($AlH_2Et_2^-$). Both the color of the final trihydride and the chemistry of compounds derived from it [e.g., $Cp_2Nb(H)(CO)$] suggest that the product is contaminated with material derived by hydrogen abstraction from carbon adjacent to oxygen in the alkoxy substituents of the reductant, even though this impurity does not show in the ¹H-NMR spectrum (142).

¹H-NMR and infrared spectra of the trihydride show absorption at $\tau = 13.72$ and $\tau = 12.73$ and $\bar\nu = 1710$ cm⁻¹, respectively, indicating the presence of M–H linkages. The latter have been directly observed in the X-ray and neutron diffraction study, which has clearly shown the planar character of the central H_3Nb fragment with the three hydrogens essentially equidistant from the metal and the central M–H bond in a bisecting position (162). The angle between the "outside" hydride ligands is 126°, in excellent agreement with the 122° predicted by Lauher and Hoffmann in their molecular-orbital study of bent Cp_2ML_m complexes ($m = 1$, 2, or 3) (79).

Many of the reactions of this and similar polyhydrides involve elimination of H_2 and the formation of monohydrides. In this respect, the niobium compound turns out to be much more reactive than its tantalum analog (162) and readily loses H_2 in the presence of other ligands to form [$Cp_2Nb(H)L$] complexes, as previously described under the appropriate listing for these monohydrides. Reactions with Lewis acids suggest a significantly higher basicity for the unique central M–H bond relative to the other two (151). Perhaps the most interesting property is its ability to catalyze H_2/D_2 exchange, possibly via a [Cp_2NbH] intermediate (145).

h. $H_5Nb(dmpe)_2$.   Reaction of $Me_5Nb(dmpe)$ with H_2 (500 atm) and (dmpe) in THF at 60°C yields the pentahydrido complex as a pale-yellow, waxy, crystalline solid, stable under N_2 for at least several days at room temperature.

A strong, broad peak in the infrared at ~1520 cm⁻¹ (Nujol) is assigned to $\bar\nu_{M-H}$, and the ¹H-NMR shows a quintet at $\tau = 12.04$ [$^2J(HP) = 39$ Hz], due to the hydrido hydrogens, both in C_6D_6 at 25°C and in toluene-d_8 at −80°C. The ¹H-NMR and ³¹P-NMR spectra suggest that the barrier to interconversion of idealized structures for [$H_5Nb(dmpe)_2$] in solution is low.

Although $H_5Nb(dmpe)_2$ decomposes upon prolonged standing at room temperature, there is no evidence for the spontaneous and reversible loss of 1 or 2 mol of H_2. However, in the presence of π-acid ligands such as CO, 2 mol H_2 are readily displaced, yielding, e.g., $HNb(dmpe)_2(CO)_2$ as previously mentioned. As with Cp_2TaH_3, the analogous reaction with tantalum is far slower (140, 141).

2. Compounds Containing Nb–H–M *bridges* (M = *Transition Element*)

a. Cp$_2$(CO)Nb$\overset{H}{\diagup\diagdown}$Fe(CO)$_4$. Addition of a slight excess of Fe(CO)$_5$ and Cp$_2$NbH$_3$ in benzene results in an essentially quantitative conversion to Cp$_2$(CO)Nb(μ-H)Fe(CO)$_4$, which can be recovered as air-sensitive dark-brown crystals. The material has been characterized by infrared $\bar{\nu}_{Nb-H-Fe}$ at 1720 cm^{-1}; ^{1}H-NMR, toluene-d_8, hydride at τ = 16.4; and by X-ray diffraction. It appears to contain both a bridging hydrogen atom and a metal–metal bond. The geometry about niobium can be viewed as a bent sandwich system, analogous to Cp$_2$NbH$_3$ with the planar H$_3$Nb group of the latter replaced by a (H,CO,Fe)Nb moiety (*163, 164*).

The reaction by which the complex is formed is thought to proceed via a formyl (HCO)Fe(CO)$_4$ species, resulting from an attack of the hydride hydrogen on CO bonded to iron.

b. Cp$_2$Nb(H)$_2$Hf(CH$_2$Ph)$_3$. Reaction of Cp$_2$NbH$_3$ with (PhCH$_2$)$_4$Hf yields toluene and the binuclear hydrido complex, which decomposes with a half-life of a few hours at room temperature (*151*). No further information on this compound is available.

c. [Cp$_2$Nb(H)$_2$]$_2$Zn. This material results from the reaction of Cp$_2$NbH$_3$ with Et$_2$Zn, and its structure has been proposed as involving two mutually staggered, bent Cp$_2$NbH$_2$ fragments (analogous to Cp$_2$NbH$_3$ without the central H) chelated to the zinc atom, so that the latter has the four hydride hydrogens tetrahedrally disposed about it.

The ^{1}H-NMR (C$_6$D$_6$) shows a hydridic resonance at τ = 19.05, and the molecular weight (cryoscopic measurements in benzene) and C,H analyses are consistent with the formula shown (*151*).

d. Cp$_2$Nb(CO)(μ-H)Zn(BH$_4$)$_2$. Reaction of NbCl$_5$, NaCp, and zinc powder in THF under CO at room temperature followed by treatment with NaBH$_4$ produces yellow crystals of Cp$_2$Nb(CO)(μ-H)Zn(BH$_4$)$_2$. X-Ray diffraction studies show that the zinc is coordinated to two bidentate BH$_4$ groups and to niobium via a bent Zn–H–Nb bond and a long direct Zn–Nb linkage (*165*).

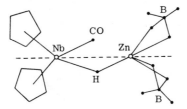

TABLE VI

Tantalum Hydride Complexes

Compound	Preparation	Properties	^{1}H NMR (τ, M–H)	Infrared (cm^{-1})	Comments	References
HTa(CO)$_2$(dmpe)$_2$	(a) TaCl$_5$/K/H$_2$/dmpe, then (2) + CO (in benzene); (b) TaCl$_5$(dmpe) + (1)CO then (2)Na/Hg and (3)H$^+$	Orange crystals, M.P. 140–141°C.	14.17 triplet of triplets (8°C); $^2J(^1$H–^{31}P) 14.25, 89.25 Hz.	$\bar{\nu}_{C-O}$ 1737; $\bar{\nu}_{Ta-H}$ 1589	X-Ray; ^{31}P NMR; mass spectrum.	171–173
HTa(PPh$_3$)$_2$(dmpe)$_2$	TaCl$_2$(dmpe)$_2$/KPPh$_2$/THF/0°C	Purple crystals.	Not found.	$\bar{\nu}_{Ta-H}$ 1648	X-Ray; ^{31}P NMR; mass spectrum.	174
HTa(η^1-C$_{10}$H$_8$)(dmpe)$_2$	[Ta(η^1-C$_{10}$H$_8$)(dmpe)$_2$]$^-$ + H$^+$	Orange crystals.	10.23 (toluene-d_8, 25°C)		X-Ray of Cl-compd.; ^{31}P NMR; mass spectrum.	175
Cp$_2$Ta(H)(CO)	(a) Cp$_2$Ta(H)(alkene)/CO; (b) Cp$_2$TaH$_3$/CO/toluene	Purple crystals.	16.75 or 16.80, 5.50 (Cp)	$\bar{\nu}_{C-O}$ 1885		145, 176
Cp$_2$Ta(H)(alkene)	(a) Cp$_2$TaCl$_2$(RMgCl(ether) R = n-C$_3$H$_7$ → C$_3$H$_6$ complex n-C$_4$H$_9$ → C$_4$H$_8$ complex n-C$_5$H$_{11}$ → C$_5$H$_{10}$ complex C$_5$H$_9$ → C$_5$H$_8$ complex; (b) Cp$_2$Ta(CHMe)(Me)/heat	Yellow → purple (depending on alkene).	13 → 13.6, depending on alkene and whether *endo* or *exo* isomer.	$\bar{\nu}_{Ta-H}$ 1750–1800; $\bar{\nu}_{C-C}$ ~1450;	^{13}C NMR	146, 177
Cp$_2^*$Ta(H)(alkyne) (Cp* = C$_5$H$_4$Me)	(a) Cp$_2^*$TaH$_3$/alkyne; (b) Cp$_2^*$TaH$_3$/alkyne/ iodobenzene, then LiAlH$_4$ alkynes: ((n-C$_3$H$_7$)$_2$C$_2$]; MeC$_2$(t-C$_4$H$_9$); MeC$_2$(i-C$_3$H$_7$.)		10.2 for (n-C$_3$H$_7$)$_2$C$_2$.	$\bar{\nu}_{Ta-H}$ 1570. $\bar{\nu}_{Ta-H}$ 1800; $\bar{\nu}_{Ta-D}$ 1275; $\bar{\nu}_{C-C}$ 1770.		178
Cp$_2$Ta(H)(PEt$_3$)	(a) Cp$_2$TaH$_3$/PEt$_3$; (b) Cp$_2$Ta(H)(alkene)/PEt$_3$	Red crystals.	19.48 or 19.56; $^2J(^1$H–^{31}P) 21 Hz; 5.68 (Cp).	$\bar{\nu}_{Ta-H}$ 1705		146, 179
[(Cp')($\eta^{1,5}$-C$_5$H$_4$)Ta(H)]$_2$	Cp$_2$TaH$_3$/heat				Similar spectral properties to "niobecene" dimer.	145
[Cp'Ta(H)(Cl)]$_2$ (Cp' = C$_5$Me$_5$R)	Cp'Ta(propylene)(Cl)$_2$/H$_2$ (40 psi, 25°C, pentane)	Green crystals.	20.45	$\bar{\nu}_{Ta-H}$ 1585 $\bar{\nu}_{Ta-D}$ 1150	Diamagnetic	180
Cp$_2$Ta$_4$(H)(Cl)$_4$(HCO)	[Cp'Ta(H)(Cl)$_2$]$_2$/CO/0°C	Orange crystals.	17.52		^{13}C NMR	180
Cp$_2$Ta$_4$(H)(Cl)$_4$(HCO)(PMe$_3$)	Previous compound/PMe$_3$	Yellow crystals.	20		X-Ray, ^{13}C NMR.	180

Compound	Preparation	Description	NMR	$\bar{\nu}_{\text{Ta-H}}$	Other methods	Ref.
[Cp₂Ta(H)Li] ?	Cp₂TaH₃/Li(n-Bu)	Orange yellow crystals, not fully characterized.		$\bar{\nu}_{\text{Ta-H}}$ 1590		181, 182
[Cp₂Ta(H)(Me)₃] ?	Ta(Me)₃Cl₂/Cp₂Pb	Air-sensitive yellow crystals. Not isolated.				183
Cp*Ta(H)(Me)₄(OCHMe₂)	Cp*Ta(Me)₄/CO/H₂	Intermediates in numerous reactions especially decomposition of Ta alkyls, etc.	Not located.	$\bar{\nu}_{\text{Ta-H}}$ 1730		184
CpTaH						185–187
Cp₂Ta(H₂SnMe₃)	Cp₂TaH₃/Me₃SnNMe₂	White crystals, m.p. 154–156°C.	14.63 (benzene-d₆); 5.53 (Cp).	$\bar{\nu}_{\text{Ta-H}}$ 1750		188
Cp₂Ta(H₂	Cp₂TaH₃ photolysis in presence of di-t-butyl peroxide.	Paramagnetic intermediates identified by ESR. Not isolable, although Me₂ analogs are.			ESR	160
Cp₂TaH₃	(a) Cp₂TaCl₂/NaAlH₂(OR)₂ (b) TaCl₅/NaCp/NaBH₄	White crystals; m.p. 187–189°C (dec.).	11.63 triplet; 13.02 doublet (benzene-d₆).	$\bar{\nu}_{\text{Ta-H}}$ 1735	X-Ray and neutron diffraction P.E.S.; mass spectrum.	162, 181, 182, 188–190
Cp*₂TaH₃ (Cp* = C₅H₄Me)	TaCl₅/LiCp*/LiAlH₄	White crystals, m.p. 89.5–91°C.	10.69 (triplet); 12.20 (doublet).	$\bar{\nu}_{\text{Ta-H}}$ 1740		191
H₃Ta(dmpe)₂	(a) TaCl₅/K/H₂/dmpe/benzene (b) Me₃Ta(dmpe)/H₂/dmpe/THF	Air-sensitive white crystals.	10.70 quintet; ²J(H–³¹P) 35 Hz.	$\bar{\nu}_{\text{Ta-H}}$ 1544 (Nujol)	Mol. weight (cryoscopy)	140, 141, 171
Cp₂Ta(CO)(μ-H)Fe(CO)₄	Cp₂TaH₃/Fe(CO)₅		16.70		Crystal structure of Nb analog.	164
(μ-H)₂(μ-Cl)₂[Ta(PMe₃)₂Cl₂]₂	(a) TaCl₅ + Na/Hg + PMe₃ (b) H₂	Green crystals	Complex multiplet centered at 18.5 attributed to Ta–H–Ta bridging hydrides.	No bands attributable to Ta–H–Ta	X-Ray structure mol weight; ³¹P NMR.	199
Cp₂TaH₃ · L	Cp₂TaH₃ + Lewis acids (L) at low temperature (L = AlEt₃, GaEt₃, ZnEt₂, CdEt₂)	Strong adducts undissociated in freezing benzene.	14.68 and 12.46 (benzene-d₆); L = AlEt₃.		Mol. weights (cryoscopy)	151
{Ta[H₂Al(OR)₂](dmpe)₂}₂ (R = —CH₂CH₂OCH₃)	TaCl₃(dmpe)₂/Na[AlH₂(OR)₂]		16.30	$\bar{\nu}_{\text{Ta-H}}$ 1605, 1640	¹H NMR of τ_TaH unchanged between +90 and −70°C; crystal structure (X-ray).	192
Ta(AlH₄)₄ ?	TaCl₅/LiAlH₄/ether	A variety of poorly characterized products.				193
[Ta(HC)CMe₃(PMe₃)Cl₃]₂	Ta(CH₂CMe₃)₂Cl₃ + PMe₃(toluene)/24 h/25°C	Red nuggets			X-Ray structure showing Ta–H–C interaction.	200

characteristic ^{1}H-NMR hydride peak at $\tau = 10.23$, compared to $\tau = 14.17$ for the carbonyl hydride HTa(dmpe)$_2$(CO)$_2$ (175). Although its structure is not known directly, that of ClTa(dmpe)$_2$(η^4-C$_{10}$H$_8$) involves an approximately pentagonal bypyramidal geometry about tantalum. The structural parameters of the chloro complex indicate considerable π-acceptor interaction by the diene portion of the arene, which may well explain the substitutional inertness of the Ta–naphthalene unit as demonstrated by the noncatalysis of D$_2$/C$_6$H$_6$ exchange by HTa(dmpe)$_2$(η^4-C$_{10}$H$_8$) in contrast to the catalytic behavior of H$_5$-Ta(dmpe)$_2$ or by other η^4-arene complexes. A very stable Ta–(η^4-C$_{10}$H$_8$) bond is indicated (175).

2. [Cp$_2$Ta(H)L] Complexes

Compounds of this general formula are known where L = CO (145, 176), various alkenes (146, 177) and alkynes (178), and also PEt$_3$ (146, 179). In addition, a tantalum analog of "niobocene" has been reported (145).

a. Cp$_2$Ta(H)(CO). This complex is made by reacting the trihydride, Cp$_2$TaH$_3$, with CO at high pressure (145), but a more convenient route is the conversion of the corresponding monohydride alkene complexes using CO at 1 atm (70°C, toluene) (176). The purple compound shows $\bar{\nu}_{\text{Ta-H}}$ at 1750 cm^{-1} and a ^{1}H-NMR resonance attributable to hydride at $\tau = 16.80$.

b. Cp$_2$Ta(H)(alkene). Treatment of Cp$_2$TaCl$_2$ with RMgCl (R = n-C$_3$H$_7$, i-C$_3$H$_7$, n-C$_4$H$_9$, n-C$_5$H$_{11}$, or C$_5$H$_9$) gives the corresponding tantalum hydride π-olefin complexes Cp$_2$Ta(H)(alkene) (alkene = C$_3$H$_6$, C$_4$H$_8$, C$_5$H$_{10}$, and C$_5$H$_8$) (146). In the C$_3$H$_6$ case, two isomers occur. Cp$_2$Ta(H)(propylene) can be produced in fair yield from the thermal decomposition of Cp$_2$Ta(CHMe)(Me) (177). Both exo and endo isomers are present, though it seems likely that the former is the initial product. The materials show infrared $\bar{\nu}_{\text{Ta-H}}$ absorption in the range 1750–1800 cm^{-1}, and ^{1}H-NMR peaks due to hydride hydrogen in the region $\tau = 13.05$ to 13.68 depending upon the alkene. They are thermally stable (to 100°C) and sublimable (e.g., C$_3$H$_6$ complex, 95°C, 10^{-2} mmHg) and may be readily converted into other tantalum hydride species by displacement of the alkene in the presence of other ligands. For example, H$_2$ (1 atm, 90°C, toluene solution) yields Cp$_2$TaH$_3$, and PEt$_3$ (90°C, toluene) gives Cp$_2$Ta(H)(PEt$_3$). Both reactions are quite slow, requiring about 14 h for completion. Reaction with CO is more complicated, in that either alkene or hydride ligands may be displaced. In the latter case, the products are of the form $Cp_2Ta(CO)R$, presumably due to the migration of the hydride onto the alkene. These alkyl carbonyl complexes show remarkable thermal stability, compared to other early transition metal alkyls, with decomposition only occurring above 150°C (146). A reaction seemingly analogous to the hydride displacement by CO occurs with aryl or alkyl isocyanides. Endo-$Cp_2Ta(H)(C_3H_6)$ and $Cp_2Ta(H)(C_4H_8)$ are converted into the adducts $Cp_2Ta(C_3H_7)(R'NC)$ and $Cp_2Ta(C_4H_9)(R'NC)$, respectively (198).

c. $[Cp_2^*Ta(H)(alkyne)]$ ($Cp^* = \eta^5\text{-}C_5H_4Me$; alkyne, $RC{\equiv}RR'$; $R = CH_3$; $R' = n\text{-}C_3H_9$, $i\text{-}C_3H_7$, or $t\text{-}C_4H_9$; $R + R' = n\text{-}C_3H_7$). Refluxing $Cp_2^*TaH_3$ with the appropriate acetylene in toluene leads to the slow formation of these monohydrido complexes (178). However, they are more readily obtained, in high yield, via the monoiodo compounds (178).

$$Cp_2^*TaH_3 + \text{alkyne} + \text{iodobenzene} \xrightarrow[\text{dioxane}]{\text{reflux}} Cp_2^*Ta(I)(\text{alkyne})$$

$$Cp_2^*Ta(I)(\text{alkyne}) \xrightarrow[\text{Et}_2\text{O}]{\text{LiAlH}_4} Cp_2^*Ta(H)(\text{alkyne})$$

The hydrides show $\bar{\nu}_{\text{Ta-H}}$ (infrared) bands around 1800 cm^{-1} (1275 cm^{-1} for the deuteride) and ^{1}H-NMR hydride absorption at $\tau = 10.2$.

Reaction of the oct-4-yne complex with HBF_4 liberates cis-oct-4-ene, and this is postulated as occurring via initial protonation of the metal (to yield $[Cp_2^*Ta(H)_2(\text{alkyne})]^+$), followed by insertion of acetylene into the Ta–H bond and C–H reductive elimination. cis-Oct-4-ene also results when H_2 is reacted with the hydrido complex at 100°C (178).

d. $Cp_2Ta(H)(PEt_3)$. Tebbe first made this compound by reacting Cp_2TaH_3 with PEt_3 (179), but an easier route employs $[Cp_2Ta(H)(\text{alkene})]$ as described above (146). The red crystalline material shows ^{1}H-NMR absorption at $\tau = 19.48$ (hydride) and $\bar{\nu}_{\text{Ta-H}}$ (infrared) at 1705 cm^{-1} but has not been otherwise characterized.

e. $[HTa(\eta^5\text{-}Cp)(\eta^{1,5}\text{-}C_5H_4)]_2$. Refluxing Cp_2TaH_3 in benzene produces "tantalocene," whose properties and structure are similar to those of its niobium analog (145).

3. $(C_5Me_4R)Ta(H)Cl_2$ and Related Compounds

Reaction of H_2 (40 psi, 25°C, pentane) with $(C_5Me_4R)Ta(\text{propylene})Cl_2$ produces a green diamagnetic material of stoichiometry $[(C_5Me_4R)Ta(H)Cl_2]_2$ (XL) showing $\bar{\nu}$ (infrared) at 1585 cm^{-1}, which moves to 1150 cm^{-1} upon deuteration, and which is presumably $\bar{\nu}_{\text{Ta-H}}$.

The ^{1}H-NMR reveals a peak at $\tau = 20.45$, attributable to Ta–H. The structures of these compounds probably contain a metal–metal single bond supported by two bridging M–H–M bonds. This reaction occurs for various R(alkyl) groups, but when R is Et the green product reacts with CO at 0°C yielding an orange compound of apparent composition $[(C_5Me_4Et)Ta(H)Cl_2]_2(CO)$ (**XLI**). However, from the ^{1}H- and ^{13}C-NMR spectra, its structure contains only one H–Ta bond, with the other hydrogen attached to carbon, in the unique manner shown below (i.e., an η^2-formyl complex).

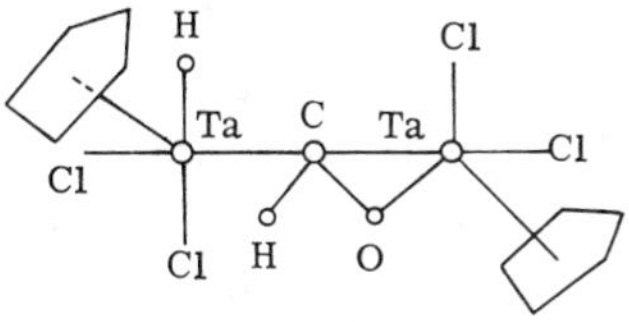

This material reacts with PMe_3, forming an adduct (**XLII**) whose structure (from X-ray data) is shown below.

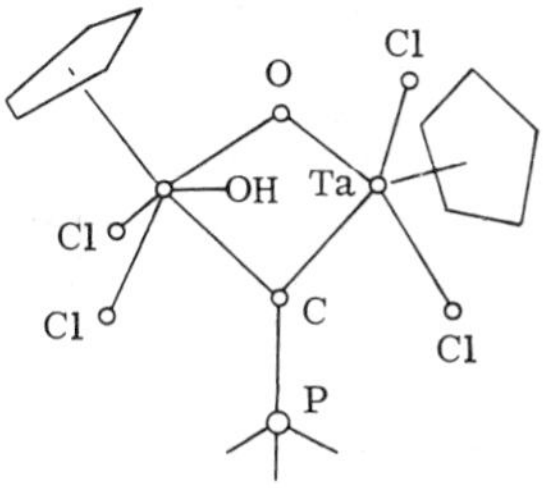

The ^{1}H-NMR spectra of these latter compounds (**XLI**) and (**XLII**) show τ_{Ta-H} at 17.52 and 20, respectively (*180*).

Compound (**XL**) can also be produced from $(C_5Me_4R)Ta(CH_2CMe_3)_2$ Cl_2 and H_2. A co-product in this reaction is red $(C_5Me_4R)_2Ta_2(H)Cl_5$, which is assumed to contain one Ta–H–Ta bridge bond. This material reacts with CO to form a complex similar to (**XLI**) but with the terminal H replaced by Cl (*180*). There is good evidence that simple formyl complexes are unstable with respect to CO–M–H rearrangement, so that compound (**XLI**) presumably owes its existence to the "stabilization" of the formyl ligand by coordination to a second metal (Lewis acid) (*180*).

4. Other "CpTa(H)" Systems

Green has isolated an orange-yellow material from the reaction of Cp_2TaH_3 with n-butyllithium in toluene, which, although not fully characterized, appears to contain the $Cp_2Ta(H)$ grouping ($\bar{\nu}_{Ta-H}$ 1590 cm^{-1}) and also a Ta–Li bond. It is, perhaps, analogous to $[Cp_2W(H)Li]_4$ (181, 182). An air-sensitive yellow compound believed to be Cp-Ta(H)(Me)$_3$ has been reported by Holliday et al., but has not been characterized. It is a co-product of the reaction between Cp_2Pb and $TaMe_3Cl_2$ (183). An apparently similar compound, $Cp^*Ta(H)(Me)_2$-(OCHMe$_2$), with $\bar{\nu}_{Ta-H}$ at 1730 cm^{-1}, has recently been reported. It is an unisolated product of the reaction of $Cp^*Ta(Me)_4$ with, successively, CO and H_2 (194). CpTa(H) moieties have been proposed as intermediates in numerous catalytic reactions. These are assumed to result from hydrogen abstractions (185–187).

a. $Cp_2Ta(H)_2(SnMe_3)$.   Dimethylamine trimethylstannane provides a convenient route to metal–metal bonds involving tin, and stoichiometric reaction of this compound with Cp_2TaH_3 in THF (reflux, 6 h) affords the white crystalline $Cp_2Ta(H)_2(SnMe_3)$ with concommittant release of amine (188).

The dihydrido complex melts without decomposition (154–156°C). It shows a hydrido absorption (^{1}H-NMR) at $\tau = 14.63$ and $\bar{\nu}_{Ta-H}$ 1750 cm^{-1}.

b. Cp_2TaH_2.   Like its niobium analog, this compound has been identified by ESR studies of photolytic reactions of Cp_2TaH_3 with di-tert-butyl peroxide. It was not isolated, although the corresponding $Cp_2Ta(Me)_2$ was (160).

c. Cp_2TaH_3 and $(C_5H_4Me)_2TaH_3$.   The Cp_2 compound can be made by reaction of Cp_2TaCl_2 with $NaAlH_2(OCH_2CH_2OCH_3)_2$ (182), or from $TaCl_5$, NaCp, and $NaBH_4$ (181). The substituted Cp complex is made from $TaCl_5$, $Li(MeC_5H_4)$, and $LiAlH_4$ (191). ^{1}H-NMR studies show two resonances attributable to the hydrido hydrogen; in the Cp_2 compound these occur at 11.63 (triplet) and 13.02 (doublet), while they are at 10.69 (triplet) and 12.20 (doublet) in the other complex (188, 191); $\bar{\nu}_{Ta-H}$ (infrared) is at 1735 and 1740 cm^{-1}, respectively.

The crystal structure of Cp_2TaH_3 has been determined by X-ray and neutron diffraction methods; it is isostructural with Cp_2NbH_3 (162). Photoelectron spectra show the relationships between $Cp_2Re(H)$, Cp_2WH_2, and Cp_2TaH_3; the varying number of lone pairs on the respective metals (2, 1, and 0) are particularly clearly indicated (189). The relationships between such hydrides have also been studied mass spectrometrically (190). The preponderance of (Cp_2TaH^+) ions over (Cp_2Ta^+), in contrast to the reverse situation for Cp_2MoH_2, is noteworthy. Clearly, the unique central M–H bond of the H_3M moiety is stronger than the other two, consistent with the reactions of Cp_2TaH_3 and its niobium analog.

The catalytic behavior of Cp_2TaH_3 and $(C_5H_4Me)_2TaH_3$ with respect to benzene–D_2 exchange has been studied. Not only do these compounds show much slower rates of exchange than Cp_2NbH_3, but they also do not discriminate among various substituted benzene substrates, while the niobium compound does. Possible reasons for these differences have been discussed in terms of a mechanism involving a series of alternating reductive-elimination and oxidative-addition steps (191).

d. $H_5Ta(dmpe)_2$. This pentahydrido complex can be made from $TaCl_5$ or $(TaPh_6^-)$, H_2 (1500 psi), and (dmpe). In the former case, potassium is also needed, and the reaction is done in benzene as opposed to THF for the hexaphenyl route. The latter is slightly more facile (45°C, 5 h, compared to 115°C, overnight) (171). The pentahydride can also be produced from its pentamethyl analog by reaction with H_2 (500 atm, 60°C in THF) in the presence of excess (dmpe) (140, 141). In each case, the air-sensitive product is white and crystalline, melting with decomposition at 133–135°C. The infrared (Nujol) $\bar{\nu}_{Ta-H}$ occurs at 1544 cm^{-1}, and the ^{1}H-NMR (toluene-d_8) hydride absorption is a quintent at $\tau = 10.70$. The compound appears to be fluxional down to at least -140°C (171). Hydride ligands are readily displaced leading to monohydrido complexes as already discussed, and D_2/benzene exchange is catalyzed by a mechanism that almost certainly involves interconversions like

$$H_5M \xrightarrow{-2H_2} HM \xrightarrow{+2H_2} H_5M$$

5. Compounds Involving Ta–H–M Bridges

M is a transition element, and the only compounds to mention here are $Cp_2(CO)Ta(H)Fe(CO)_4$, which is produced in a fashion analogous to the niobium compound and has a similar structure (164), and the very unusual $(\mu\text{-}H)_2$-$(\mu\text{-}Cl)_2$-ditantalum species $[Ta_2Cl_6(PMe_3)_4H_2]$, whose X-ray structure has recently been determined (199). The bridging hydrogens were not directly located, but their positions were clearly indicated by the disposition of the other atoms, and they were identified in a 360 MHz ^{1}H-NMR study.

6. Ta–H–Al Bridges

a. $Cp_2TaH_3AlEt_3$ and related compounds. Reaction of Cp_2TaH_3 with Et_3Al produces a bridge-bonded adduct similar to the niobium compound. ^{1}H-NMR absorptions due to the hydride hydrogens occur at $\tau = 12.46$ and 14.68.

Similar adducts are formed with $GaEt_3$, $ZnEt_2$, and $CdEt_2$. These compounds are more stable than their niobium counterparts (*151*).

b. (dmpe)$_2$Ta[H$_2$Al(OCH$_2$CH$_2$OCH$_3$)$_2$]. This compound, in which Ta–H–Al linkages are implied (infrared $\bar{\nu}$ 1605 and 1640 cm^{-1}; $\tau = 16.30$, ^{1}H-NMR) can be isolated from the reaction of TaCl$_2$(dmpe)$_2$ with Na[H$_2$Al(OCH$_2$CH$_2$OCH$_3$)$_2$]. The crystal structure has been determined by X-ray diffraction methods, but the bridging hydrogens were not located (*192*).

c. [H$_n$Ta(AlH$_4$)$_{4-n}$]($n = 0$–4). Compounds of this general formula apparently occur in the reaction products from reduction of TaCl$_5$ by LiAlH$_4$ (in Et$_2$O). They do not appear to have been characterized (*193*).

7. A Ta–H–C *Bridge?*

A neutron diffraction study of [Ta(CHCMe$_3$)(PMe$_3$)Cl$_3$]$_2$ has revealed an interaction between Ta and H of the adjacent (CH) group, which may be considered as a Ta–H–C linkage. The importance of such an interaction in facilitating removal of the α hydrogen (e.g., by an alkyl ligand to give an alkane) was noted (*200*).

REFERENCES

1. Wailes, P. C., Coutts, R. S. P., and Weigold, H. "Organometallic Chemistry of Titanium, Zirconium and Hafnium." Academic Press, New York, 1974.
2. Fischer, A. K., and Wilkinson, G. *J. Inorg. Nucl. Chem.* **2,** 149 (1956).
3. Clauss, K., and Bestian, H., *Justus Liebigs Ann. Chem.* **654,** 8 (1962).
4. Yokokawa, K., and Azuma, K., *Bull. Chem. Soc. Jpn.* **38,** 859 (1965).
5. Shikata, K., Yokokawa, K., Nakao, S., and Azuma, K., *J. Chem. Soc. Jpn., Ind. Chem. Sect.* **68,** 1248 (1965).
6. Watt, G. W., Baye, L. J., and Drummond, F. O., *J. Am. Chem. Soc.* **88,** 1138 (1966).
7. Watt, G. W., and Drummond, F. O., *J. Am. Chem. Soc.* **92,** 826 (1970).
8. Salzmann, J. J., and Mosimann, P., *Helv. Chim. Acta* **50,** 1831 (1967).
9. Calderazzo, F., Salzmann, J: J., and Mosimann, P., *Inorg. Chim. Acta* **1,** 65 (1967).
10. van Tamelen, E. E., Fechter, R. B., Schneller, S. W., Boche, G., Greeley, R. H., and Akermark, B., *J. Am. Chem. Soc.* **91,** 1551 (1969).
11. van Tamelen, E. E., Seeley, D., Schneller, S., Rudler, H., and Cretney, W., *J. Am. Chem. Soc.* **92,** 5251 (1970).
12. Klei, B., and Teuben, J. J., *J. Chem. Soc., Chem. Commun.* p. 659 (1978).
13. Fieselmann, B. F., and Stucky, G. D., *Inorg. Chem.* **17,** 2074 (1978).
14. Brintzinger, H., *J. Am. Chem. Soc.* **88,** 4305 (1966).
15. Brintzinger, H., *J. Am. Chem. Soc.* **89,** 6871 (1967).

16. Martin, H. A., and Jellinek, F., *J. Organomet. Chem.* **6**, 293 (1966).
17. Westerhoff, A., and de Liefe Meyer, H. J., *J. Organomet. Chem.* **139**, 71 (1977).
18. Henrici-Olivé, G., and Olivé, S., *J. Organomet. Chem.* **19**, 309 (1969).
19. Henrici-Olivé, G., and Olivé, S., *J. Organomet. Chem.* **23**, 155 (1970).
20. Kenworthy, J. G., Myatt, J., and Symons, M. C. R., *J. Chem. Soc. A* p. 3428 (1971).
21. Kenworthy, J. G., Myatt, J., and Symons, M. C. R., *J. Chem. Soc. A* p. 1020 (1971).
22. Bercaw, J. E., and Brintzinger, H. H., *J. Am. Chem. Soc.* **91**, 7301 (1969).
23. Bercaw, J. E., Marvich, R. H., Bell, L. G., and Brintzinger, H. H., *J. Am. Chem. Soc.* **94**, 1219 (1972).
24. Brintzinger, H. H., and Bercaw, J. E., *J. Am. Chem. Soc.* **92**, 6182 (1970).
25. Guggenberger, L. J., and Tebbe, F. N., *J. Am. Chem. Soc.* **98**, 4137 (1976).
26. Guggenberger, L. J., and Tebbe, F. N., *J. Am. Chem. Soc.* **95**, 7870 (1973).
27. Wailes, P. C., and Weigold, H., *J. Organomet. Chem.* **24**, 713 (1970).
28. Guggenberger, L. J., *Inorg. Chem.* **12**, 294 (1973).
29. Davison, A., and Wreford, S. S., *J. Am. Chem. Soc.* **96**, 3017 (1974).
30. Marvich, R. H., and Brintzinger, H. H., *J. Am. Chem. Soc.* **93**, 2046 (1971).
31. van Tamelen, E. E., Cretney, W., Klaentschi, N., and Miller, J. S., *J. Chem. Soc., Chem. Commun.* p. 481 (1972).
32. Rausch, M. D., Boon, W. H., and Mintz, E. A., *J. Organomet. Chem.* **160**, 81 (1978).
33. Peng, M., and Brubaker, C. H., *Inorg. Chim. Acta* **26**, 231 (1978).
34. El Murr, N., Chaloyard, A., and Tirouflet, J., *J. Chem. Soc., Chem. Commun.* p. 446 (1980).
35. Chaloyard, A., Dormond, A., Tirouflet, J., and El Murr, N., *J. Chem. Soc., Chem. Commun.* p. 214 (1980).
36. Lappert, M. F., Riley, P. E., and Yarrow, P. I. W., *J. Chem. Soc., Chem. Commun.* p. 305 (1979).
37. Volpin, M. E., and Shur, V. B., *Nature (London)* **209**, 1236 (1966).
38. van Tamelen, E. E., *Acc. Chem. Res.* **3**, 361 (1970).
39. Bercaw, J. E., *J. Am. Chem. Soc.* **96**, 5087 (1974), and references therein.
40. Henrici-Olivé, G., and Olivé, S., *Angew. Chem.* **81**, 679 (1969).
41. Pez, G., *J. Am. Chem. Soc.* **98**, 8072 (1976).
42. Pez, G., *J. Am. Chem. Soc.* **98**, 8079 (1976).
43. McDermott, J. X., Wilson, M. E., and Whitesides, G. M., *J. Am. Chem. Soc.* **98**, 6529 (1976).
44. Manriquez, J. M., McAlister, D. R., Rosenberg, E., Shiller, A. M., Williamson, K. L., Chan, S. I., and Bercaw, J. E., *J. Am. Chem. Soc.* **100**, 3078 (1978).
45. Vol'pin, M. E., Belyi, A. A., Shur, V. B., Lyakhovetsky, Yu. L., Kudryav, R. V., and Bubnov, N. N., *J. Organomet. Chem.* **27**, C5 (1971).
46. Razuvaev, G. A., Latyaeva, V. N., Vishinskaya, L. I., and Samarina, T. P., *J. Organomet. Chem.* **164**, 41 (1979).
47. Schmidt, F. A., Ryntina, N. M., Saraev, V. V., Gruznykh, V. A., and Makarov, V. A., *React. Kinet. Catal. Lett.* **5**, 101 (1976).
48. Shikata K., Nakao, K., and Azuma, K., *J. Chem. Soc. Jpn., Ind. Chem. Sect.* **68**, 1251 (1965); *Chem. Abstr.* **63**, 16470 (1965).
49. Tebbe, F. N., U.S. Patent 3,933,876 (1976).
50. Flamini, A., Cole-Hamilton, D. J., and Wilkinson, G., *J. Chem. Soc., Dalton Transl.* p. 454 (1978).
51. Marks, T. J., and Kolb, J. R., *Chem. Rev.* **77**, 263 (1977).
52. Mamaeva, G. I., Hargittai, I., and Spiridonov, V. P. *Inorg. Chim. Acta* **25**, L123 (1977).

53. Fachinetti, G., Floriani, C., Mellini, M., and Merlino, S., *J. Chem. Soc. Chem. Commun.* p. 300 (1976).
54. Volkov, V. V., and Myakishev, K. G., *Izv. Sib. Otd. Akad. Nauk SSSR, Ser. Khim. Nauk* **1**, 77 (1977); *Chem. Abstr.* **86**, 132712 (1977).
55. Tebbe, F. N., and Guggenberger, L. J., *J. Chem. Soc., Chem. Commun.* p. 227 (1973).
56. Natta, G., Mazzanti, G., Corradini, P., Giannini, U., and Cesca, S., *Atti Accad. Naz. Lincei, Cl. Sci. Fiz. Mat. Nat., Rend.* **26**, 150 (1959).
57. Natta, G., Pino, P., Mazzanti, G., and Giannini, U., *J. Am. Chem. Soc.* **79**, 2975 (1957).
58. Corradini, P., and Sirigu, A., *Inorg. Chem.* **6**, 601 (1967).
59. Bulychev, B. M., Tokareva, S. E., Soloveichik, G. L., and Evdokimova, E. V., *J. Organomet. Chem.* **179**, 263 (1979).
60. Dozzi, G., Cucinella, S., and Mazzei, A., *J. Organomet. Chem.* **164**, 1 (1979).
61. Sato, F., Sato, S., Kodama, H., and Sato, M., *J. Organomet. Chem.* **142**, 71 (1977).
62. Ashby, E. C., and Lin, J. J., *Tetrahedron Lett.* **51**, 4481 (1977).
63. Brunet, J. J., Vanderesse, R., and Caubere, P., *J. Organomet. Chem.* **157**, 125 (1978).
64. McMurry, J. E., *Acc. Chem. Res.* **7**, 281 (1974).
65. Mukaiyama, T., *Angew. Chem., Int. Ed. Engl.* **17**, 817 (1977).
66. Isagawa, K., Tatsumi, K., Kosugi, H., and Otsuji, Y., *J. Chem. Lett.* **9**, 1017 (1977).
67. Ashby, E. C., and Noding, S. A., *J. Org. Chem.* **45**, 1035 (1980).
68. Ashby, E. C., and Noding, S. A., *J. Org. Chem.* **45**, 1041 (1980).
69. Ivin, K. J., Rooney, J. J., Stewart, C. D., Green, M. L. H., and Mahtab, R., *J. Chem. Soc., Chem. Commun.* p. 604 (1978).
70. Evitt, E. R., and Bergman, R. G., *J. Am. Chem. Soc.* **101**, 3973 (1979).
71. Pez, G. P., *J. Chem. Soc., Chem. Commun.* p. 560 (1977).
72. Bonds, W. D., Brubaker, C. H., Chandrasekaran, E. S., Gibbons, C., Grubbs, R. H., and Kroll, L. C., *J. Am. Chem. Soc.* **97**, 2128 (1975).
73. James, B. D., Nanda, R. K., and Wallbridge, M. G. H., *Inorg. Chem.* **6**, 1979 (1967).
74. Nöth, H., and Hartwimmer, R., *Chem. Ber.* **93**, 2238 (1960).
75. Labinger, J. A., and Komadina, K. H., *J. Organomet. Chem.* **155**, C25 (1978).
76. Stevenson, P. E., and Lipscomb, W. N., *J. Chem. Phys.* **50**, 3306 (1969).
77. Chaillet, M., Arriau, J., Leclerc, D., Marey, T., and Tirouflet, J., *J. Organomet. Chem.* **117**, 27 (1976).
78. Lorenz, D. R., and Wasson, J. R., *J. Inorg. Nucl. Chem.* **37**, 2265 (1975).
79. Lauher, J. W., and Hoffmann, R., *J. Am. Chem. Soc.* **98**, 1729 (1976).
80. Schwartz, J., and Labinger, J. A., *Angew. Chem., Int. Ed. Engl.* **15**, 333 (1976).
81. Bercaw, J. E., "Transition Metal Hydrides" (R. Bau, ed.), p. 136. American Chemical Society, Washington, D.C., 1978.
82. Wolczanski, P. T., and Bercaw, J. E., *Acc. Chem. Res.* **13**, 121 (1980).
83. Datta, S., Wreford, S. S., Beatty, R. P., and NcNeese, T. J., *J. Am. Chem. Soc.* **101**, 1053 (1979).
84. Beatty, R. P., Datta, S., and Wreford, S. S., *Inorg. Chem.* **18**, 3139 (1979).
85. Gell, K. I., and Schwartz, J., *J. Am. Chem. Soc.* **100**, 3246 (1978).
86. Jeffrey, J., Lappert, M. F., Luong-Thi, N. T., Atwood, J. L., and Hunter, W. E., *J. Chem. Soc., Chem. Commun.* p. 1081 (1978).
87. Kautzner, B., Wailes, P. C., and Weigold, H., *J. Chem. Soc., Chem. Commun.* p. 1105 (1969).
88. Wailes, P. C., and Weigold, H., *J. Organomet. Chem.* **24**, 405 (1970).

89. Evans, J., Okrasinski, S. J., Pribula, A. J., and Norton, J. R., *J. Am. Chem. Soc.* **98**, 4000 (1976).
90. Manriquez, J. M., McAlister, D. R., Sanner, R. D., and Bercaw, J. E., *J. Am. Chem. Soc.* **100**, 2716 (1978).
91. McAlister, D. R., Erwin, D. K., and Bercaw, J. E., *J. Am. Chem. Soc.* **100**, 5966 (1978).
92. Britzinger, H. H., *J. Organomet. Chem.* **171**, 337 (1979).
93. Gell, K. I., and Schwartz, J., *J. Chem. Soc., Chem. Commun.* p. 244 (1979).
94. Hart, D. W., and Schwartz, J., *J. Am. Chem. Soc.* **96**, 8115 (1974).
95. Wailes, P. C., and Weigold, H., *Inorg. Synth.* **19**, 223 (1979).
96. Wailes, P. C., Weigold, H., and Bell, A. P., *J. Organomet. Chem.* **27**, 373 (1971).
97. Carr, D. B., and Schwartz, J., *J. Am. Chem. Soc.* **101**, 3521 (1979).
98. Sorrell, T. N., *Tetrahedron Lett.* **50**, 4985 (1978).
99. Manriquez, J. M., McAlister, D. R., Sanner, R. D., and Bercaw, J. E., *J. Am. Chem. Soc.* **98**, 6733 (1976).
100. Fachinetti, G., Floriani, C., Roselli, A., and Pucci, S., *J. Chem. Soc., Chem. Commun.* p. 269 (1978).
101. Wolczanski, P. T., and Bercaw, J. E., *J. Am. Chem. Soc.* **101**, 6450 (1979).
102. Wolczanski, P. T., Threlkel, R. S., and Bercaw, J. E., *J. Am. Chem. Soc.* **101**, 218 (1979).
103. Pez, G. P., Putnik, C. F., Suib, S. L., and Stucky, G. D., *J. Am. Chem. Soc.* **101**, 6933 (1979).
104. Remick, R. J., Asunta, T. A., and Skell, P. S., *J. Am. Chem. Soc.* **101**, 1321 (1979).
105. Samuel, E., Maillard, P., and Gianotti, G., *J. Organomet. Chem.* **142**, 289 (1977).
106. Wailes, P. C., Weigold, H., and Bell, A. P., *J. Organomet. Chem.* **43**, C32 (1972).
107. Couturier, S., and Gautheron, B., *J. Organomet. Chem.* **157**, C61 (1978).
108. Erker, G., and Rosenfeld, F., *Angew. Chem.* **90**, 648 (1978).
109. Weigold, H., Bell, A. P., and Willing, R. I., *J. Organomet. Chem.* **73**, C23 (1974).
110. Kablitz, H., and Wilke, G., *J. Organomet. Chem.* **51**, 241 (1973).
111. Tebbe, F. N., and Parshall, G. W., *J. Am. Chem. Soc.* **93**, 3793 (1971).
112. Benfield, F. W. S., and Green, M. L. H., *J. Chem. Soc., Dalton Trans.* p. 1324 (1974).
113. Nakamoto, K., "Infra-red Spectra of Inorganic and Co-ordination Compounds." Wiley, New York, 1970.
114. Chum, P. W., and Wilson, S. E., *Tetrahedron Lett.* p. 15 (1976).
115. Sato, F., Sato, S., and Sato, M., *J. Organomet. Chem.* **122**, C25 (1976).
116. Wailes, P. C., Weigold, H., and Bell, A. P., *J. Organomet. Chem.* **43**, C29 (1972).
117. Shoer, L. I., Gell, K. I., and Schwartz, J., *J. Organomet. Chem.* **136**, C19 (1977).
118. Shoer, L. I., and Schwartz, J., *J. Am. Chem. Soc.* **99**, 5831 (1977).
119. Tolstikov, G. A., Miftakhov, M. S., and Valeev, F. A., *Izv. Akad. Nauk SSSR, Ser. Khim.* p. 2576 (1979); *Chem. Abstr.* **92**, 146863 (1980).
120. Hieber, W., Peterhans, E., and Winter, E., *Chem. Ber.* **94**, 2572 (1961).
121. Hieber, W., Winter, E., and Schubert, E., *Chem. Ber.* **95**, 3070 (1962).
122. Kruck, T., and Hempel, H. U., *Angew. Chem., Int. Ed. Engl.* **13**, 201 (1974).
123. Puttfarcken, U., and Rehder, D., *J. Organomet. Chem.* **157**, 321 (1978).
124. Puttfarcken, U., and Rehder, D., *J. Organomet. Chem.* **185**, 219 (1980).
125. Davison, A., and Ellis, J. E., *J. Organomet. Chem.* **36**, 131 (1972).
126. Ellis, J. E., and Faltynek, R. A., *J. Organomet. Chem.* **93**, 205 (1975).
127. Franke, U., and Weiss, E., *J. Organomet. Chem.* **152**, C19 (1978).
128. Davison, A., and Reger, D. L., *J. Organomet. Chem.* **23**, 491 (1970).
129. Kinney, R. J., Jones, W. D., and Bergman, R. G., *J. Am. Chem. Soc.* **100**, 635 (1978).

130. Kinney, R. J., Jones, W. D., and Bergman, R. G., *J. Am. Chem. Soc.* **100**, 7902 (1978).
131. Murr, N. E. L., Moise, C., Riveccie, M., and Tirouflet, J., *Inorg. Chim. Acta* **32**, 189 (1979).
132. Evans, A. G., Evans, J. C., and Mortimer, J., *J. Am. Chem. Soc.* **101**, 3204 (1979).
133. Evans, A. G., Evans, J. C., and Moon, E. H., *J. Chem. Soc., Dalton Trans.* p. 2390 (1974).
134. Fischer, E. O., and Schneider, R. J. J., *Chem. Ber.* **103**, 3684 (1970).
135. Ellis, J. E., Faltynek, R. A., and Hentges, S. G., *J. Am. Chem. Soc.* **99**, 626 (1977).
136. Marks, T. J., and Kennelly, W. J., *J. Am. Chem. Soc.* **97**, 1439 (1975).
137. Marks, T. J., and Kennelly, W. J., *Chem. Rev.* **77**, 265 (1977).
138. Nöth, H., *Angew. Chem.* **73**, 371 (1961).
139. Nöth, H., private communication (1980).
140. Schrock, R. R., *J. Organomet. Chem.* **121**, 373 (1976).
141. Byrne, J. W., Blaser, H. U., and Osborn, J. A., *J. Am. Chem. Soc.* **97**, 3871 (1975).
142. Labinger, J. A., *in* "Transition Metal Hydrides" (R. Bau, ed.) p. 149. American Chemical Society, Washington, D.C. 1978.
143. Labinger, J. A., and Wong, K. S., *J. Organomet. Chem.* **170**, 373 (1979).
144. Kirillova, N. I., Gusev, A. I., and Struchkov, Yu. T., *Zh. Strukt. Khim.* **13**, 473 (1972). *Chem. Abstr.* **77**, 119409v (1972).
145. Tebbe, F. N., and Parshall, G. W., *J. Am. Chem. Soc.* **93**, 3793 (1971).
146. Klazinga, A. H., and Teuben, J. H., *J. Organomet. Chem.* **157**, 413 (1978).
147. Labinger, J. A., and Schwartz, J., *J. Am. Chem. Soc.* **97**, 1596 (1975).
148. Lucas, C. R., and Green, M. L. H., *J. Chem. Soc., Chem. Commun.* p. 1005 (1972).
149. Verkade, C. P., Westerhof, A., and Liefde Meijer, H. J., *J. Organomet. Chem.* **154**, 317 (1978).
150. Otto, E. E. H., and Brintzinger, H. H., *J. Organomet. Chem.* **170**, 209 (1979).
151. Tebbe, F. N., *J. Am. Chem. Soc.* **95**, 5412 (1973); also private communication in Ref. *142*.
152. Otto, E. E. H., and Brintzinger, H. H., *J. Organomet. Chem.* **148**, 29 (1978).
153. Guggenberger, L. J., Meakin, P., and Tebbe, F. N., *J. Am. Chem. Soc.* **96**, 5420 (1974).
154. Guggenberger, L. J., *Inorg. Chem.* **12**, 294 (1973).
155. Guggenberger, L. J., and Tebbe, F. N., *J. Am. Chem. Soc.* **93**, 5924 (1971).
156. Nesmeyanov, A. N., Lemenovskii, D. A., Fedin, V. P., and Perevalova, E. G., *Dokl. Akad. Nauk SSSR.* **245**, 609 (1979); *Chem. Abstr.* **91**, 20652w (1979).
157. Struchkov, Yu. T., *Pure Appl. Chem.* **52**, 741 (1980).
158. Bunker, M. J., Green, M. L. H., Couldwell, C., and Prout, K., *J. Organomet. Chem.* **192**, C6 (1980).
159. Kirillova, N. I., Gusev, A. I., Pasynskii, A. A., and Struchkov, Yu. T., *Zh. Strukt. Khim.* **15**, 288 (1974); *Chem. Abstr.* **81**, 17847b (1974).
160. Elson, I. H., Kochi, J. K., Klabunde, U., Manzer, L. F., Parshall, G., and Tebbe, F. N., *J. Am. Chem. Soc.* **96**, 7374 (1974).
161. Elson, I. H., and Kochi, J. K., *J. Am. Chem. Soc.* **97**, 1262 (1975).
162. Wilson, R. D., Koetzle, T. F., Hart, D. W., Kvick, A., Tipton, D. L., and Bau, R., *J. Am. Chem. Soc.* **99**, 1775 (1977).
163. Labinger, J. A., Wong, K. S., and Scheidt, W. R., *J. Am. Chem. Soc.* **100**, 3254 (1978).
164. Labinger, J. A., Wong, K. S., and Scheidt, W. R., *Inorg. Chem.* **18**, 136 (1979).

165. Porai-Koshits, M. A., Antsyshkina, A. S., Pasynskii, A. A., Sadikov, G. G., Skripkin, Yu. V., and Ostrikova, V. N., *Inorg. Chim. Acta* **34**, L285; *Koord. Khim.* **5**, 1103 (1979); *Chem. Abstr.* **91**, 149659h (1979).
166. Lucas, C. R., *Inorg. Synth.* **16**, 107 (1976).
167. Kirillova, N. I., Gusev, A. I., and Struchkov, Yu. T., *Zh. Strukt. Khim.* **15**, 718 (1974); *Chem. Abstr.* **81**, 128044g (1974).
168. King, R. B., *Z. Naturforsch., B: Anorg. Chem., Org. Chem., Biochem., Biophys., Biol.* **18B**, 157 (1963).
169. Imoto, H., and Corbett, J. D., *Inorg. Chem.* **19**, 1241 (1980), and references therein.
170. Westerhof, A., and Liefde Meijer, H. J., *J. Organomet. Chem.* **139**, 71 (1977).
171. Tebbe, F. N., *J. Am. Chem. Soc.* **95**, 5823 (1973).
172. Datta, S., and Wreford, S. S., *Inorg. Chem.* **16**, 1134 (1977).
173. Meakin, P., Guggenberger, L. J., Tebbe, F. N., and Jesson, J. P., *Inorg. Chem.* **13**, 1025 (1974).
174. Domaille, P. J., Foxman, B. M., McNeese, T. J., and Wreford, S. S., *J. Am. Chem. Soc.* **102**, 4114 (1980).
175. Albright, J. O., Datta, S., Dezube, B., Kouba, J. K., Marynick, D. S., Wreford, S. S., and Foxman, B. M., *J. Am. Chem. Soc.* **101**, 611 (1979); see also Albright, J. O., Brown, L. D., Datta, S., Kouba, J. K., Wreford, S. S., and Foxman, B. M., *ibid.* **99**, 5518 (1977).
176. Klazinga, A. H., and Teuben, J. H., *J. Organomet. Chem.* **165**, 31 (1979).
177. Sharp, P. R., and Schrock, R. R., *J. Organomet. Chem.* **171**, 43 (1979).
178. Labinger, J. A., Schwartz, J., and Townsend, J. M., *J. Am. Chem. Soc.* **96**, 4009 (1974).
179. Barefield, E. K., Parshall, G. W., and Tebbe, F. N., *J. Am. Chem. Soc.* **92**, 5234 (1970).
180. Belmonte, P., Schrock, R. R., Churchill, M. R., and Youngs, W. J., *J. Am. Chem. Soc.* **102**, 2858 (1980).
181. Green, M. L. H., McCleverty, J. A., Pratt, L., and Wilkinson, G., *J. Chem. Soc.* p. 4854 (1961).
182. Green, M. L. H., and Moreau, J. J. E., *J. Organomet. Chem.* **161**, C25 (1978).
183. Holliday, A. K., Makin, P. H., and Puddephatt, R. J., *J. Chem. Soc., Dalton Trans.* p. 228 (1979).
184. Wood, C. D., McLain, S. J., and Schrock, R. R., *J. Am. Chem. Soc.* **101**, 3210 (1979).
185. McLain, S. J., and Schrock, R. R., *J. Am. Chem. Soc.* **100**, 1315 (1978).
186. Schrock, R. R., and Fellmann, J. D., *J. Am. Chem. Soc.* **100**, 3359 (1978).
187. Schrock, R. R., *J. Organomet. Chem.* **122**, 209 (1976).
188. Harris, D. H., Keppie, S. A., and Lappert, M. F., *J. Chem. Soc., Dalton Trans.* p. 1653 (1973).
189. Green, J. C., Jackson, S. E., and Higginson, B., *J. Chem. Soc., Dalton Trans.* p. 403 (1975).
190. Macašek, F., Mikulaj, V., and Drienovsky, P., *Collect. Czech. Chem. Commun.* **35**, 993 (1970).
191. Klabunde, U., and Parshall, G. W., *J. Am. Chem. Soc.* **94**, 9081 (1972).
192. McNeese, T. J., Wreford, S. S., and Foxman, B. M., *J. Chem. Soc., Chem. Commun.* p. 500 (1978).
193. Golovanova, A. I., Kost, M. E., and Mikheeva, V. I., *Izv. Akad. Nauk SSSR. Ser. Khim.* 1448 (1973); *Chem. Abstr.* **79**, 132477b (1973).
194. Wood, C. D., and Schrock, R. R., *J. Am. Chem. Soc.* **101**, 5421 (1979).

195. Norton, J. R., *Acc. Chem. Res.* **12**, 139 (1979).

196. Bond, A. M., Bixler, J. W., Mocellin, E., Datta, S., James, E. J., and Wreford, S. S., *Inorg. Chem.* **19**, 1760 (1980).

197. Klingler, R. J., Huffman, J. C., and Kochi, J. K., *J. Am. Chem. Soc.* **102**, 208 (1980).

198. Klazinga, A. H., and Teuben, J. H., *J. Organomet. Chem.* **192**, 75 (1980).

199. Sattelberger, A. P., Wilson, R. B., and Huffman, J. C., *J. Am. Chem. Soc.* **102**, 7113 (1980).

200. Schultz, A. J., Williams, J. M., Schrock, R. R., Rupprecht, G. A., and Fellmann, J. D., *J. Am. Chem. Soc.* **101**, 1593 (1979).

INDEX

A

Aluminum borates, 207, 222
  structure, 190, 198
Ammonia, as solvent for metals, 135–185
  electronic properties of dilute solutions,
    142–148
    concentration dependence, 142–144,
      168
    electrical conductivity, 142–143
    electron–cation interactions, 142–
      145
    electron–electron interaction, 145–
      148
    electron spin pairing, 146
    ESR studies, 146–147
    magnetic susceptibility, 143, 145
  for lanthanides, 148
  nonmetal-to-metal transition, 168–176
Ammonium borates, 192, 216

B

Band structure, of cesium auride, 240–
  241
Barium borates, 191, 220
Beryllium borates, 216–217
Borate complexes, 187–237, see also specific metals
  in aqueous solution, 200–210
  hydrolysis, 200
  B–O bond lengths, 198–199
  and Hückel theory, 199
  formation constants and pH, 204
  hexaborates, 195
  metaborates, 211–213, 215, 219
  mobilities, 209
  nomenclature, 188
  nonaborate, 195
  pentaborates, 197–198, 211, 213–216
  polyborate equilibria in water, 201–
    207
  of rare earths, 196
  rate of polyborate formation, 206–207
  solubility, 207–209
    in alkalis, 208
    in aqueous metal salt solutions, 208–
      209
    in seawater, 207

  in strong acid, 208
  structures, 188–200
    anhydrous compounds, 195–198
    basic units, 189
    and decomposition, 209
    hydrated compounds, 189–195
    infrared spectra, 200–201
    NMR studies, 299–200
  synthesis, 209–225
    cation effects, 209
    and pH, 210
    and temperature, 210
  tetraborates, 195, 197, 211, 213–216
  triborates, 194–196
  of two metals, 221–222
Boric acid
  acidity, 200–201
    effect of neutral salts, 202
  in aqueous solution, 201
  in seawater, 208

C

Cadmium borates, 207, 221
  structure, 197
Calcium borates, 218–219
  stability ranges, 209–210
  structures, 190–195, 197–199
  in water, 218
Carbonylate anions, reaction with silicon
  halides, 3–10
Carborane complexes, of gold, 252
Cesium
  in amine solvents, 153–159
    ESR studies, 153–157
    magnetic data, 156–157
    optical spectra, 157–159
  anion, 165–166
  auride, 240–243
    band structure, 240–241
    crystal structure, 240, 242
    electrical conductivity, 240–241
    dissociation energy, 242
    phase change, 240
  borates, 215–216
    structures, 194, 197
  hexafluoroaurate, 259
Chemical vapor deposition, and silicide
  synthesis, 109–111

Chromium compounds
  borate, 223
  germylenes, 117
  stannyl carbene, 77
  stannylene, 117
Chromium silyl complexes
  anionic pentacarbonyl, 11–12
  tricarbonylcyclopentadienyl deriva-
    tives, 4, 7, 58
  cleavage reactions, 40–41, 43
  Lewis base adducts, 64
  ligand exchange at silicon, 60
  mass spectra, 99
Cluster complexes
  of gold, 243–249, 261–262
  silyl derivatives, 111–116
Cobalt compounds
  borates, 207, 224
    structures, 190
  carbonyl thallium derivative, 11
  cyclopentadienyl hydride in Ziegler–
    Natta catalysis, 282–283
  germyl clusters, 115–116
  octacarbonyl, aluminum tribromide
    adduct, 77
  silicides, 110
  stannyl complexes, 80, 116
Cobalt silyl complexes
  clusters, 17–18, 20, 111–112, 114–116
  cyclopentadienyl derivative, 23
  phosphine hydride, 32
    reductive elimination, 81
  structures, 84, 88–89
  tetracarbonyl derivatives, 6–9, 11, 13–
    14, 17–19, 51
  carbonyl spectra, 66
  carbonyl substitution, 70, 72–75
  cleavage reactions, 41, 43–44
  disproportionation, 80
  Lewis base adducts, 63, 65–67
  ligand exchange at silicon, 61, 63
  mass spectra, 99–100
  metal exchange, 57
  NMR spectra and fluxionality, 70,
    94–96
  optical activity, 102–105
  photoelectron spectra, 106–107
  pyrolysis, 107–108
  silicide formation, 110

silicon migration, 78–79
siloxy complex formation, 10
Copper complexes
  borates, 191
  hexanuclear cluster, 245
  silyls, 12, 47, 55

D

Dinitrogen complexes, of titanium
  cyclopentadienyls, 269, 272, 275–277
  phenoxides, 279

E

Electrons, solvation in dilute solution,
    138–142
  bimolecular reactions, 139
Electron spin resonance (ESR) spectra
  of dicyclopentadienyl complexes
    niobium hydrides, 320–321
    photolysis of zirconium dimethyl,
      295
    tantalum dihydride, 332
    of vanadium dichloride and alkyl-
      aluminum chloride, 309, 312–
      313
  of dilute metal–ammonia solutions,
      146–147
  of gold (II) complexes, 251–254
  of lithium in frozen methylamine, 177
  of paramagnetic states in amines and
      ethers, 152–157, 159, 163
  of phenoxytitanium complex, 279
  of rubidium in hexamethylphosphor-
      amide, 161–165
  of solvated electron, 152
  of titanium dicyclopentadienyls
    dichloride and LiAlH$_4$, 281–282
    dihydride anion, 269
    hydride-bridged dimeric anion, 270
    hydride phosphine complexes, 278
    with main group hydrides, 270
ESCA, of gold complexes
  binding energy and oxidation state,
      241–242
  binuclear compounds, 255, 257
  cesium auride, 241

Ethylenediamine as metal solvent
and metal anions, 165–166
optical spectra of alkali metal solutions, 158–159
and solvated electron, 144

F

Fulvalene complexes
as alkene disproportionation catalysts, 283
cobalt dication, 272
titanium hydrides, 271–272, 288

G

Germanium transition-metal complexes, 29, 56, 75–76
clusters, 111, 114
germylenes, 117, 120
Gold, electron affinity, 240
Gold complexes, 239–266
aurides, 240–243
in liquid ammonia, 166, 242–243
clusters
analogy with boranes, 243
with eleven gold atoms, 247–249, 262
hexanuclear compounds, 244–245, 261
MO calculations, 245, 247, 249
Mössbauer spectra, 244, 247, 249
nonanuclear compounds, 246–248, 262
octanuclear compounds, 245–246, 261–262
pentanuclear compounds, 243–244
with thirteen gold atoms, 262
divalent compounds, 251–257
carbamate derivatives, 251, 253–254, 256
carborane, 252
dithiolates, 252–254
with gold–gold bonds, 254–257
phthalocyanines, 252
as reaction intermediates, 251
tetrachloro anion, 251
ylides, 255–257
mixed-valence compounds, 249–251

chlorides, 250–251
pentavalent derivatives, 257–261
hexafluoroaurates, 257–260
pentafluoride, 257, 260–261
silyl, 12
trifluoride and xenon difluoride, 258

H

Hafnium complexes
hydrides, 303–305
dicyclopentadienyls, 304–305
ditertiaryphosphine trihydride, 304–305
silyl, 12
Hexamethylphosphoramide, as metal solvent
ESR spectra of frozen rubidium solution, 161–165
and solvated electrons, 139–141
Hydride complexes, 267–340, see also specific metals
Hydrosilation, of alkynes, 90, 106

I

Infrared spectra
of borates, 200–201, 203, 205–206, 211
of silyl complexes, 91–93
Iridium, silyl complexes
bis(phosphine), 13, 23–24, 28, 43
cluster, 6, 120
trifluorophosphine derivative, 19–21
Iron compounds
borates, 190, 207, 224
germyl clusters, 114
silicides, 110
silylenes, 37–38, 117–118
stannyl clusters, 114
Iron silyl complexes
anionic tetracarbonyl, 37
binuclear carbonyls, 3, 5, 16, 33, 44–45, 116
bis(ditertiaryphosphine), 32
clusters, 112–114
dicarbonylcyclopentadienyl, 2, 5, 8–9, 12, 16
carbonyl substitution, 71

Iron silyl complexes, dicarbonylcyclopentadienyl (*cont'd*)
  cleavage reactions, 40–44, 51–53,
    106
  Lewis base adducts, 63, 65–67
  ligand exchange at silicon, 60–62
  mass spectra, 99–100
  optical activity, 101–102
  silicon migration, 78
  ferrasilacyclopentane, 9, 22, 34
  fluxionality, 93, 97
  infrared spectra, 91–92
  Mössbauer spectra, 107
  structures, 83–84, 87–88, 120
  tetracarbonyl bis(silyl), 13, 15–16, 22–
    23, 43, 93
    carbonyl substitution, 70, 72
    cis–trans equilibria, 96
    cleavage reactions, 43–44, 50, 54,
      118
    Lewis base adducts, 64, 68–69
    metal exchange reactions, 57
    NMR spectra, 93–95
    pyrolysis, 107
    silicide formation, 110
  tetracarbonylsilyl hydride
    reaction with isoprene, 75
    reductive elimination, 81

L

Lead complexes
  borates, 198, 223
  cobalt carbene, 76
Lithium
  borates, 211
    structures, 190–192, 197
  in frozen methylamine, 177
    ESR study, 177
    magnetic susceptibility, 180
  in liquid ammonia
    effect of temperature, 175–176
    magnetic susceptibility, 179–180
    phase diagram, 176
    tetraammine complex, 176

M

Magnesium borates, 217–218
  structures, 190–195
  in water, 217

Manganese compounds
  allyl, 7
  borates, 223–224
    structures, 190–191, 194
  germylene, 120
  germyls
    carbene, 76
    clusters, 111
  silicides, 110
  stannyl clusters, 111
  thallium carbonyls, 11
Manganese silyl complexes
  clusters, 112
  dicarbonylcyclopentadienyls, 22, 34,
    119
    cleavage reactions, 43–44, 48–49, 53
    hydrogen substitution, 73
    optical activity, 101–102, 104
    reductive elimination, 8
    stereochemistry, 29
  infrared spectra, 91–92
  pentacarbonyl derivatives, 4, 5, 7, 12–
    13, 15, 33, 118
    carbonyl substitution, 71, 75
    cleavage reactions, 40, 42–43, 48,
      50–51
    kinetics of cleavage, 106
    Lewis base adducts, 63–64, 66–69
    ligand exchange, 58–60
    mass spectra, 98–100
    metal exchange, 57
    photoelectron spectra, 106–107
    pyrolysis, 107–109
    silicide formation, 109–111
  silacyclobutane, 10
  siloxane formation, 10
  structures, 82–83, 86–87, 90–91
Mercury compounds
  bis(silyls), 11, 13
  borates, 221
Metal solutions in nonaqueous solvents,
    135–185, *see also* specific solvents
  concentrated solutions, 168–178
    expanded-metal compounds, 176–178
    liquid–liquid phase separation, 174–
      176
  diamagnetic states, 165–168
    clusters, 167
    optical spectra, 165–166
    electrolytic nature, 138

nonmetal-to-metal transitions, 168–178
  Herzfeld theory, 169–170
  inhomogeneity of transition, 173–174
  Mott transition, 170–172
  paramagnetic states, 148–161, 165–169
    continuum model, 159–161
    ESR studies, 152–157
    multistate model, 159
    optical spectra, 157–159
  and solvated electrons, 138–142
  quantitative theory, 138–142
Methylamine, as solvent for metals
  lithium complex, 177, 180
  and solvated electrons, 139–141
Molecular orbital (MO) calculations
  dicyclopentadienyl metal dihydrides, 318–322
  titanium, 284
  zirconium, 290–291
  gold clusters, 245, 247, 249
Molybdenum silyl complexes
  pentacarbonyl, 120
    anion, 11–12
  tricarbonylcyclopentadienyl, 4, 11, 13, 22, 33
    cleavage reactions, 43, 51–52
    Lewis base adducts, 64
    ligand exchange at silicon, 60
    mass spectra, 99
    tropylium derivative, 33
Mössbauer ^{197}Au spectra
  of binuclear gold(II) compounds, 255, 257
  of cesium auride, 241
  of gold clusters, 244, 247, 249
  of gold pentafluoride, 261

N

Nickel complexes
  borates, 207, 224–225
    structure, 190
  silyls
    anions, 37
    binuclear bis(phosphine), 24
    bipyridyl complex, 32, 46, 54
    cyclopentadienyls, 32–33, 43
    dicarbonyl, 24, 46

Niobium hydrides, 314–325
  aluminum alkyl derivatives, 317, 324
  borohydride complex, 317, 324
  bridged to transition metals, 317, 323
  σ,η^5-C_5H_4 derivative, 319–320
  dicyclopentadienyls
    carbonyl hydride, 318, 320
    dihydride, 316, 320–321
    monohydrides, 315–316, 318–319, 322
    trihydride, 298, 317–323
  ditertiaryphosphine derivatives
    bis(ethylene), 315
    dicarbonyls, 315, 320
    pentahydride, 315, 317, 322
  hexanuclear cluster, 317, 325
Nitrogen complexes, of zirconium, 296
Nuclear magnetic resonance (NMR) spectra
  ^{11}B NMR spectra, of borates, 199–200, 203
  ^{13}C NMR spectra, of transition-metal silyls, 93–96
    fluxional cobalt complex, 70
  ^{133}Cs NMR spectra, of cesium auride, 247
  of Group IA anions, 166
    sodium, 178–179
  ^{1}H NMR spectra
    of borates, 199
    of fluxional Group IVB-bridged complexes, 96–97
    and ligand exchange of manganese silyl, 59
    of metal hydrides, 298, 302
    niobium, 318
    of metals in ammonia, 142–143
  ^{29}Si NMR spectra, of silyl complexes, 93–94
  ^{51}V NMR spectra, of vanadium carbonyl hydrides, 306–308, 311
Nucleophilicity, of metal carbonyl anions, 3

O

Osmium silyl complexes, 17, 20
  cis–trans isomerization, 96
  clusters, 17, 111, 114, 119

Oxidative addition
  of alkanes to zirconium atoms, 295
  of halogens to gold(II), 256–257
  to platinum silyls, 76
  of silanes, 14, 21–30, 119
    adduct stability, 28
    kinetics of addition to Ir(I), 30, 106
    optical activity, 101

P

Palladium silyl complexes, 25, 35
pH effects, and borates, 203, 210
Phosphine methylenes
  reaction with metal hydrides
    of chromium, 7
    of silicon, 52
  silicon adduct, 63
Phthalocyanine complexes, 252
Platinum silyl complexes
  in alkyne hydrosilation, 106
  bis(phosphine) derivatives, 12, 13, 21,
    25–27, 29, 31–32, 40
  binuclear compounds, 25, 27
  cleavage reactions, 41–44, 50, 55, 68
  ligand exchange at silicon, 61, 63, 65
  optical activity, 103–105
  clusters, 112, 116
  platinadisilacyclopentene, 54–55
  silicon tetrafluoride adduct, 35–36
  structures, 84, 89–90
Potassium
  in amine solvents
    ESR spectra, 153–155
    optical spectra, 157–159, 165–166
  aurides, 242–243
  borates, 213–215
    solubility, 213–214
    structures, 192–194, 198–199
  in liquid ammonia, 144

R

Raman spectra
  of borates, 200–201, 205–206
  of hexafluoroaurate salts, 258–260
Rhenium silyl complexes, 15
  clusters, 112
  dicarbonylcyclopentadienyls, 21–22
  infrared spectra, 91–92

pentacarbonyl derivatives
  cleavage reactions, 42, 49
  Lewis base adducts, 63–64, 66–67
  mass spectra, 98–100
  photoelectron spectra, 106
  pyrolysis, 107
  from sila-acyl, 38
  structures, 82–83, 85–87
Rhodium silyl complexes, 32
  cyclopentadienyls, 19, 23
  of trifluorophosphine, 19–21
  structures, 84
Rubidium
  anion, 165–166
    electron photodetachment, 166
    optical spectra, 166
  auride, 241
  borates, 215
    structures, 194
  in nonaqueous solvents
    ESR studies, 153–155, 161–165
    paramagnetic states, 150–152
Ruthenium silyl complexes, 17
  clusters, 47, 88, 114
    anion, 119
  phosphine hydrides, 23, 43
  hydrogen substitution, 73, 75
  reductive elimination, 81
  structures, 84, 87–88
  tetracarbonylbis(silyl)
    carbonyl exchange, 74
    carbonyl substitution, 70, 72
    cis–trans equilibration, 96
    cleavage reactions, 45–46
    and cycloalkenes, 55–56
    fluxionality, 93–95

S

Silatrane complex, 89
Silicides, from molecular silyl complexes,
  107–111
Silicon–transition-metal complexes, 1–
  133, see also specific metals
  cleavage of silicon–metal bond, 39–57
    alkenes, 44–46, 54–56
    alkynes, 44–47, 54–55
    amines, 43, 50
    benzaldehyde, 42, 49–50
    dihydrogen, 40, 48

Grignards, 44, 53
Group IVB compounds, 43, 52
halogen compounds, 40–41, 48–49
lithium alkyls, 44, 53
lithium aluminum hydride, 43–44, 52
mercury reagents, 53
metal carbonyls, 56–57
nitriles, 45, 54
phosphorus donors, 51–52
sodium borohydride, 43, 53
thiophenol, 42, 50
water and alcohols, 41–42, 49
clusters, 111–116, 119–120
disproportionation, 79–80
base catalysis, 80
at silicon, 79–80
electronegativity of metal group, 98
in hydrosilation catalysis, 90, 106
kinetics, 106
Lewis base adducts at silicon, 63–70
amines, 63–70
infrared spectra, 66–68
NMR spectra, 67–68
phosphines, 63–65, 67
reactivity, 69
as salts of carbonyl anions, 66–69
ligand exchange at silicon, 58–63
of alkenyl group, 62
of alkoxy group, 62
of amino group, 62
of halogen, 58–59, 62
of hydrogen, 58
mass spectra, 98–100
appearance potentials, 100
bond dissociation energies, 100
migration, 77–79
effect of polar solvents, 77–79
via thermolysis, 79
Mössbauer spectra, 107
NMR spectra, 93–98
fluxionality, 93–97
isomerization, 93–96
stereochemistry, 93–98
optical activity, 100–105
and electrophilic attack, 101
and nucleophilic attack, 104–105
photoelectron spectra, 106–107
pyrolysis, 107–111
chemical vapor deposition, 109–111

product distribution, 108–109
silicide formation, 109–111
reaction at carbonyl ligand, 76–77
of organolithium reagents, 76–77
reductive elimination, 80–81
σ-donor ability of SiMe$_3$ group, 87
sila-acyl derivative, 38, 49
stereochemistry, 29
structural studies, 82–91
bond lengths, 84–85, 120
effect of carbonyl substitution, 87
hydrogen bridging, 85–87
neutron diffraction, 82–84
X-ray diffraction, 82–91, 120
substitution at transition metal, 70–76
of carbon monoxide, 70–75
of hydrogen, 73, 75
synthesis, 3–38
acid–base adducts, 31, 35–36, 51–52
addition to strained heterocycles, 31, 34
from alkali metal silyls and metal halides, 11–12, 55
via aluminum silyl, 119
via elimination reactions, 30–33
via mercury bis(silyl), 11, 13, 118
by oxidative addition, 21–30, 101, 119
rearrangement reactions, 36–37
from silanes and metal carbonyls, 13–21
from silicon halides and metal anions, 3–11
via silyl anions, 37
solvent effects, 9–10
thermal stability, 107
trans influence of silicon, 13, 89
vibrational spectra, 91–93
Silver borates, 216
Silylenes
complexes, 37, 51, 116, 118
as catalyst intermediates, 118
extrusion from disilanes, 114, 118
from hydridosilanes, 14
Sodium
in amine solvents
metal association, 165–166
structure of complex, 168
anion, 178–179
aurides, 242–243

Sodium (cont'd)
  borates, 212–213
    borax, 213
    phase equilibria, 212
  solution infrared spectra, 205
  stability range, 209–210
  structures, 190–194, 199–200
  in liquid ammonia, 135
    and metal concentration, 135, 143,
      168
    optical properties, 143–144
    temperature dependence, 174
Strontium borates, 193, 220

T

Tantalum hydrides, 325–334
  of aluminum, 327, 333–334
  bridged to carbon, 327, 334
  bridged to iron tricarbonyl, 327, 333
  σ,η⁵-C₅H₄ derivative, 326, 330
  cyclopentadienyl chloride dimer, 326,
    330–331
    CO adduct, 331
  dicyclopentadienyl derivatives
    dihydrides, 327, 332
    monohydrides, 326, 329–330
    trihydrides, 298, 327, 330, 332–334
  ditertiaryphosphine complexes
    dicarbonyl, 325–326, 328
    dicarbonyl cation, 328
    diphenylphosphide, 328
    naphthalene derivative, 328–329
    pentahydride, 327, 329, 333
  hexacarbonyl derivatives, 311
  trimethylphosphine dimer, 327, 333
Thallium
  borates, 190, 220
  reaction with metal carbonyls, 11
Tin
  pentamethylcyclopentadienyl cation,
    99
  transitio-metal complexes, 56, 75, 76
  clusters, 111, 114
Titanium complexes, 267–284
  binuclear fulvalene derivative, 271
    hydrolysis, 271
    reaction with diphenyl zinc, 278
  bis(pentamethylcyclopentadienyl), 275
    dihydride, 276, 278, 289, 298

dinitrogen complex, 275
  hydrogen shift, 277
  monohydride, 277–278, 289
borohydrides, 279–280
  dicyclopentadienyl, 279, 283
  Schiff base derivative, 280
  σ,η⁵-C₅H₄ derivatives, 274, 288, 294–
    295
dicyclopentadienyls
  alkylaluminum derivatives, 280–
    282, 287–288
  borohydride, 279, 283
  dicarbonyl, 269
  dihydride anion, 269, 287
  dimethyl, 269–270
  diphenyl and diphenyl zinc, 278, 289
  electrochemical reduction of halides,
    273–274
  monohydride, 270
  phosphine hydride, 278, 289
  polymeric hydride, 270–271, 274
  1,4-tetramethylene, 275
  silyls, 12, 22, 40, 83, 85, 119–120
hydrides
  in alkene disproportionation, 283
  in alkene hydrometallation, 282
  in alkene isomerization, 282
  hydridic character, 283
  as hydrogenation catalysts, 278, 289
  phenoxides, 279, 289
  polymer-attached catalysts, 283
  polymeric trihydrides, 278–279, 289
  reactions, 275–276
  theoretical studies, 284
  as Ziegler–Natta catalysts, 282
titanocene, 268–271
  dinitrogen derivatives, 269, 273
Titanium silyl complexes, 12, 22, 40, 83,
  85, 119–120
Trans influence, of silyl group, 13, 89
Tungsten complexes
  dicyclopentadienyl hydride lithium tet-
    ramer, 330
  silyls
    anionic pentacarbonyl, 11–12
    clusters, 38, 112, 117
    dicyclopentadienyl, 32, 73
    structures, 83, 86
    tricarbonylcyclopentadienyl, 40, 43,
      52

U

UV photoemission spectrum, of cesium auride, 241

V

Vanadium complexes
  carbonyl carbide, 79
  hydrides, 305–314
    aluminum halide, 312–313
    borohydrides, 309, 314
    hexacarbonyl derivatives, 306–308, 310–312
    silyl, 21–22
    tetracarbonyldiarsine cation, 309, 313
    tricarbonylcyclopentadienyl anions, 308, 312–313
    tricarbonylcyclopentadienyl dihydride, 309, 313
    tricarbonyldiarsine trihydride, 309, 313
    tricarbonylmesitylene derivative, 308, 312
  silyls
    carbonyls, 3, 64, 69–70, 79, 107
    paramagnetic dicyclopentadienyls, 21–22

X

X-ray photoelectron spectra, of gold clusters, 247, 249

Z

Ziegler–Natta catalysis, by hydride complexes
  cobalt, 282–283
  titanium, 282
Zinc borates, 207, 220–221
  structures, 190, 192, 197

Zirconium hydride complexes, 284–285, 290–303
  bis(pentamethylcyclopentadienyl) dihydride, 284–285, 292, 300
    addition of isonitriles, 294, 301
    ^{1}H NMR spectra, 298
    hydridic nature, 297
    insertion reactions, 292–293, 301
    and metal carbonyls, 293–294, 301
    phosphine adducts, 296–297
  bis(pentamethylcyclopentadienyl) monohydrides
    alkoxy complexes, 292–293, 301
    alkyl dimer, 285, 290, 293
    σ,η^5-C_5H_4 derivative, 295
    cyclohexadienylphosphine complex, 285, 299
  cyclooctatetraene dihydride, 297, 300
    alkylaluminum derivative, 297–298
    as hydrogenation catalyst, 297
  dicyclopentadienyls
    alkyl hydride dimer, 285, 290–291, 299
    aluminum hydride derivative, 291, 302
    borohydrides, 291–292, 299, 302, 305
    bridged to alkylaluminum group, 302–303
    dihydride, 291, 296, 298, 300, 305
    hydridochloride, 284, 291–294, 299
    as hydrogenation catalyst, 296
    in hydrozirconation, 291–292
    ditertiaryphosphine trihydride, 298, 300
    formyl complex, 293–294
    ^{1}H NMR spectra, 298
    naphthyl-bridged derivative, 295, 299
    neopentyl complex, 295, 299
    tetrahydroindenyl dihydride, 297, 300
    zirconium–hydrogen stretching frequency, 298
Zirconium silyl complexes, 12, 40, 118
  structure, 83, 85
Zirconoxycarbenes, 294

CONTENTS OF PREVIOUS VOLUMES

VOLUME 1

Mechanisms of Redox Reactions of
Simple Chemistry
H. Taube

Compounds of Aromatic Ring Systems
and Metals
E. O. Fischer and H. P. Fritz

Recent Studies of the Boron Hydrides
William N. Lipscomb

Lattice Energies and Their Significance
in Inorganic Chemistry
T. C. Waddington

Graphite Intercalation Compounds
W. Rüdorff

The Szilard-Chambers Reactions in Solids
Garman Harbottle and Norman Sutin

Activation Analysis
D. N. F. Atkins and A. A. Smales

The Phosphonitrilic Halides and Their
Derivatives
N. L. Paddock and H. T. Searle

The Sulfuric Acid Solvent System
R. J. Gillespie and E. A. Robinson

AUTHOR INDEX—SUBJECT INDEX

VOLUME 2

Stereochemistry of Ionic Solids
J. D. Dunitz and L. E. Orgel

Organometallic Compounds
John Eisch and Henry Gilman

Fluorine-Containing Compounds of
Sulfur
George H. Cady

Amides and Imides of the Oxyacids of
Sulfur
Margot Becke-Goehring

Halides of the Actinide Elements
Joseph J. Katz and Irving Sheft

Structure of Compounds Containing
Chains of Sulfur Atoms
Olav Foss

Chemical Reactivity of the Boron
Hydrides and Related Compounds
F. G. A. Stone

Mass Spectrometry in Nuclear Chemistry
*H. G. Thode, C. C. McMullen, and
K. Fritze*

AUTHOR INDEX—SUBJECT INDEX

VOLUME 3

Mechanisms of Substitution Reactions of
Metal Complexes
Fred Basolo and Ralph G. Pearson

Molecular Complexes of Halogens
L. J. Andrews and R. M. Keefer

Structure of Interhalogen Compounds
and Polyhalides
*E. H. Wiebenga, E. E. Havinga, and
K. H. Boswijk*

Kinetic Behavior of the Radiolysis
Products of Water
Christiane Ferradini

The General, Selective, and Specific
Formation of Complexes by Metallic
Cations
G. Schwarzenbach

Atmosphere Activities and Dating
Procedures
A. G. Maddock and E. H. Willis

Polyfluoroalkyl Derivatives of Metalloids
and Nonmetals
R. E. Banks and R. N. Haszeldine

AUTHOR INDEX—SUBJECT INDEX

VOLUME 4

Condensed Phosphates and Arsenates
Erich Thilo

Olefin, Acetylene, and π-Allylic
Complexes of Transition Metals
R. G. Guy and B. L. Shaw

Recent Advances in the Stereochemistry
of Nickel, Palladium, and Platinum
J. R. Miller

The Chemistry of Polonium
K. W. Bagnall

The Use of Nuclear Magnetic Resonance
in Inorganic Chemistry
E. L. Muetterties and W. D. Phillips

Oxide Melts
J. D. Mackenzie

AUTHOR INDEX—SUBJECT INDEX

VOLUME 5

The Stabilization of Oxidation States of
the Transition Metals
R. S. Nyholm and M. L. Tobe

Oxides and Oxyfluorides of the Halogens
M. Schmeisser and K. Brandle

The Chemistry of Gallium
N. N. Greenwood

Chemical Effects of Nuclear Activation
in Gases and Liquids
I. G. Campbell

Gaseous Hydroxides
O. Glenser and H. G. Wendlandt

The Borazines
E. K. Mellon, Jr., and J. J. Lagowski

Decaborane-14 and Its Derivatives
M. Frederick Hawthorne

The Structure and Reactivity of
Organophosphorus Compounds
R. F. Hudson

AUTHOR INDEX—SUBJECT INDEX

VOLUME 6

Complexes of the Transition Metals with
Phosphines, Arsines, and Stibines
G. Booth

Anhydrous Metal Nitrates
C. C. Addison and N. Logan

Chemical Reactions in Electric
Discharges
Adli S. Kana'an and John L. Margrave

The Chemistry of Astatine
A. H. W. Aten, Jr.

The Chemistry of Silicon–Nitrogen
Compounds
U. Wannagat

Peroxy Compounds of Transition Metals
J. A. Connor and E. A. V. Ebsworth

The Direct Synthesis of Organosilicon
Compounds
J. J. Zuckerman

The Mössbauer Effect and Its Application
in Chemistry
E. Fluck

AUTHOR INDEX—SUBJECT INDEX

VOLUME 7

Halides of Phosphorus, Arsenic,
Antimony, and Bismuth
L. Kolditz

The Phthalocyanines
A. B. P. Lever

Hydride Complexes of the Transition
Metals
M. L. H. Green and D. L. Jones

Reactions of Chelated Organic Ligands
Quintus Fernando

Organoaluminum Compounds
Roland Köster and Paul Binger

Carbosilanes
G. Fritz, J. Grobe, and D. Kummer

AUTHOR INDEX—SUBJECT INDEX

VOLUME 8

Substitution Products of the Group VIB
Metal Carbonyls
Gerard R. Dobson, Ingo W. Stolz, and
Raymond K. Sheline

Transition Metal Cyanides and Their
Complexes
B. M. Chadwick and A. G. Sharpe

Perchloric Acid
G. S. Pearson

Neutron Diffraction and Its Application
in Inorganic Chemistry
G. E. Bacon

352 CONTENTS OF PREVIOUS VOLUMES

Nuclear Quadrupole Resonance and Its
Application in Inorganic Chemistry
Masaji Kubo and Daiyu Nakamura

The Chemistry of Complex
Aluminohydrides
E. C. Ashby

AUTHOR INDEX—SUBJECT INDEX

VOLUME 9

Liquid–Liquid Extraction of Metal Ions
D. F. Peppard

Nitrides of Metals of the First Transition
Series
R. Juza

Pseudohalides of Group IIIB and IVB
Elements
M. F. Lappert and H. Pyszora

Stereoselectivity in Coordination
Compounds
J. H. Dunlop and R. D. Gillard

Heterocations
A. A. Woolf

The Inorganic Chemistry of Tungsten
R. V. Parish

AUTHOR INDEX—SUBJECT INDEX

VOLUME 10

The Halides of Boron
A. G. Massey

Further Advances in the Study of
Mechanisms of Redox Reactions
A. G. Sykes

Mixed Valence Chemistry—A Survey and
Classification
Melvin B. Robin and Peter Day

AUTHOR INDEX—SUBJECT INDEX—
VOLUMES 1–10

VOLUME 11

Technetium
*K. V. Kotegov, O. N. Pavlov, and
V. P. Shvedov*

Transition Metal Complexes with Group
IVB Elements
J. F. Young

Metal Carbides
William A. Frad

Silicon Hydrides and Their Derivatives
B. J. Aylett

Some General Aspects of Mercury
Chemistry
H. L. Roberts

Alkyl Derivatives of the Group II Metals
B. J. Wakefield

AUTHOR INDEX—SUBJECT INDEX

VOLUME 12

Some Recent Preparative Chemistry of
Protactinium
D. Brown

Vibrational Spectra of Transition Metal
Carbonyl Complexes
Linda M. Haines and M. H. Stiddard

The Chemistry of Complexes Containing
2,2'-Bipyridyl, 1,10-Phenanthroline,
or 2,2',6',2''-Terpyridyl as Ligands
W. R. McWhinnie and J. D. Miller

Olefin Complexes of the Transition
Metals
H. W. Quinn and J. H. Tsai

Cis and Trans Effects in Cobalt(III)
Complexes
J. M. Pratt and R. G. Thorp

AUTHOR INDEX—SUBJECT INDEX

VOLUME 13

Zirconium and Hafnium Chemistry
E. M. Larsen

Electron Spin Resonance of Transition
Metal Complexes
B. A. Goodman and J. B. Raynor

Recent Progress in the Chemistry of
Fluorophosphines
John F. Nixon

Transition Metal Cluster with π-Acid
　　Ligands
　　R. D. Johnston

AUTHOR INDEX—SUBJECT INDEX

VOLUME 14

The Phosphazotrihalides
　　M. Bermann

Low Temperature Condensation of High
　　Temperature Species as a Synthetic
　　Method
　　P. L. Timms

Transition Metal Complexes Containing
　　Bidentate Phosphine Ligands
　　W. Levason and C. A. McAuliffe

Beryllium Halides and Pseudohalides
　　N. A. Bell

Sulfur–Nitrogen–Fluorine Compounds
　　O. Glemser and R. Mews

AUTHOR INDEX—SUBJECT INDEX

VOLUME 15

Secondary Bonding to Nonmetallic
　　Elements
　　N. W. Alcock

Mössbauer Spectra of Inorganic
　　Compounds: Bonding and Structure
　　G. M. Bancroft and R. H. Platt

Metal Alkoxides and Dialkylamides
　　D. C. Bradley

Fluoroalicyclic Derivatives of Metals and
　　Metalloids
　　W. R. Cullen

The Sulfur Nitrides
　　H. G. Heal

AUTHOR INDEX—SUBJECT INDEX

VOLUME 16

The Chemistry of Bis(trifluoromethyl)-
　　amino Compounds
　　H. G. Ang and Y. C. Syn

Vacuum Ultraviolet Photoelectron
　　Spectroscopy of Inorganic Molecules
　　R. L. DeKock and D. R. Lloyd

Fluorinated Peroxides
　　*Ronald A. De Marco and Jean'ne
　　M. Shreeve*

Fluorosulfuric Acid, Its Salts, and
　　Derivatives
　　Albert W. Jache

The Reaction Chemistry of Diborane
　　L. H. Long

Lower Sulfur Fluorides
　　F. Seel

AUTHOR INDEX—SUBJECT INDEX

VOLUME 17

Inorganic Compounds Containing the
　　Trifluoroacetate Group
　　C. D. Garner and B. Hughes

Homopolyatomic Cations of the Elements
　　R. J. Gillespie and J. Passmore

Use of Radio-Frequency Plasma in
　　Chemical Synthesis
　　S. M. L. Hamblyn and B. G. Reuben

Copper(I) Complexes
　　F. H. Jardine

Complexes of Open-Chain Tetradenate
　　Ligands Containing Heavy Donor
　　Atoms
　　C. A. McAuliffe

The Functional Approach to Ionization
　　Phenomena in Solutions
　　U. Mayer and V. Gutmann

Coordination Chemistry of the Cyanate,
　　Thiocyanate, and Selenocyanate Ions
　　A. H. Norbury

SUBJECT INDEX

VOLUME 18

Structural and Bonding Patterns in
　　Cluster Chemistry
　　K. Wade

Coordination Number Pattern Recognition Theory of Carborane Structures
Robert E. Williams

Preparation and Reactions of Perfluorohalogenoorganosulfenyl Halides
A. Haas and U. Niemann

Correlations in Nuclear Magnetic Shielding. Part I
Joan Mason

Some Applications of Mass Spectroscopy in Inorganic and Organometallic Chemistry
Jack M. Miller and Gary L. Wilson

The Structures of Elemental Sulfur
Beat Meyer

Chlorine Oxyfluorides
K. O. Christe and C. J. Schack

SUBJECT INDEX

VOLUME 19

Recent Chemistry and Structure Investigation of Nitrogen Triiodide, Tribromide, Trichloride, and Related Compounds
Jochen Jander

Aspects of Organo-Transition-Metal Photochemistry and Their Biological Implications
Ernst A. Koerner von Gustorf, Luc H. G. Leenders, Ingrid Fischler, and Robin N. Perutz

Nitrogen–Sulfur-Fluorine Ions
R. Mews

Isopolymolybdates and Isopolytungstates
Karl-Heinz Tytko and Oskar Glemser

SUBJECT INDEX

VOLUME 20

Recent Advances in the Chemistry of the Less-Common Oxidation States of the Lanthanide Elements
D. A. Johnson

Ferrimagnetic Fluorides
Alain Tressaud and Jean Michel Dance

Hydride Complexes of Ruthenium, Rhodium, and Iridium
G. L. Geoffroy and J. R. Lehman

Structures and Physical Properties of Polynuclear Carboxylates
Janet Catterick and Peter Thornton

SUBJECT INDEX

VOLUME 21

Template Reactions
Maria De Sousa Healy and Anthony J. Rest

Cyclophosphazenes
S. S. Krishnamurthy, A. C. Sau, and M. Woods

A New Look at Structure and Bonding in Transition Metal Complexes
Jeremy K. Burdett

Adducts of the Mixed Trihalides of Boron
J. Stephen Hartman and Jack M. Miller

Reorganization Energies of Optical Electron Transfer Processes
R. D. Cannon

Vibrational Spectra of the Binary Fluorides of the Main Group Elements
N. R. Smyrl and Gleb Mamantov

The Mössbauer Effect in Supported Microcrystallites
Frank J. Berry

SUBJECT INDEX

VOLUME 22

Lattice Energies and Thermochemistry of Hexahalometallate(IV) Complexes, A_2MX_6, which Possess the Antifluorite Structure
H. Donald B. Jenkins and Kenneth F Pratt

Reaction Mechanisms of Inorganic Nitrogen Compounds
G. Stedman

Thio-, Seleno-, and Tellurohalides of the Transition Metals
M. J. Atherton and J. H. Holloway

Correlations in Nuclear Magnetic Shielding, Part II
Joan Mason

Cyclic Sulfur–Nitrogen Compounds
H. W. Roesky

1,2-Dithiolene Complexes of Transition Metals
R. P. Burns and C. A. McAuliffe

Some Aspects of the Bioinorganic Chemistry of Zinc
Reg H. Prince

SUBJECT INDEX

VOLUME 23

Recent Advances in Organotin Chemistry
Alwyn G. Davies and Peter J. Smith

Transition Metal Vapor Cryochemistry
William J. Power and Geoffrey A. Ozin

New Methods for the Synthesis of Trifluoromethyl Organometallic Compounds
Richard J. Lagow and John A. Morrison

1,1-Dithiolato Complexes of the Transition Elements
R. P. Burns, F. P. McCullough, and C. A. McAuliffe

Graphite Intercalation Compounds
Henry Selig and Lawrence B. Ebert

Solid-State Chemistry of Thio-, Seleno-, and Tellurohalides of Representative and Transition Elements
J. Fenner, A. Rabenau, and G. Trageser

SUBJECT INDEX

VOLUME 24

Thermochemistry of Inorganic Fluorine Compounds
A. A. Woolf

Lanthanide, Yttrium, and Scandium Trihalides: Preparation of Anhydrous Materials and Solution Thermochemistry
J. Burgess and J. Kijowski

The Coordination Chemistry of Sulfoxides with Transition Metals
J. A. Davies

Selenium and Tellurium Fluorides
A. Engelbrecht and F. Sladky

Transition-Metal Molecular Clusters
B. F. G. Johnson and J. Lewis

INDEX